COMMUNICATION SYSTEMS

COMMUNICATION SYSTEMS

MARCELO S. ALENCAR, PH.D., P.E.
Federal University of Campina Grande
Brazil

VALDEMAR C. DA ROCHA JR., PH.D., P.E.
Federal University of Pernambuco
Brazil

Springer

Marcelo S. Alencar
Federal University of Campina Grande
Brazil

Valdemar C. da Rocha, Jr.
Federal University of Pernambuco
Brazil

Alencar, Marcelo S., 1957-
Communication systems / Marcelo S. Alencar, Valdemar C. da Rocha.
p. cm.
Includes bibliographical references and index.
ISBN 0-387-25481-1 (alk. paper) -- ISBN 0-387-27097-3 (e-book)
1. Signal processing--Digital techniques. 2. Digital communications. I. Rocha, Valdemar C. da, 1947- II. Title.

TK5102.9.A39 2005
621.382--dc22 2005049937

ISBN 0-387-25481-1 e-ISBN 0-387-27097-3 Printed on acid-free paper.
ISBN 978-0387-25481-4

Printed in the United States of America.

9 8 7 6 5 4 3 2 1 SPIN 11307617

springeronline.com

This book is dedicated
to our families.

SOLUTIONS MANUAL

As an aid to teachers interested in teaching this book as a course, a solutions manual is available by contacting the Publisher.

Contents

Appendices

Preface

This book is the result of several years of teaching and research at the Federal University of Campina Grande and Federal University of Pernambuco, Brazil. It is intended to serve as an introductory textbook for courses dealing with Communication Systems or Modulation Theory. The modulation theory is dealt with using stochastic processes, which is novel for undergraduate texts. The book is suitable for the undergraduate as well as the initial graduate levels of Electrical Engineering courses.

Chapter 1 covers signal and Fourier analysis and presents an introduction to Fourier transform, convolution and definitions of autocorrelation and power spectral density. Chapter 2 introduces the concepts of probability, random variables and stochastic processes and their applications to the analysis of linear systems.

Speech coding is dealt with in Chapter 3, which also deals with digitizing of analog signals, quantization and coding for compression. Chapter 4 presents amplitude modulation with random signals, including digital signals and performance evaluation methods.

Quadrature amplitude modulation using random signals, including SSB, QUAM, QAM and QPSK is the subject of Chapter 5. Chapter 6 explains angle modulation with random modulating signals, including frequency and phase modulation, FSK and PSK.

Channel modeling is the subject of Chapter 7, which includes channel characteristics and propagation. Chapter 8 deals with transmission and reception of the modulated carrier, and presents several features of the transmitting and receiving equipment. Chapter 9 presents the main concepts of mobile communication systems, both analog and digital.

The book has five appendices. Appendix A covers Fourier series and transforms, Hilbert transform, and their properties. Appendix B presents formulas used in the text. Appendix C shows tables of the radio-frequency spectrum. Appendix D presents the CDMA cellular system. Appendix E presents the GSM cellular system. An important feature is the many examples and problems to be found throughout the book.

MARCELO S. ALENCAR, VALDEMAR C. DA ROCHA JR.

Acknowledgments

The authors are grateful to Professor Elvino S. Sousa, University of Toronto, Canada, Professor Paddy Farrell and Professor Bahram Honary, Lancaster University, UK, Professor Michael Darnell and Professor Garik Markarian, Leeds University, UK, for technical communications and useful discussions related to communication systems.

The authors are also grateful to all the members of the Communications Research Groups, certified by the National Council for Scientific and Technological Development (CNPq), at both the Federal University of Campina Grande and the Federal University of Pernambuco, for their collaboration in many ways, helpful discussions and friendship, as well as our colleagues at the Institute for Advanced Studies in Communications.

The authors wish also to acknowledge the contribution of Francisco Madeiro, from the Catholic University of Pernambuco, and Waslon T. A. Lopes, from ÁREA1 Salvador, who wrote the chapter on speech coding, and the support of Thiago T. Alencar, who helped with translation of parts of the text.

The authors are indebted to their families for their patience and support during the course of the preparation of this book.

Finally, the authors are thankful to Alex Greene and Melissa Guasch, from Springer, who strongly supported this project from the beginning and helped with the reviewing process.

Chapter 1

SIGNAL ANALYSIS

1.1 Introduction

The objective of this chapter is to provide the reader with the necessary mathematical basis for understanding probability theory and stochastic processes. The reader will become familiar with concepts and equations involving Fourier series, which have a significant historical relevance for the theory of communications. Furthermore, both the theory and properties of the Fourier transform will be presented, which constitute powerful tools for spectral analysis (Alencar, 1999).

1.2 Fourier Analysis

The basic Fourier theory establishes fundamental conditions for the representation of an arbitrary function in a finite interval as a sum of sinuoids. In fact this is just an instance of the more general Fourier representation of signals in which a periodic signal $f(t)$, under fairly general conditions, can be represented by a *complete set* of orthogonal functions. By a *complete set* $\mathcal{S}$ of orthogonal functions it is understood that, except for those orthogonal functions already in $\mathcal{S}$, there are no other orthogonal functions not belonging to $\mathcal{S}$ to be considered. It is assumed in the sequel that a periodic signal $f(t)$ satisfies the Dirichlet conditions, i.e., that $f(t)$ is a bounded function which in any one period has at most a finite number of local maxima and minima and a finite number of points of discontinuity (Wylie, 1966). The representation of signals by orthogonal functions has very often an error, which diminishes as the number of component terms in the corresponding series is increased.

The fact that a periodic signal $f(t)$ can in general be expanded as a sum of mutually orthogonal functions, demands for a closer look at the concepts of periodicity and orthogonality.

Periodicity relates to the repetitive character of the function. A function $f(t)$ is defined to be a *periodic function* of period T, if and only if, T is the smallest positive number for which $f(t+T) = f(t)$. In other words, $f(t)$ is periodic if its domain contains $t+T$ whenever it contains t, and $f(t+T) = f(t)$. It follows from the definition of a periodic function that if T represents the period of $f(t)$ then $f(t) = f(t+nT)$, for $n = 1, 2, \ldots$, i.e., $f(t)$ will repeat its values when integer multiples of T (Wozencraft and Jacobs, 1965) are added to its argument.

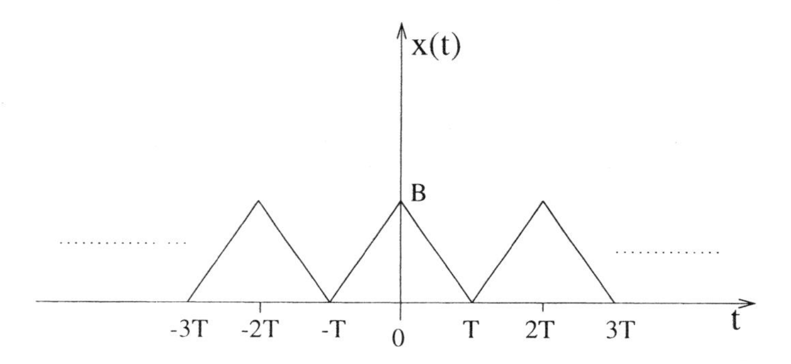

Figure 1.1. Example of a periodic signal.

If $f(t)$ and $g(t)$ are two periodic functions with the same period T, then their sum $f(t) + g(t)$ will also be a periodic function with period T. We prove this result by making $h(t) = f(t) + g(t)$ and noticing that $h(t+T) = f(t+T) + g(t+T) = f(t) + g(t) = h(t)$.

We shall now investigate the concept of *orthogonality*. Orthogonality provides the tool for introducing the concept of a *basis*, i.e., of a *minimum set of functions* that can be used to generate other functions. However, orthogonality by itself does not guarantee that a complete vector space is generated.

Two real functions $u(t)$ and $v(t)$, defined in the interval $\alpha \leq t \leq \beta$, are orthogonal if their inner product is null, i.e., if

$$(u(t), v(t)) = \int_{\alpha}^{\beta} u(t)v(t)dt = 0 \tag{1.1}$$

The set of functions $f_n(t)$, as illustrated in Figure 1.2, can be used for representing signals in the time domain. This set of functions constitutes an orthogonal set in the interval $(0, 1)$.

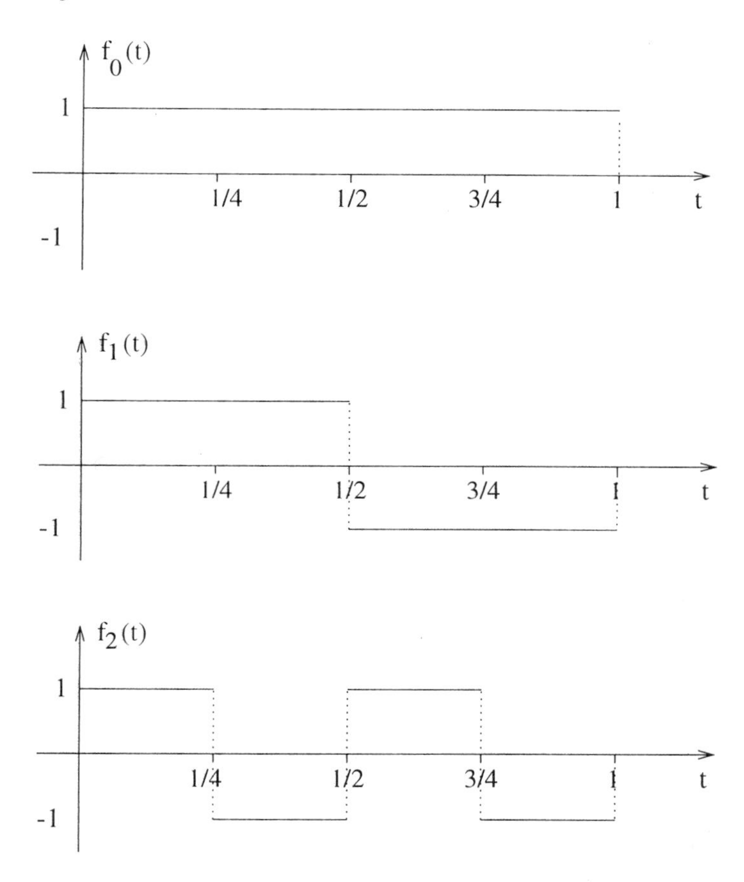

Figure 1.2. Set of orthogonal functions.

1.2.1 The Trigonometric Fourier Series

The trigonometric Fourier series representation of the signal $f(t)$ can be written as

$$f(t) = a_0 + \sum_{n=1}^{\infty} [a_n \cos(n\omega_0 t) + b_n \sin(n\omega_0 t)], \tag{1.2}$$

where the term a_0 (the average value of the function $f(t)$) indicates whether or not the signal contains a DC value and the terms a_n and b_n are denominated the Fourier series coefficients, where n is a positive integer. The equality sign holds in (1.2) for all values of t only when $f(t)$ is periodic. However, the Fourier series representation is a useful tool for any type of signal as long as that signal representation is required only in the $[0, T]$ interval. Outside that interval the Fourier series representation will always be periodic, even if the signal $f(t)$ is not periodic (Knopp, 1990).

The sine and cosine functions are examples of orthogonal functions because they satisfy the following equations, for integer values of n and m, denominated orthogonality relations:

$$\int_0^T \cos(n\omega_o t)\,\sin(m\omega_o t)dt = 0, \text{ for all integers } n, m, \tag{1.3}$$

$$\int_0^T \cos(n\omega_o t)\cos(m\omega_o t)dt = \begin{cases} 0 & \text{if } n \neq m \\ \frac{T}{2} & \text{if } n = m \end{cases} \tag{1.4}$$

$$\int_0^T \sin(n\omega_o t)\,\sin(m\omega_o t)dt = \begin{cases} 0 & \text{if } n \neq m \\ \frac{T}{2} & \text{if } n = m \end{cases} \tag{1.5}$$

where $\omega_0 = 2\pi/T$.

As a consequence of the orthogonality conditions, explicit expressions for the coefficients a_n and b_n of the Fourier trigonometric series can be computed. By integrating both sides in expression (1.2) in the interval $[0, T]$, it follows that (Oberhettinger, 1990)

$$\int_0^T f(t)dt = \int_0^T a_o dt + \sum_{n=1}^{\infty} \int_0^T a_n \cos(n\omega_o t)dt + \sum_{n=1}^{\infty} \int_0^T b_n \sin(n\omega_o t)dt$$

and since

$$\int_0^T a_n \cos(n\omega_o t)dt = \int_0^T b_n \sin(n\omega_o t)dt = 0,$$

it follows that

$$a_o = \frac{1}{T}\int_0^T f(t)dt. \tag{1.6}$$

Now, by multiplying both sides in expression (1.2) by $\cos(m\omega_o t)$ and integrating in the interval $[0, T]$, it follows that

$$\int_0^T f(t)\cos(m\omega_o t)dt = \int_0^T a_o \cos(m\omega_o t)dt + \sum_{n=1}^{\infty}\int_o^T a_n \cos(n\omega_o t)\cos(m\omega_o t)dt + \sum_{n=1}^{\infty}\int_o^T b_n \cos(m\omega_0 t)\sin(n\omega_o t)dt, \quad (1.7)$$

which after simplification produces

$$a_n = \frac{2}{T}\int_0^T f(t)\cos(n\omega_o t)dt, \text{ for } n = 1, 2, 3, \ldots \quad (1.8)$$

In a similar manner b_n is found by multiplying both sides in expression (1.2) by $\sin(n\omega_o t)$ and integrating in the interval $[0, T]$, i.e.,

$$b_n = \frac{2}{T}\int_0^T f(t)\sin(n\omega_o t)dt, \quad (1.9)$$

for $n = 1, 2, 3, \ldots$.

1.2.2 Even Functions and Odd Functions

A function is called an odd function if it is anti-symmetric with respect to the ordinate axis, i.e., if $f(-t) = -f(t)$, where $-t$ and t are assumed to belong to the function domain. Examples of odd functions are provided by the functions t, t^3, $\sin t$ and $t^{|2n+1|}$.

Similarly, a function is called an even function if it is symmetric with respect to the ordinate axis, i.e., if $f(-t) = f(t)$, where t and $-t$ are assumed to belong to the function domain. Examples of even functions are provided by the functions 1, t^2, $\cos t$, $|t|$, $\exp(-|t|)$ and $t^{|2n|}$.

Some Elementary Properties

a) The sum (difference) and the product (quotient) of two even functions is an even function;

b) The sum (difference) of two odd functions is an odd function;

c) The product (quotient) of two odd functions is an even function;

d) The sum (difference) of an even function and an odd function is neither an even function nor an odd function;

e) The product (quotient) between an even function and an odd function is an odd function.

Two other important properties are the following.

f) If $f(t)$ is an even periodic function of period T, then

$$\int_{-T/2}^{T/2} f(t)dt = 2\int_{0}^{T/2} f(t)dt. \tag{1.10}$$

g) If $f(t)$ is an odd periodic function of period T, then

$$\int_{-T/2}^{T/2} f(t)dt = 0. \tag{1.11}$$

Properties (f) and (g) allow for a considerable simplification when computing coefficients of a trigonometric Fourier series:

h) If $f(t)$ is an even function then $b_n = 0$, and

$$a_n = \frac{2}{T}\int_{0}^{T} f(t)\cos(n\omega_o t)dt, \text{ for } n = 1,2,3, \ldots . \tag{1.12}$$

i) If $f(t)$ is an odd function then $a_n = 0$ and

$$b_n = \frac{2}{T}\int_{0}^{T} f(t)\sin(n\omega_o t)dt, \text{ for } n = 1,2,3, \ldots . \tag{1.13}$$

Example: Compute the coefficients of the trigonometric Fourier series for the waveform $f(t) = A[u(t+\tau) - u(t-\tau)]$, which repeats itself with period T, where $u(t)$ denotes the unit step function and $2\tau \leq T$.

Solution: Since the given signal is symmetric with respect to the ordinate axis, it follows that $f(t) = f(-t)$ and the function is even. Therefore $b_n = 0$, and all that is left for computing is a_o, and a_n for $n = 1, 2, \ldots$. The expression for computing the average value a_0 is given by

$$a_o = \frac{1}{T}\int_{-\frac{T}{2}}^{\frac{T}{2}} f(t)dt = \frac{1}{T}\int_{-\tau}^{\tau} Adt = \frac{2A\tau}{T}.$$

In the previous equation the maximum value of τ is $T/2$. The coefficients a_n for $n = 1, 2, \ldots$ are computed as

$$a_n = \frac{2}{T}\int_{0}^{T} f(t)\cos(n\omega_o t)dt = \frac{2}{T}\int_{-\tau}^{\tau} A\cos(n\omega_o t)dt,$$

$$a_n = \frac{4A}{T}\int_0^{\tau}\cos(n\omega_o t)dt = \frac{4A}{Tn\omega_o}\sin(n\omega_o t)\Big|_0^{\tau} = (4A\tau/T)\frac{\sin(n\omega_o\tau)}{n\omega_0\tau}.$$

The signal $f(t)$ is then represented by the following trigonometric Fourier series

$$f(t) = \frac{2A\tau}{T} + \left(\frac{4A\tau}{T}\right)\sum_{n=1}^{\infty}\frac{\sin(n\omega_o\tau)}{n\omega_0\tau}\cos(n\omega_o t).$$

1.2.3 The Compact Fourier Series

It is also possible to represent the Fourier series in a form known as the *compact* Fourier series as follows

$$f(t) = C_0 + \sum_{n=1}^{\infty} C_n\cos(n\omega_o t + \theta_n). \tag{1.14}$$

By expanding the expression $C_n\cos(n\omega_o t+\theta)$ as $C_n\cos(n\omega_o t)\cos\theta_n - C_n\sin(n\omega_o t)\sin\theta_n$ and comparing this result with (1.2) it follows that $a_o = C_o$, $a_n = C_n\cos\theta_n$ and $b_n = -C_n\sin\theta_n$. It is now possible to compute C_n as a function of a_n and b_n. For that purpose it is sufficient to square a_n and b_n and add the result, i.e.,

$$a_n^2 + b_n^2 = C_n^2\cos^2\theta_n + C_n^2\sin^2\theta_n = C_n^2. \tag{1.15}$$

From Equation (1.15) the modulus of C_n can be written as

$$C_n = \sqrt{a_n^2 + b_n^2}. \tag{1.16}$$

In order to determine θ_n it suffices to divide b_n by a_n, i.e.,

$$\frac{b_n}{a_n} = -\frac{\sin\theta_n}{\cos\theta_n} = -\tan\theta_n, \tag{1.17}$$

which when solved for θ_n produces

$$\theta_n = -\arctan\left(\frac{b_n}{a_n}\right). \tag{1.18}$$

1.2.4 The Exponential Fourier Series

Since the set of exponential functions $e^{jn\omega_o t}$, $n = 0, \pm1, \pm2, \ldots$, is a complete set of orthogonal functions in an interval of magnitude T, where $T = 2\pi/\omega_o$, then it is possible to represent a function $f(t)$ by a linear combination of exponential functions in an interval T.

$$f(t) = \sum_{-\infty}^{\infty} F_n e^{jn\omega_0 t} \tag{1.19}$$

where

$$F_n = \frac{1}{T} \int_{\frac{-T}{2}}^{\frac{T}{2}} f(t) e^{-jn\omega_0 t} dt. \tag{1.20}$$

Equation (1.19) represents the exponential Fourier series expansion of $f(t)$ and equation (1.20) is the expression to compute the associated series coefficients. The exponential Fourier series is also known as the *complex Fourier series.* It is immediate to show that equation (1.19) is just another way of expressing the Fourier series as given in (1.2). Replacing $\cos(n\omega_o t) + j\sin(n\omega_0 t)$ for $e^{n\omega_0 t}$ (Euler's identity) in (1.19), it follows that

$$\begin{aligned} f(t) &= F_o + \sum_{n=-\infty}^{-1} F_n[\cos(n\omega_o t) + j\sin(n\omega_o t)] \\ &+ \sum_{n=1}^{\infty} F_n[\cos(n\omega_o t) + j\sin(n\omega_o t)], \end{aligned}$$

or

$$f(t) = F_o + \sum_{n=1}^{\infty} F_n[\cos(n\omega_o t) + j\sin(n\omega_o t)] + F_{-n}[\cos(n\omega_o t) - j\sin(n\omega_o t)].$$

Grouping the coefficients of the sine and cosine terms, it follows that

$$f(t) = F_o + \sum_{n=1}^{\infty} (F_n + F_{-n})\cos(n\omega_o t) + j(F_n - F_{-n})\sin(n\omega_o t). \tag{1.21}$$

Comparing the above expression with (1.2) it follows that

$$a_o = F_o,\ a_n = (F_n + F_{-n}) \text{ and } b_n = j(F_n - F_{-n}), \tag{1.22}$$

and that

$$F_o = a_o, \tag{1.23}$$

$$F_n = \frac{a_n - jb_n}{2}, \tag{1.24}$$

and

$$F_{-n} = \frac{a_n + jb_n}{2}. \tag{1.25}$$

In case the function $f(t)$ is even, i.e., if $b_n = 0$, then

$$a_o = F_o, \; F_n = \frac{a_n}{2} \text{ and } F_{-n} = \frac{a_n}{2}. \tag{1.26}$$

Example: Compute the exponential Fourier series for the train of impulses

$$\delta_T(t) = \sum_{n=-\infty}^{\infty} \delta(t - nT).$$

Solution: The complex coefficients are given by

$$F_n = \frac{1}{T} \int_{\frac{-T}{2}}^{\frac{T}{2}} \delta_T(t) e^{-jn\omega_o t} dt = \frac{1}{T}, \tag{1.27}$$

since

$$\int_{-\infty}^{\infty} \delta(t - t_o) f(t) dt = f(t_o), \text{ (Impulse filtering)}. \tag{1.28}$$

It follows that $f(t)$ can be written as

$$f(t) = \frac{1}{T} \sum_{n=-\infty}^{\infty} e^{-jn\omega_o t}. \tag{1.29}$$

In practice, in order to obtain an impulse train it is sufficient to pass a binary digital signal through a differentiator circuit and then pass the resulting waveform through a half-wave rectifier.

The Fourier series expansion of a periodic function is equivalent to its decomposition in frequency components. In general, a periodic function with period T has frequency components $0, \pm\omega_o, \pm 2\omega_o, \pm 3\omega_o, \ldots, \pm n\omega_o$, where $\omega_o = 2\pi/T$ is the fundamental frequency and the multiples of ω_0 are called harmonics. Notice that the spectrum exists only for discrete values of ω and that the spectral components are spaced by at least ω_o.

1.3 Fourier Transform

It was shown earlier that an arbitrary function can be represented in terms of an exponential (or trigonometric) Fourier series in a finite interval. If such a function is periodic this representation can be extended for the entire interval $(-\infty, \infty)$. However, it is interesting to observe the spectral behavior of a function in general, periodic or not, in the entire interval $(-\infty, \infty)$. In order to do that we truncate the function $f(t)$ in the interval $[-T/2, T/2]$, obtaining $f_T(t)$. It is possible then to represent this function as a sum of exponentials in the entire interval $(-\infty, \infty)$ by making T approach infinity. In other words

$$\lim_{T\to\infty} f_T(t) = f(t).$$

The $f_T(t)$ signal can be represented by the exponential Fourier series as

$$f_T(t) = \sum_{n=-\infty}^{\infty} F_n e^{jn\omega_o t}, \tag{1.30}$$

where $\omega_o = 2\pi/T$ and

$$F_n = \frac{1}{T}\int_{-\frac{T}{2}}^{\frac{T}{2}} f_T(t)e^{-jn\omega_o t}dt. \tag{1.31}$$

F_n represents the spectral amplitude associated to each component of frequency $n\omega_o$. As T increases, the amplitudes diminish but the spectrum shape is not altered. This increase in T forces ω_o to diminish and the spectrum to become denser. In the limit, as $T \to \infty$, ω_o becomes infinitesimally small, being represented by $d\omega$. On the other hand, there are now infinitely many components and the spectrum is no longer a discrete one, becoming a continuous spectrum in the limit.

For convenience, we will write $TF_n = F(\omega)$, i.e., the product TF_n becomes a function of the variable ω, since $n\omega_o \to \omega$. Replacing $\frac{F(\omega)}{T}$ for F_n in 1.30, we obtain

$$f_T(t) = \frac{1}{T}\sum_{n=-\infty}^{\infty} F(\omega)e^{j\omega t}. \tag{1.32}$$

Replacing $\omega_0/2\pi$ for $1/T$ it follows that

$$f_T(t) = \frac{1}{2\pi}\sum_{n=-\infty}^{\infty} F(\omega)e^{j\omega t}\omega_0. \tag{1.33}$$

In the limit, as T approaches infinity, we have

$$f(t) = \frac{1}{2\pi}\int_{-\infty}^{\infty} F(\omega)e^{j\omega t}d\omega \tag{1.34}$$

which is known as the inverse Fourier transform.

Similarly, from 1.31, as T approaches infinity, one obtains

$$F(\omega) = \int_{-\infty}^{\infty} f(t)e^{-j\omega t}dt \tag{1.35}$$

which is known as the direct Fourier transform, sometimes denoted in the literature as $F(\omega) = \mathcal{F}[f(t)]$. A Fourier transform pair is often denoted as $f(t) \longleftrightarrow F(\omega)$

In the sequel we present some important Fourier transforms (Haykin, 1988).

Bilateral Exponential Signal

If $f(t) = e^{-a|t|}$ it follows from (1.35) that

$$\begin{aligned} F(\omega) &= \int_{-\infty}^{\infty} e^{-a|t|}e^{-j\omega t}dt & (1.36)\\ &= \int_{-\infty}^{0} e^{at}e^{-j\omega t}dt + \int_{0}^{\infty} e^{-at}e^{-j\omega t}dt \\ &= \frac{1}{a-j\omega} + \frac{1}{a+j\omega}, \\ F(\omega) &= \frac{2a}{a^2+\omega^2}. & (1.37)\end{aligned}$$

Gate Function

The gate function is defined by the expression $p_T(t) = A[u(t+T/2) - u(t-T/2)]$, or

$$p_T(t) = \begin{cases} A & \text{if } |t| \leq T/2 \\ 0 & \text{if } |t| > T/2 \end{cases} \tag{1.38}$$

where $u(t)$ denotes the *unit step* function, defined as

$$u(t) = \begin{cases} 1 & \text{if } t \geq 0 \\ 0 & \text{if } t < 0 \end{cases} \tag{1.39}$$

The gate function is illustrated in Figure 1.3. The Fourier transform of the gate function can be calculated as

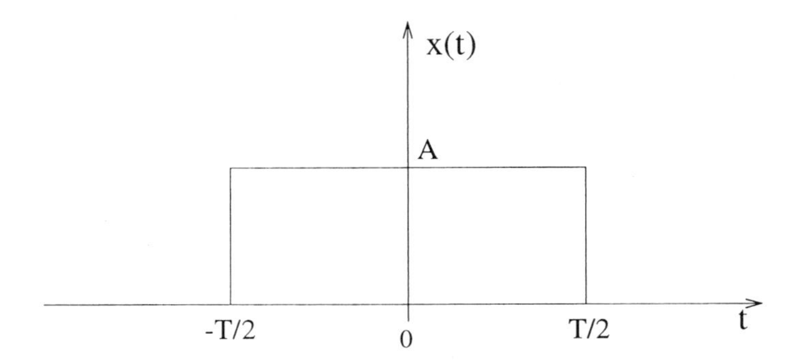

Figure 1.3. Gate function.

$$
\begin{aligned}
F(\omega) &= \int_{-\frac{T}{2}}^{\frac{T}{2}} Ae^{-j\omega t}dt && (1.40)\\
&= \frac{A}{j\omega}(e^{j\omega\frac{T}{2}} - e^{-j\omega\frac{T}{2}})\\
&= \frac{A}{j\omega}2j\sin(\omega T/2),
\end{aligned}
$$

which can be rearranged as

$$F(\omega) = AT\left(\frac{\sin(\omega \mathrm{T}/2)}{\omega T/2}\right),$$

and finally

$$F(\omega) = AT\mathrm{Sa}\left(\frac{\omega T}{2}\right), \quad (1.41)$$

where $\mathrm{Sa}\,(x) = \frac{\sin x}{x}$ is the *sampling* function. This function converges to one, as x goes to zero. The sampling function, the magnitude of which is illustrated in Figure 1.4, is of great relevance in communication theory.

The *sampling* function obeys the following important relationship

$$\int_{-\infty}^{\infty} \frac{k}{\pi}\mathrm{Sa}\,(kt)dt = 1. \quad (1.42)$$

The area under this curve is equal to 1. As k increases, the amplitude of the sampling function increases, the spacing between zero crossings diminishes and most of the signal energy concentrates near the origin. For $k \to \infty$ the function converges to an impulse function, i.e.

$$\delta(t) = \lim_{k\to\infty} \frac{k}{\pi}\mathrm{Sa}\,(kt). \quad (1.43)$$

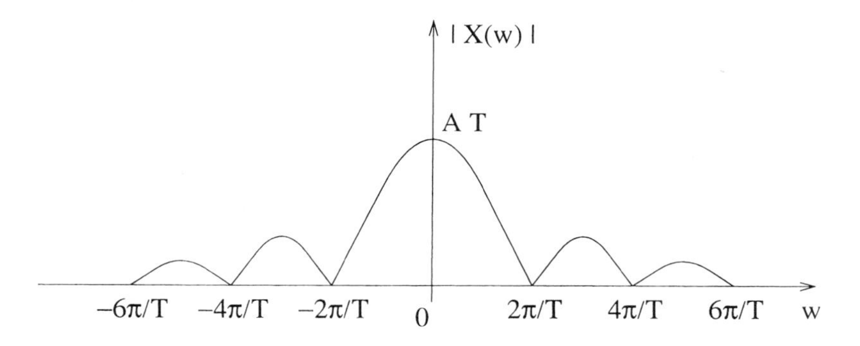

Figure 1.4. Magnitude plot of the Fourier transform of the gate function.

In this manner, in the limit it is true that $\int_{-\infty}^{\infty} \delta(t)dt = 1$. Since the function concentrates its non zero values near the origin, it follows that $\delta(t) = 0$ for $t \neq 0$. Therefore

$$\int_{-\infty}^{\infty} f(t)\delta(t)dt = f(0) \int_{-\infty}^{\infty} \delta(t)dt = f(0). \tag{1.44}$$

In general we can write (1.44) as

$$\int_{-\infty}^{\infty} f(t)\delta(t - t_o) = f(t_o). \tag{1.45}$$

This important relationship, mentioned earlier in (1.28), is known as the filtering property of the impulse function.

Impulse Function or Dirac's Delta Function

By making $f(t) = \delta(t)$ in (1.35) it follows that

$$F(\omega) = \int_{-\infty}^{\infty} \delta(t)e^{-j\omega t}dt. \tag{1.46}$$

Using the impulse filtering property it follows that $F(\omega) = 1$. Thus, we conclude that the impulse function contains a continuum of equal amplitude spectral components.

Alternatively, by making $F(\omega) = 1$ in 1.34 and simplifying, the impulse function can be written as

$$\frac{1}{\pi}\int_{0}^{\infty} \cos \omega t d\omega.$$

The Constant Function

If $f(t)$ is a constant function then its Fourier transform in principle would not exist since this function does not satisfy the absolute integrability criterion. In general $F(\omega)$, the Fourier transform of $f(t)$, is expected to be finite, i.e.,

$$|F(\omega)| \leq \int_{-\infty}^{\infty} |f(t)||e^{-j\omega t}|dt < \infty, \tag{1.47}$$

since $|e^{-j\omega t}| = 1$, then

$$\int_{-\infty}^{\infty} |f(t)|dt < \infty. \tag{1.48}$$

However that is just a sufficiency condition and not a necessary condition for the existence of the Fourier transform, since there exist functions that although do not satisfy the condition of absolute integrability, in the limit have a Fourier transform (Carlson, 1975). This is a very important observation since this approach is often used in the computation of Fourier transforms of many functions. Returning to the constant function, we remark that it can be approximated by a gate function with amplitude A and width τ, and then making τ approach very large values,

$$\mathcal{F}[A] = \lim_{\tau\to\infty} A\tau \mathrm{Sa}\,(\frac{\omega\tau}{2}) \tag{1.49}$$

$$= 2\pi A \lim_{\tau\to\infty} \frac{\tau}{2\pi} \mathrm{Sa}\,(\frac{\omega\tau}{2})$$

$$\mathcal{F}[A] = 2\pi A\delta(\omega). \tag{1.50}$$

This result is not only a very interesting one but also somehow intuitive since a constant function in time represents a DC level and, as was to be expected, contains no spectral component except for the one at $\omega = 0$.

The Sine and the Cosine Fourier Transforms

Since both the sine and the cosine functions are periodic functions, they do not satisfy the condition of absolute integrability. However their respective Fourier transforms exist in the limit when τ goes to infinity. Assuming the function to exist only in the interval $(\frac{-\tau}{2}, \frac{\tau}{2})$ and to be zero outside this interval, and considering the limit of the expression when τ goes to infinity,

$$\mathcal{F}(\sin \omega_0 t) = \lim_{\tau \to \infty} \int_{\frac{-\tau}{2}}^{\frac{\tau}{2}} \sin \omega_0 t \; e^{-j\omega t} dt \tag{1.51}$$

$$= \lim_{\tau \to \infty} \int_{\frac{-\tau}{2}}^{\frac{\tau}{2}} \frac{e^{-j(\omega-\omega_0)t}}{2j} - \frac{e^{-j(\omega+\omega_0)t}}{2j} dt$$

$$= \lim_{\tau \to \infty} \left[\frac{j\tau \sin(\omega+\omega_0)\frac{\tau}{2}}{2(\omega+\omega_0)\frac{\tau}{2}} - \frac{j\tau \sin(\omega-\omega_0)\frac{\tau}{2}}{2(\omega-\omega_0)\frac{\tau}{2}} \right]$$

$$= \lim_{\tau \to \infty} \left\{ j\frac{\tau}{2}\mathrm{Sa}\left[\frac{(\omega+\omega_0)}{2}\right] - j\frac{\tau}{2}\mathrm{Sa}\left[\frac{\tau(\omega+\omega_0)}{2}\right] \right\}.$$

Therefore,

$$\mathcal{F}(\sin\omega_0 t) = j\pi[\delta(\omega+\omega_0) - \delta(\omega-\omega_0)].$$

Applying a similar reasoning it follows that

$$\mathcal{F}(\cos \omega_0 t) = \pi[\delta(\omega-\omega_0) + \delta(\omega+\omega_0)]. \tag{1.52}$$

The Fourier Transform of $e^{j\omega_0 t}$

Using Euler's identity, $e^{j\omega_0 t} = \cos \omega_0 t + j\sin\omega_0 t$, it follows that

$$\mathcal{F}[e^{j\omega_0 t}] = \mathcal{F}[\cos \omega_0 t + j\sin\omega_0 t]. \tag{1.53}$$

Substituting in (1.53) the Fourier transforms of the sine and of the cosine functions, respectively, it follows that

$$\mathcal{F}[e^{j\omega_0 t}] = 2\pi\delta(\omega-\omega_0). \tag{1.54}$$

The Fourier Transform of a Periodic Function

We consider next the exponential Fourier series representation of a periodic function $f_T(t)$ of period T

$$f_T(t) = \sum_{n=-\infty}^{\infty} F_n e^{jn\omega_0 t}. \tag{1.55}$$

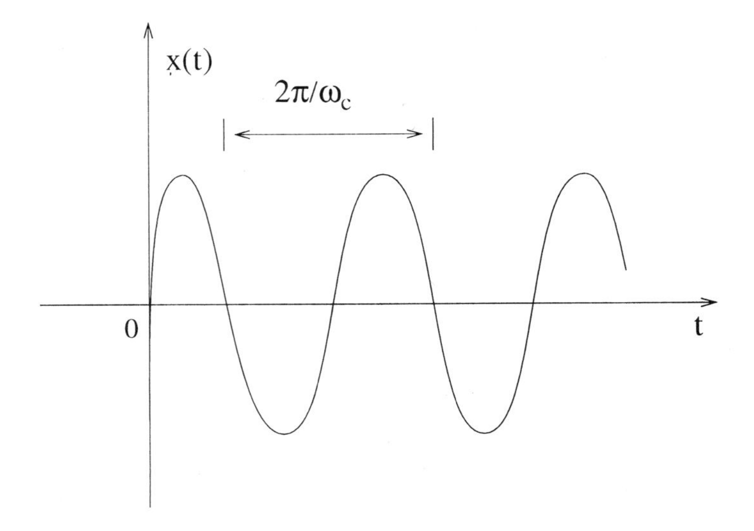

Figure 1.5. The sine function.

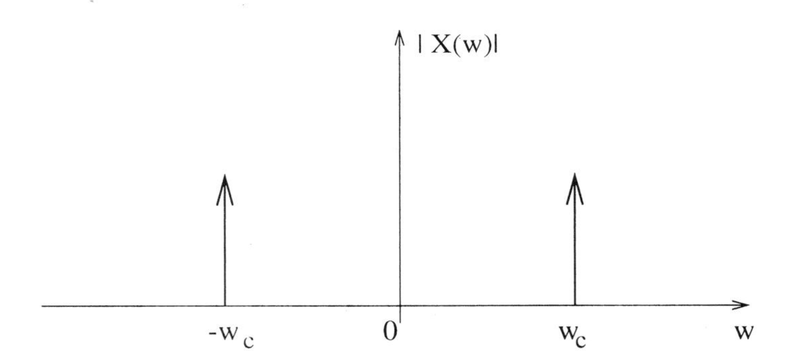

Figure 1.6. Magnitude plot of the Fourier transform of the sine function.

Applying the Fourier transform Fourier to both sides in (1.55) it follows that

$$\mathcal{F}[f_T(t)] = \mathcal{F}\left[\sum_{n=-\infty}^{\infty} F_n e^{jn\omega_0 t}\right] \tag{1.56}$$

$$= \sum_{n=-\infty}^{\infty} F_n \mathcal{F}[e^{jn\omega_0 t}]. \tag{1.57}$$

Now, applying in (1.57) the result from (1.54) it follows that

$$F(\omega) = 2\pi \sum_{n=-\infty}^{\infty} F_n \delta(\omega - n\omega_0). \tag{1.58}$$

1.4 Some Properties of the Fourier Transform

Linearity

Linearity is an important property when studying communication systems. A system is defined to be a linear system if satisfies the properties of homogeneity and additivity.

1 Homogeneity – If the application of the signal $x(t)$ at the system input produces $y(t)$ at the system output, then the application of the input $\alpha x(t)$, where α is a constant, produces $\alpha y(t)$ at the output;

2 Additivity – If the application of the signals $x_1(t)$ and $x_2(t)$ at the system input produces respectively $y_1(t)$ and $y_2(t)$ at the system output, then the application of the input $x_1(t) + x_2(t)$ produces $y_1(t) + y_2(t)$ at the output.

By applying the tests for homogeneity and additivity it is immediate to check that the process that generates the signal $s(t) = A\cos(\omega_c t + \Delta m(t) + \theta)$ from an input signal $m(t)$ is non linear. By applying the same test to the signal $r(t) = m(t)\cos(\omega_c t + \theta)$ it is immediate to show that the process generating $r(t)$ is linear.

The Fourier transform is a linear operator, i.e., if a function can be written as a linear combination of other (well behaved) functions, the corresponding Fourier transform will be given by a linear combination of the corresponding Fourier transforms of each one of the functions involved in the linear combination (Gagliardi, 1988).

If $f(t) \longleftrightarrow F(\omega)$ and $g(t) \longleftrightarrow G(\omega)$ it then follows that

$$\alpha f(t) + \beta g(t) \longleftrightarrow \alpha F(\omega) + \beta G(\omega). \tag{1.59}$$

Proof: Let $h(t) = \alpha f(t) + \beta g(t) \rightarrow$, then it follows that

$$\begin{aligned} H(\omega) &= \int_{-\infty}^{\infty} h(t) e^{-j\omega t} dt \\ &= \alpha \int_{-\infty}^{\infty} f(t) e^{-j\omega t} dt + \beta \int_{-\infty}^{\infty} g(t) e^{-j\omega t} dt, \end{aligned}$$

and finally

$$H(\omega) = \alpha F(\omega) + \beta G(\omega). \tag{1.60}$$

Scaling

$$\mathcal{F}[f(at)] = \int_{-\infty}^{\infty} f(at) e^{-j\omega t} dt. \tag{1.61}$$

Initially let us consider $a > 0$ in (1.61). By letting $u = at$ it follows that $dt = (1/a)du$. Replacing u for at in (1.61) it follows that

$$\mathcal{F}[f(at)] = \int_{-\infty}^{\infty} \frac{f(u)}{a} e^{-j\frac{\omega}{a}u} du$$

which simplifies to

$$\mathcal{F}[f(at)] = \frac{1}{a} F\left(\frac{\omega}{a}\right).$$

Let us consider now the case where $a < 0$. By a similar procedure it follows that

$$\mathcal{F}[f(at)] = -\frac{1}{a} F\left(\frac{\omega}{a}\right).$$

Therefore, finally

$$\mathcal{F}[f(at)] = \frac{1}{|a|} F\left(\frac{\omega}{a}\right). \tag{1.62}$$

This result points to the fact that if a signal is compressed in the time domain by a factor a, then its Fourier transform will expand in the frequency domain by that same factor.

Symmetry

This is an interesting property which can be fully observed in even functions. The symmetry property states that if

$$f(t) \longleftrightarrow F(\omega), \tag{1.63}$$

then it follows that

$$F(t) \longleftrightarrow 2\pi f(-\omega). \tag{1.64}$$

Proof: By definition,

$$f(t) = \frac{1}{2\pi} \int_{-\infty}^{+\infty} F(\omega) e^{j\omega t} d\omega,$$

which after multiplication of both sides by 2π becomes

$$2\pi f(t) = \int_{-\infty}^{+\infty} F(\omega) e^{j\omega t} d\omega.$$

By letting $u = -t$ it follows that

$$2\pi f(-u) = \int_{-\infty}^{+\infty} F(\omega) e^{-j\omega u} d\omega,$$

and now by making $t = \omega$, we obtain

$$2\pi f(-u) = \int_{-\infty}^{+\infty} F(t) e^{-jtu} dt.$$

Finally, by letting $u = \omega$ it follows that

$$2\pi f(-\omega) = \int_{-\infty}^{+\infty} F(t) e^{-j\omega t} dt. \tag{1.65}$$

Example: The Fourier transform of a constant function can be easily derived by use of the symmetry property. Since

$$A\delta(t) \longleftrightarrow A,$$

it follows that

$$A \longleftrightarrow 2\pi A\delta(-\omega) = 2\pi A\delta(\omega).$$

Time Domain Shift

Given that $f(t) \longleftrightarrow F(\omega)$, it then follows that $f(t-t_0) \longleftrightarrow F(\omega)e^{-j\omega t_0}$. Let $g(t) = f(t - t_0)$. In this case it follows that

$$G(\omega) = \mathcal{F}[g(t)] = \int_{-\infty}^{\infty} f(t - t_0) e^{-j\omega t} dt. \tag{1.66}$$

By making $\tau = t - t_0$ it follows that

$$G(\omega) = \int_{-\infty}^{\infty} f(\tau) e^{-j\omega(\tau + t_0)} d\tau \tag{1.67}$$

$$= \int_{-\infty}^{\infty} f(\tau) e^{-j\omega\tau} e^{-j\omega t_0} d\tau, \quad (1.68)$$

and finally

$$G(\omega) = e^{-j\omega t_0} F(\omega). \quad (1.69)$$

This result shows that whenever a function is shifted in time its frequency domain amplitude spectrum remains unaltered. However the corresponding phase spectrum experiences a rotation proportional to ωt_0.

Frequency Domain Shift

Given that $f(t) \longleftrightarrow F(\omega)$ it then follows that $f(t)e^{j\omega_0 t} \longleftrightarrow F(\omega - \omega_0)$.

$$\mathcal{F}[f(t)e^{j\omega_0 t}] = \int_{-\infty}^{\infty} f(t) e^{j\omega_0 t} e^{-j\omega t} dt \quad (1.70)$$

$$= \int_{-\infty}^{\infty} f(t) e^{-j(\omega-\omega_0)t} dt,$$

$$\mathcal{F}[f(t)e^{j\omega_0 t}] = F(\omega - \omega_0). \quad (1.71)$$

Differentiation in the Time Domain

Given that

$$f(t) \longleftrightarrow F(\omega), \quad (1.72)$$

it then follows that

$$\frac{df(t)}{dt} \longleftrightarrow j\omega F(\omega). \quad (1.73)$$

Proof: Let us consider the expression for the inverse Fourier transform

$$f(t) = \frac{1}{2\pi} \int_{-\infty}^{\infty} F(\omega) e^{j\omega t} d\omega. \quad (1.74)$$

Differentiating in time it follows that

$$\begin{aligned}
\frac{df(t)}{dt} &= \frac{1}{2\pi} \frac{d}{dt} \int_{-\infty}^{\infty} F(\omega) e^{j\omega t} d\omega \\
&= \frac{1}{2\pi} \int_{-\infty}^{\infty} \frac{d}{dt} F(\omega) e^{j\omega t} d\omega \\
&= \frac{1}{2\pi} \int_{-\infty}^{\infty} j\omega F(\omega) e^{j\omega t} d\omega,
\end{aligned}$$

and then

$$\frac{df(t)}{dt} \longleftrightarrow j\omega F(\omega). \tag{1.75}$$

In general it follows that

$$\frac{d^n f(t)}{dt} \longleftrightarrow (j\omega)^n f(\omega). \tag{1.76}$$

By computing the Fourier transform of the signal $f(t) = \delta(t) - \alpha e^{-\alpha t}u(t)$, it is immediate to show that, by applying the property of differentiation in time, this signal is the time derivative of the signal $g(t) = e^{-\alpha t}u(t)$.

Integration in the Time Domain

Let $f(t)$ be a signal with zero average value, i.e., let $\int_{-\infty}^{\infty} f(t)dt = 0$. By defining

$$g(t) = \int_{-\infty}^{t} f(\tau)d\tau, \tag{1.77}$$

it follows that

$$\frac{dg(t)}{dt} = f(t),$$

and since

$$g(t) \longleftrightarrow G(\omega), \tag{1.78}$$

then

$$f(t) \longleftrightarrow j\omega G(\omega),$$

and

$$G(\omega) = \frac{F(\omega)}{j\omega}. \tag{1.79}$$

In this manner it follows that for a signal with zero average value

$$f(t) \longleftrightarrow F(\omega)$$

$$\int_{-\infty}^{t} f(\tau)d\tau \longleftrightarrow \frac{F(\omega)}{j\omega}. \tag{1.80}$$

Generalizing, for the case where $f(t)$ has a non-zero average value, it follows that

$$\int_{-\infty}^{t} f(\tau)d\tau \longleftrightarrow \frac{F(\omega)}{j\omega} + \pi\delta(\omega)F(0). \tag{1.81}$$

The Convolution Theorem

The convolution theorem is a powerful tool for analyzing the frequency contents of a signal, allowing obtention of many relevant results. One instance of the use of the convolution theorem, of fundamental importance in communication theory, is the sampling theorem which will be the subject of the next section.

The convolution between two time functions $f(t)$ and $g(t)$ is defined by the following integral

$$\int_{-\infty}^{\infty} f(\tau)g(t-\tau)d\tau, \tag{1.82}$$

which is often denoted as $f(t) * g(t)$.

Let $h(t) = f(t) * g(t)$ and let $h(t) \longleftrightarrow H(\omega)$. It follows that

$$H(\omega) = \int_{-\infty}^{\infty} h(t)e^{-j\omega t}dt = \int_{-\infty}^{\infty}\int_{-\infty}^{\infty} f(\tau)g(t-\tau)e^{-j\omega t}dtd\tau. \tag{1.83}$$

$$H(\omega) = \int_{-\infty}^{\infty} f(\tau)\int_{-\infty}^{\infty} g(t-\tau)e^{-j\omega t}dtd\tau, \tag{1.84}$$

$$H(\omega) = \int_{-\infty}^{\infty} f(\tau)G(\omega)e^{-j\omega\tau}d\tau \tag{1.85}$$

and finally,

$$H(\omega) = F(\omega)G(\omega). \tag{1.86}$$

The convolution of two time functions is equivalent in the frequency domain to the product of their respective Fourier transforms. For the case where $h(t) = f(t)\cdot g(t)$, proceeding in a similar manner we obtain

$$H(\omega) = \frac{1}{2\pi}[F(\omega) * G(\omega)]. \tag{1.87}$$

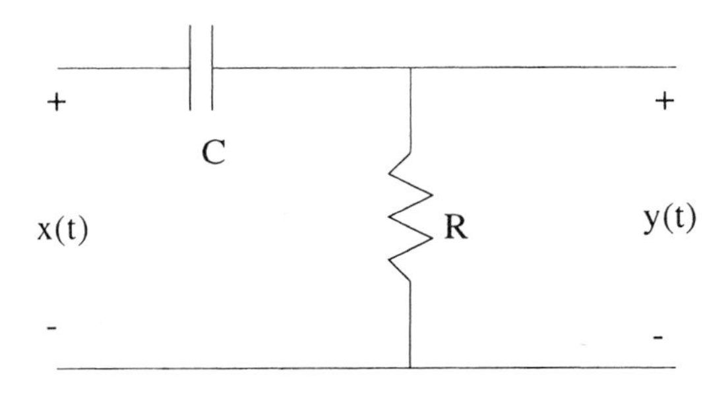

Figure 1.7. RC circuit.

In other words, the product of two time functions has a Fourier transform given by the convolution of their respective Fourier transforms. The convolution operation is often used when computing the response of a linear circuit, given its impulse response and an input signal.

Example: The circuit in Figure 1.7 has the impulse response $h(t)$ given by

$$h(t) = \frac{1}{RC} e^{-\frac{t}{RC}} u(t).$$

The application of the unit impulse $x(t) = \delta(t)$ as the input to this circuit causes an output $y(t) = h(t) * x(t)$. In the frequency domain, by the convolution theorem it follows that $Y(\omega) = H(\omega)X(\omega) = H(\omega)$, i.e., the Fourier transform of the impulse response of a linear system is the system transfer function.

Using the frequency convolution theorem it can be shown that

$$\cos(\omega_c t)u(t) \longleftrightarrow \frac{\pi}{2}[\delta(\omega + \omega_c) + \delta(\omega - \omega_c)] + j\frac{\omega}{\omega_c^2 - \omega^2}.$$

1.5 The Sampling Theorem

A bandlimited signal $f(t)$, having no frequency components above $\omega_M = 2\pi f_M$, can be reconstructed from its samples, collected at uniform time intervals $T_s = 1/f_s$, i.e., at a sampling rate f_s, where $f_s \geq 2f_M$.

By a bandlimited signal $f(t) \longleftrightarrow F(\omega)$ it is understood that there is a frequency ω_M above which $F(\omega) = 0$, i.e., that $F(\omega) = 0$ for $|\omega| > \omega_M$. Nyquist concluded that all the information about $f(t)$, as illustrated in Figure 1.8, is contained in the samples of this signal, collected at regular time intervals T_s. In this manner the signal can be completely recovered from its samples. For a bandlimited signal $f(t)$, i.e., such that $F(\omega) = 0$

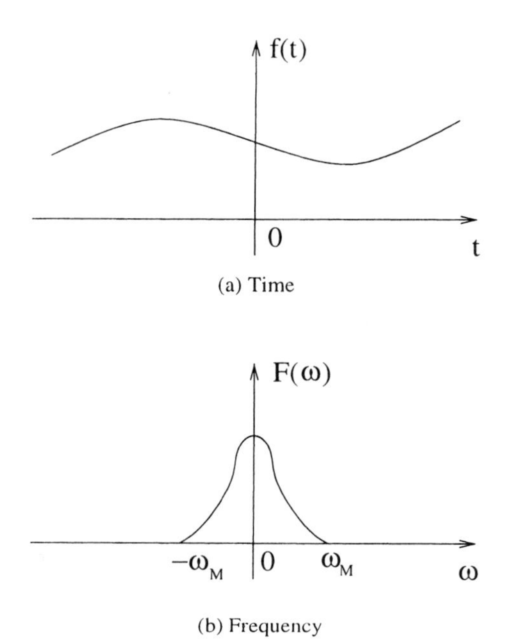

Figure 1.8. Bandlimited signal $f(t)$ and its spectrum.

for $|\omega| > \omega_M$, it follows that

$$f(t) * \frac{\sin(at)}{\pi t} = f(t), \text{ if } a > \omega_M,$$

because in the frequency domain this corresponds to the product of $F(\omega)$ by a gate function of width greater than $2\omega_M$.

The function $f(t)$ is sampled once every T_s seconds or, equivalently, sampled with a sampling frequency f_s, where $f_s = 1/T_s \geq 2f_M$.

Consider the signal $f_s(t) = f(t)\delta_T(t)$, where

$$\delta_T(t) = \sum_{n=-\infty}^{\infty} \delta(t - nT) \longleftrightarrow \omega_o \delta_{\omega_o} = \omega_o \sum_{n=-\infty}^{\infty} \delta(\omega - n\omega_o). \quad (1.88)$$

The signal $\delta_T(t)$ is illustrated in Figure 1.9. The signal $f_s(t)$ represents $f(t)$ sampled at uniform time intervals T_s seconds. From the frequency convolution theorem it follows that the Fourier transform of the product of two functions in the time domain is given by the convolution of their respective Fourier transforms. It now follows that

$$f_s(t) \longleftrightarrow \frac{1}{2\pi}[F(\omega) * \omega_0 \delta_{\omega_0}(\omega)] \quad (1.89)$$

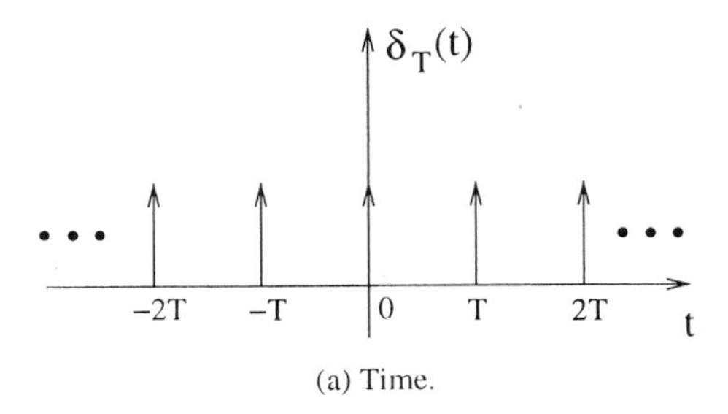

(a) Time.

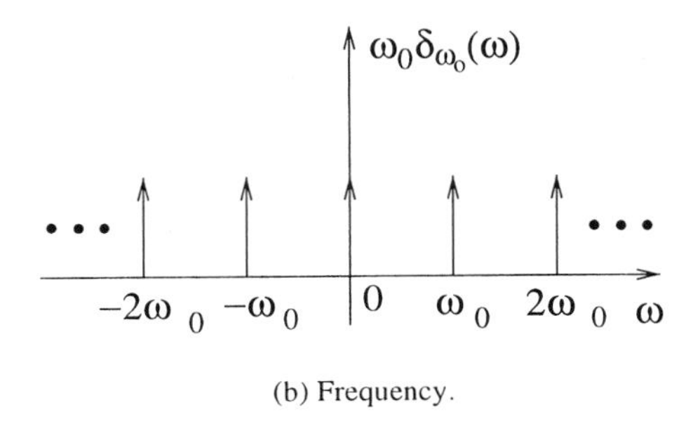

(b) Frequency.

Figure 1.9. Impulse train used for sampling.

and thus

$$f_s(t) \longleftrightarrow \frac{1}{T}[F(\omega) * \delta_{\omega_0}(\omega)] = \frac{1}{T} \sum_{n=-\infty}^{\infty} F(\omega - n\omega_o). \qquad (1.90)$$

It can be observed from Figure 1.10 that if the sampling frequency ω_s is less than $2\omega_M$, there will be an overlap of spectral components. This will cause a loss of information because the original signal can no longer be fully recovered from its samples. As ω_s becomes smaller than $2\omega_M$, the sampling rate diminishes causing a partial loss of information. Therefore the minimum sampling frequency that allows perfect recovery of the signal is $\omega_s = 2\omega_M$, and is known as the Nyquist sampling rate. In order to recover the original spectrum $F(\omega)$, it is enough to pass the sampled signal through a low-pass filter with cut-off frequency ω_M.

For applications in telephony, the sampling frequency is $f_S = 8,000$ samples per second, or 8 k samples/s. Then the speech signal is quantized, as will be discussed later, for 256 distinct levels. Each level corresponds to an 8-bit code ($2^8 = 256$).

After encoding, the signal is transmitted at a rate of 8,000 samples/s $\times$ 8 bits/sample = 64 kbits/s and occupies a bandwidth of approximately 64 kHz.

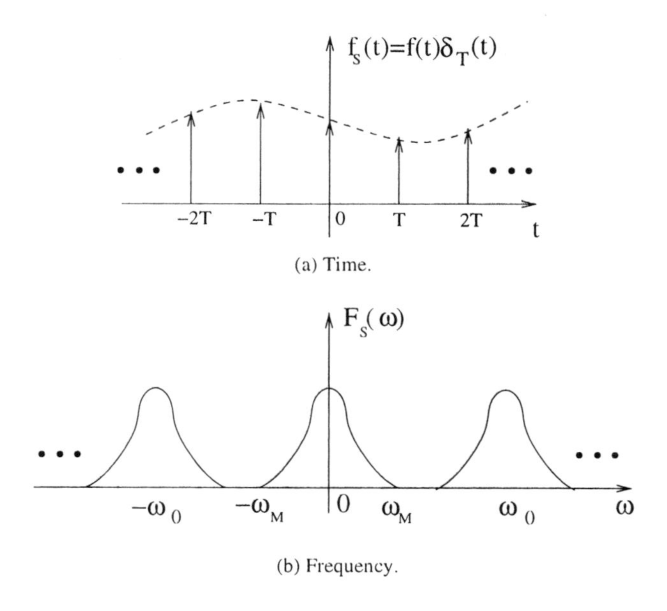

Figure 1.10. Sampled signal and its spectrum.

If the sampling frequency w_S is lower than $2\pi B$, there will be spectra overlap and, as a consequence, information loss. As long as w_S becomes lower than $2\pi B$, the sampling rate becomes lower, leading to partial loss of information. Therefore, the sampling frequency for a baseband signal to be recovered without loss is $w_S = 2\pi B$, known as the Nyquist sampling frequency.

As mentioned, if the sampling frequency is lower than the Nyquist frequency, the signal will not be completely recovered, since there will be spectral superposition, leading to distortion in the highest frequencies. This phenomenon is known as *aliasing*. On the other hand, increasing the sampling frequency for a value higher than the Nyquist frequency leads to spectra separation higher than the minimum necessary to recover the signal.

1.6 Parseval's Theorem

For a real signal $f(t)$ of finite energy, often called simply a real energy signal, the energy E associated with $f(t)$ is given by

$$E = \int_{-\infty}^{\infty} f^2(t)dt$$

and can equivalently be calculated by the formula

$$E = \frac{1}{2\pi} \int_{-\infty}^{\infty} |F(\omega)|^2 d\omega$$

It follows that

$$\int_{-\infty}^{\infty} f^2(t)dt = \frac{1}{2\pi} \int_{-\infty}^{\infty} |F(\omega)|^2 d\omega. \tag{1.91}$$

The relationship given in (1.91) is known as Parseval's theorem or Parseval's identity. For a real signal $x(t)$ with energy E it can be shown, by using Parseval's identity, that the signals $x(t)$ and $y(t) = x(t - \tau)$ have the same energy E.

Another way of expressing Parseval's identity is as follows.

$$\int_{-\infty}^{\infty} f(x)G(x)dx = \int_{-\infty}^{\infty} F(x)g(x)dx. \tag{1.92}$$

1.7 Average, Power and Autocorrelation

As mentioned earlier, the average value of a real signal $x(t)$ is given by

$$\overline{x}(t) = \lim_{T\to\infty} \frac{1}{T} \int_{\frac{-T}{2}}^{\frac{T}{2}} x(t)dt. \tag{1.93}$$

The instantaneous power of $x(t)$ is given by

$$p_X(t) = x^2(t). \tag{1.94}$$

If the signal $x(t)$ exists for the whole interval $(-\infty, +\infty)$, the total power $\overline{P}_X$ is defined for a real signal $x(t)$ as the power dissipated in a 1 ohm resistor, when a voltage $x(t)$ is applied to this resistor (or a current $x(t)$ flows through the resistor) (Lathi, 1989). Thus,

$$\overline{P}_X = \lim_{T\to\infty} \frac{1}{T} \int_{\frac{-T}{2}}^{\frac{T}{2}} x^2(t)dt. \tag{1.95}$$

From the previous definition the unit to measure $\overline{P}_X$ corresponds to the square of the units of the signal $x(t)$ ($volt^2$, amp^2). These units will only be converted to *watts* if they are normalized by units of impedance (ohm). It is common use to express the power in *decibell* (dB). The power in *decibell* is given by the espression (Gagliardi, 1988)

$$\overline{P}_{X,dB} = 10 \log \overline{P}_X. \tag{1.96}$$

The total power ($\overline{P}_X$) contains two components: one DC component, due to a nonzero average value of the signal $x(t)$ ($\overline{P}_{DC}$), and an AC component ($\overline{P}_{AC}$). The DC power of the signal is given by

$$\overline{P}_{DC} = (\overline{x}(t))^2. \tag{1.97}$$

It follows that the AC power can be determined by removing the DC power from the total power, i.e.,

$$\overline{P}_{AC} = \overline{P}_X - \overline{P}_{DC}. \tag{1.98}$$

Time Autocorrelation of Signals

The average time autocorrelation $\overline{R}_X(\tau)$, or simply autocorrelation, of a real signal $x(t)$ is defined as follows

$$\overline{R}_X(\tau) = \lim_{T \to \infty} \frac{1}{T} \int_{\frac{-T}{2}}^{\frac{T}{2}} x(t)x(t+\tau)dt. \tag{1.99}$$

The change of variable $y = t + \tau$ allows equation (1.99) to be written as

$$\overline{R}_X(\tau) = \lim_{T \to \infty} \frac{1}{T} \int_{\frac{-T}{2}}^{\frac{T}{2}} x(t)x(t-\tau)dt. \tag{1.100}$$

From equations (1.99) and (1.100), it follows that $\overline{R}_X(\tau)$ is an even function of τ, and thus (Lathi, 1989)

$$\overline{R}_X(-\tau) = \overline{R}_X(\tau). \tag{1.101}$$

From the definition of autocorrelation and power it follows that

$$\overline{P}_X = \overline{R}_X(0) \tag{1.102}$$

and

$$\overline{P}_{DC} = \overline{R}_X(\infty), \tag{1.103}$$

i.e., from its autocorrelation function it is possible to obtain information about the power of a signal. The autocorrelation function can also be considered in the frequency domain by taking its Fourier transform, i.e.,

$$\mathcal{F}\{\overline{R}_X(\tau)\} = \int_{-\infty}^{+\infty} \lim_{T\to\infty} \frac{1}{T} \int_{\frac{-T}{2}}^{\frac{T}{2}} x(t)x(t+\tau)e^{-j\omega\tau} dt\, d\tau = \tag{1.104}$$

$$= \lim_{T\to\infty} \frac{1}{T} \int_{\frac{-T}{2}}^{\frac{T}{2}} x(t) \int_{-\infty}^{+\infty} x(t+\tau) d\tau dt$$

$$= \lim_{T\to\infty} \frac{1}{T} \int_{\frac{-T}{2}}^{\frac{T}{2}} x(t)X(\omega)e^{j\omega t} dt$$

$$= X(\omega) \lim_{T\to\infty} \frac{1}{T} \int_{\frac{-T}{2}}^{\frac{T}{2}} x(t)e^{j\omega t} dt$$

$$= \lim_{T\to\infty} \frac{X(\omega)X(-\omega)}{T}$$

$$= \lim_{T\to\infty} \frac{|X(\omega)|^2}{T} \tag{1.105}$$

The *power spectral density* $\overline{S}_X$ of a signal $x(t)$ is defined as the Fourier transform of the autocorrelation function $\overline{R}_X(\tau)$ of $x(t)$, i.e., as

$$\overline{S}_X = \int_{-\infty}^{\infty} \overline{R}_X(\tau)e^{-j\omega\tau} d\tau. \tag{1.106}$$

Example: Find the power spectral density of the sinusoidal signal $x(t) = A\cos(\omega_0 t + \theta)$ illustrated in Figure 1.11a.

Solution:

$$\overline{R}_X(\tau) = \lim_{T\to\infty} \frac{1}{T} \int_{\frac{-T}{2}}^{\frac{T}{2}} A^2 \cos(\omega_0 t + \theta) \cos\left[\omega_0(t+\tau) + \theta\right] dt$$

$$= \frac{A^2}{2} \lim_{T\to\infty} \frac{1}{T} \left[\int_{\frac{-T}{2}}^{\frac{T}{2}} \cos\omega_0\tau dt + \int_{\frac{-T}{2}}^{\frac{T}{2}} \cos\left(2\omega_0 t + \omega_0\tau + 2\theta\right) dt \right]$$

$$= \frac{A^2}{2} \cos\omega_0\tau.$$

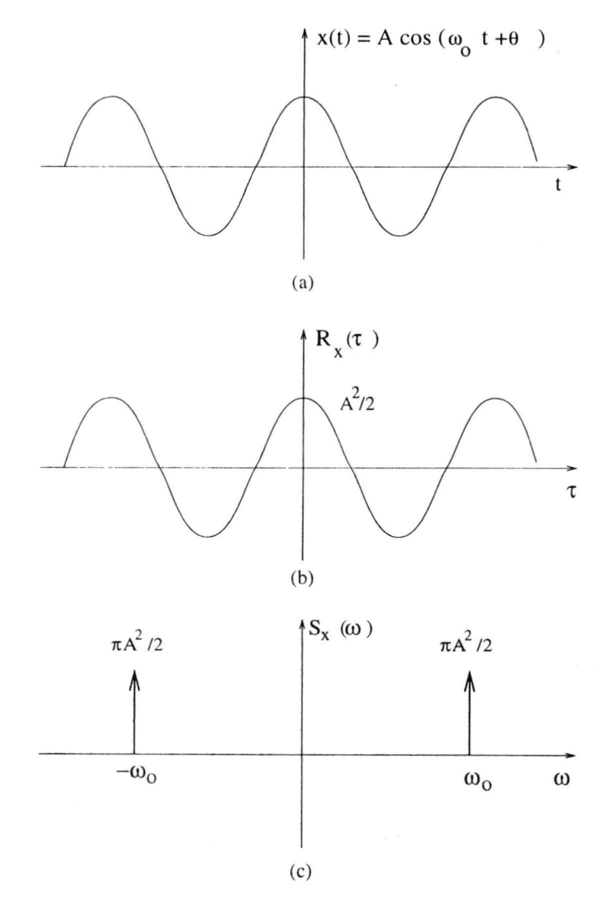

Figure 1.11. Sinusoidal signal and its autocorrelation and power spectral density.

Notice that the autocorrelation function (Figure 1.11b) is independent of the phase θ. The power spectral density (Figure 1.11c) is given by

$$\overline{S}_X(\omega) = \mathcal{F}\left[R_X(\tau)\right]$$

$$\overline{S}_X(\omega) = \frac{\pi A^2}{2}\left[\delta(\omega + \omega_0) + \delta(\omega - \omega_0)\right].$$

The power or mean square average of $x(t)$ is given by

$$\overline{P}_X = \overline{R}_X(0) = \frac{A^2}{2}.$$

1.8 Problems

1) Consider the signal

$$x(t) = \begin{cases} 1, & 0 \leq t < \pi \\ -1, & \pi \leq t \leq 2\pi \end{cases}$$

which is approximated as $\tilde{x}(t) = \frac{4}{\pi}\sin(t)$, in the time interval considered.

   (a) Show that the error in the approximation is orthogonal to the function $\tilde{x}(t)$;

   (b) Show that the energy of $x(t)$ is the sum of the energy in the error signal with the energy of the signal $\tilde{x}(t)$.

2) Calculate the instantaneous power and the average power of the following signals:

   (a) $x(t) = \cos(2\pi t)$

   (b) $y(t) = \sin(2\pi t)$

   (c) $z(t) = x(t) + y(t)$

3) Determine the constant A such that $f_1(t)$ and $f_2(t)$ are orthogonal for all t, where: $f_1(t) = e^{-|t|}$ and $f_2(t) = 1 - Ae^{-2|t|}$.

4) Given the set of functions $f_n(t)$, as illustrated in Figure 1.2, show that:

   (a) This set of functions constitutes an orthogonal set in the interval $(0, 1)$. Is the set orthonormal?

   (b) Represent a given signal $f(t) = 2t$ in the interval $(0, 1)$, using this set of orthogonal functions.

   (c) Plot the function $f(t)$ and its approximate representation $\tilde{f}(t)$ in the same graph.

   (d) Determine the energy of the error signal resulting from the approximation.

5) Represent the functions shown in Figure 1.12, using the unit step.

6) Represent analytically the graph of Figure 1.13, by using generalized functions.

7) Calculate the following integrals

   (a) $\int_{-\infty}^{\infty} e^{-\alpha t} u(t) dt$,

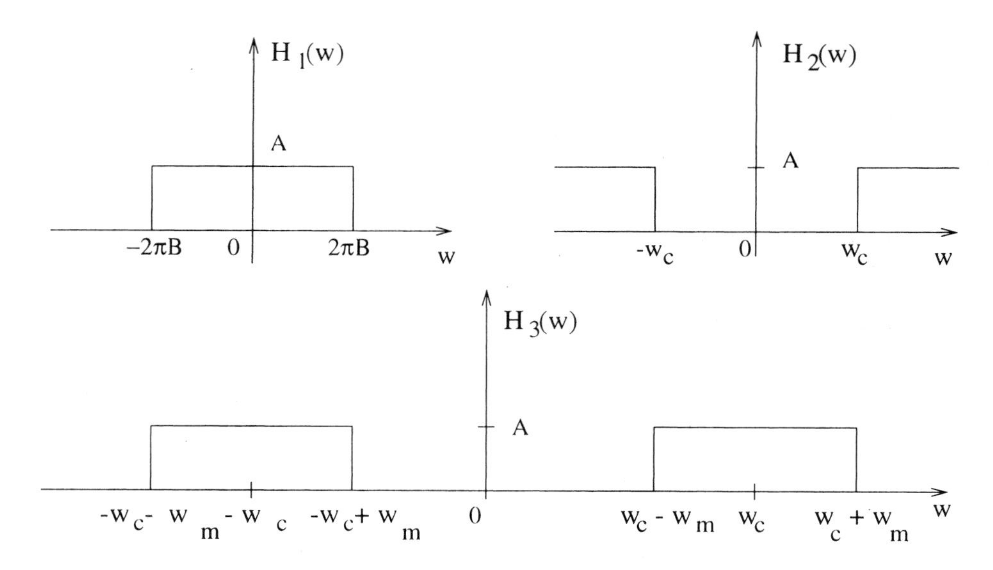

Figure 1.12. Functions to be represented by means of the unit step.

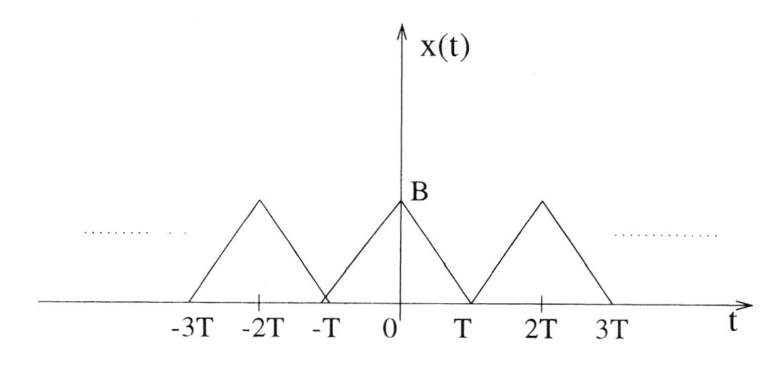

Figure 1.13. Function to be represented with a unit step, a ramp or an impulse.

(b) $\int_{-\infty}^{\infty} e^{-\alpha t}\delta(t)dt$,

(c) $\int_{-\infty}^{\infty} e^{-\alpha t}r(t)dt$.

8) Calculate the Fourier series for the function $f(t) = \sin^3(\omega t)$.

9) Calculate the Fourier transform of the impulse function assuming the Fourier transform of the unit step function is known.

10) Calculate the inverse Fourier transform of the function

$$F(\omega) = A[u(\omega + \omega_0) - u(\omega - \omega_0)].$$

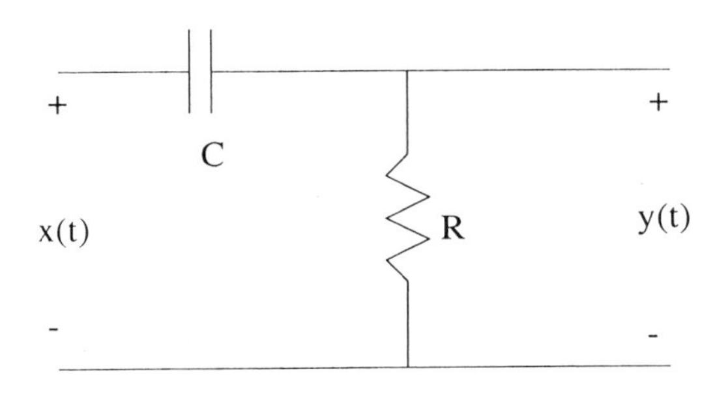

Figure 1.14. RC circuit.

11) Calculate the Fourier transform of the function $f(t) = Ae^{-\alpha t}u(t)$, and plot the corresponding magnitude and phase diagrams.

12) Plot the magnitude and phase diagrams of the Fourier transform of the function $\delta(t + t_0)$.

13) For the circuit of Figure 1.14, with impulse response $h(t)$, find the response for the excitation $x(t)$ given by

$$x(t) = te^{-\frac{t}{RC}}u(t),$$

where

$$h(t) = \frac{1}{RC}e^{-\frac{t}{RC}}u(t)$$

14) Find the Fourier transform of the function $g(t) = f(t)\cos(\omega_c t)$, given the Fourier transform of $f(t)$.

15) Show that for a function $f(t)$ in general:

$$\int_{-\infty}^{t} f(t)dt \longleftrightarrow \frac{F(\omega)}{j\omega} + \pi F(0)\delta(\omega).$$

16) Prove that, for a real energy signal $f(t)$, the energy associated to $f(t)$,

$$\int_{-\infty}^{\infty} f^2(t)dt$$

can be calculated by the formula

$$\frac{1}{2\pi}\int_{-\infty}^{\infty} |F(\omega)|^2 d\omega.$$

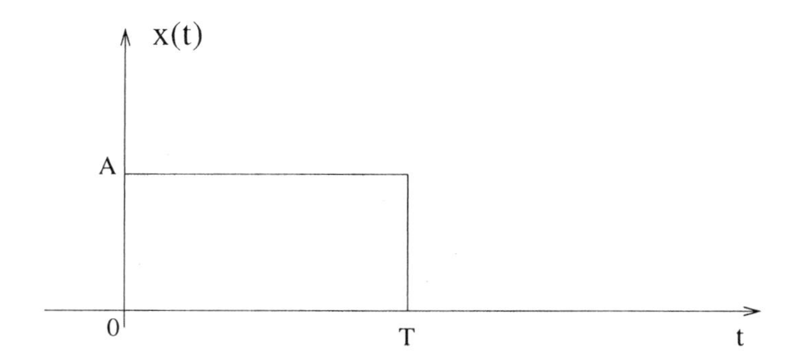

Figure 1.15. Shifted gate function.

17) Use the property of the convolution in the frequency domain to show that

$$\cos(\omega_c t)u(t) \longleftrightarrow \frac{\pi}{2}[\delta(\omega + \omega_c) + \delta(\omega - \omega_c)] + j\frac{\omega}{\omega_c^2 - \omega^2}.$$

18) A signal $x(t)$ has the exponential Fourier series expansion as given below. Find its corresponding trigonometric Fourier series expansion.

$$x(t) = -\frac{2A}{\pi} \sum_{n=-\infty}^{\infty} \frac{1}{4n^2 - 1} e^{j2\pi nt}.$$

19) By defining the cutoff frequency as the smallest frequency for which the first spectral zero occurs, determine the cutoff frequency (ω_0) of the signal $x(t)$ in Figure 1.15.

20) Calculate the Fourier transform of the signals represented in Figures 1.16 and 1.17.

21) Calculate the frequency response of a linear system the transfer function of which is given below, when the input is the pulse $x(t) = A[u(t + T/2) - u(t - T/2)]$. Plot the corresponding magnitude and phase diagrams of the frequency response.

$$H(\omega) = ju(-\omega) - ju(\omega).$$

22) A signal $x(t)$ is given by the expression

$$x(t) = \frac{\sin(At)}{\pi t}.$$

Determine the Nyquist frequency for sampling this signal.

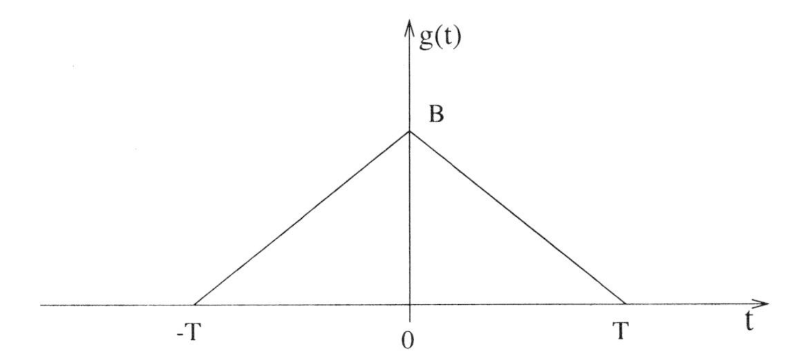

Figure 1.16. Triangular waveform.

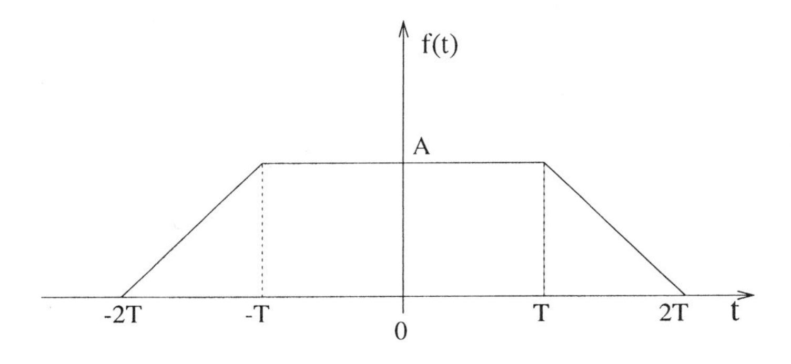

Figure 1.17. Trapezoidal waveform.

23) What is the least sampling rate required to sample the signal $f(t) = \sin^3(\omega_0 t)$? Show graphically the effect caused by a reduction of the sampling rate, falling below the Nyquist rate.

24) Calculate the Fourier transform of the signal

$$g(t) = Ae^{-t}u(t)$$

and then apply the property of integration in the time domain to obtain the Fourier transform of $f(t) = A(1 - e^{-t})u(t)$.

25) Determine the magnitude and phase spectra of the signal $f(t) = te^{-at}u(t)$.

26) A voltage signal $v_0(t) = V_0 + \sum_{n=1}^{\infty} V_n \cos(n\omega_0 t + \theta_n)$ is applied to the input of a circuit, producing the current $i_0(t) = I_0 + \sum_{m=1}^{\infty} I_m \cos(m\omega_0 t + \phi_m)$. By using the orthogonality concept calculate the power (P) ab-

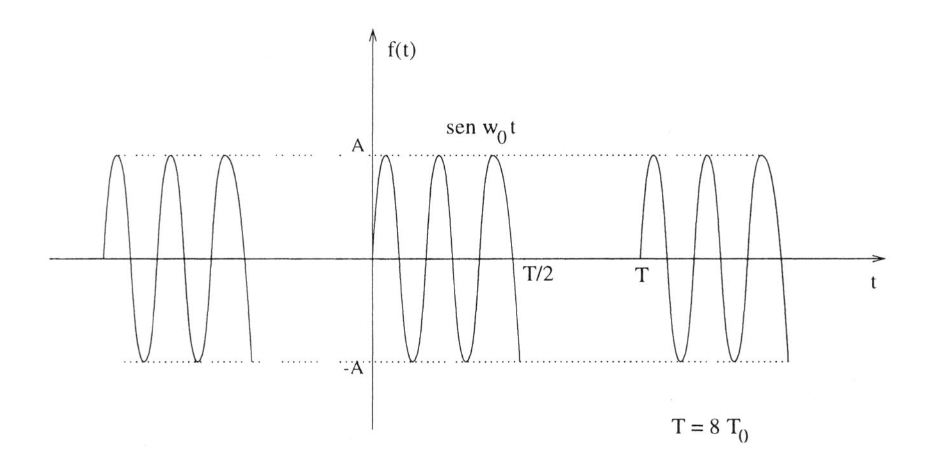

Figure 1.18. ASK signal.

sorbed by the circuit, considering

$$P = \frac{1}{T}\int_{-T/2}^{T/2} v(t)i(t)dt.$$

What is the power (P) for the case where $\theta_n = \phi_n$?

27) Calculate the Fourier transform of the signal $f(t)$ given in Figure 1.18 (ASK signal).

28) Define an ideal low-pass filter and explain why it is not physically realizable. Indicate the corresponding filter transfer function and the filter impulse response.

29) Given the linear system shown in Figure 1.19, where T represents a constant delay, determine:

    (a) The system transfer function $H(\omega)$,

    (b) The system impulse response $h(t)$.

30) Given the spectrum of the signal in Figure 1.20, determine:

    (a) The Nyquist sampling frequency;

    (b) The spectrum of the sampled signal for the following sampling frequencies: $\omega_A = \omega$, $\omega_A = 2\omega$ e $\omega_A = 6\omega$;

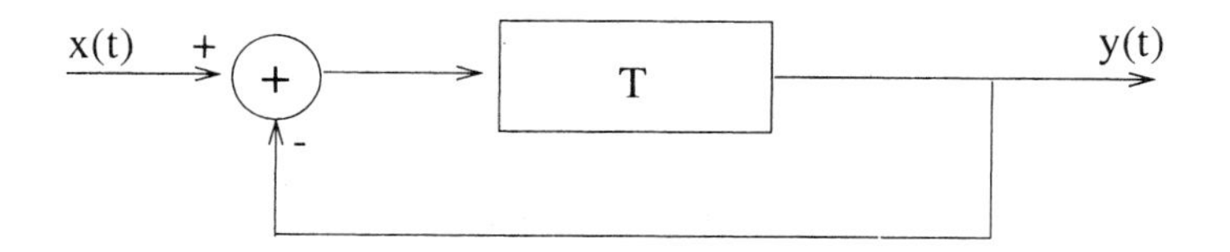

Figure 1.19. Linear system with feedback.

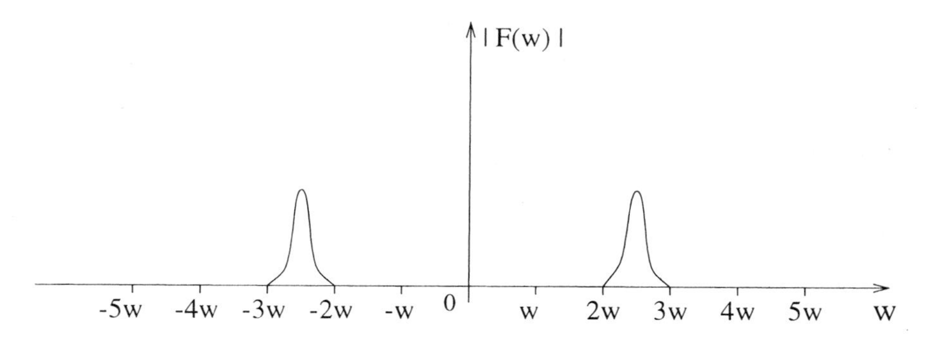

Figure 1.20. Signal spectrum.

(c) Whether it is possible to reconstruct the spectrum using a low-pass filter.

31) Let $f(t)$ be the signal with spectrum $F(\omega)$ as follows

$$F(\omega) = \frac{AT}{1 + jwT}, \quad \text{where : } T = 0.5366\mu s, \;\; A = 4V.$$

(a) Calculate and plot $|F(\omega)|$.

(b) Calculate the frequency for which $|F(\omega)|$ corresponds to a value 3dB below the maximum amplitude value in the spectrum.

(c) Calculate the energy of the signal $f(t)$.

32) Represent the following signals using the unit step

a) $r(t + T)$;

b) $\delta(t - T) + \delta(t + T)$;

c) a periodic sawtooth waveform with period T and peak amplitude A.

Plot the corresponding graphs.

33) Calculate the Fourier transform of each one of the following signals

a) $x(t) = e^{-t+t_o}u(t - t_o)$;

b) $y(t) = tu(t)$;

c) $z(t) = y'(t)$.

Draw the time signals as well as the associated magnitude spectra.

34) Verify, by applying the properties of homogeneity and additivity, whether the process generating the signal $s(t) = A\cos(\omega_c t + \Delta m(t) + \theta)$ from the input signal $m(t)$ is linear. Perform the same test for the signal $r(t) = m(t)\cos(\omega_c t + \theta)$.

35) A linear system has impulse response $h(t) = 2[u(t) - u(t-T)]$. Using convolution, determine the system response to the input signal $x(t) = u(t) - u(t - T)$.

36) A digital signal $x(t)$ has autocorrelation function

$$\bar{R}_X(\tau) = A^2[1 - \frac{|\tau|}{T_b}][u(\tau + T_b) - u(\tau - T_b)],$$

where T_b is the bit duration. Determine the total power, the AC power and the DC power of the given signal. Calculate the signal power spectral density. Plot the diagrams representing these functions.

37) Calculate the Fourier transform of a periodic signal $x(t)$, represented analytically as

$$x(t) = \sum_{-\infty}^{\infty} F_n e^{jn\omega_0 t}.$$

38) Prove the following property of the Fourier transform

$$x(\alpha t) \longleftrightarrow \frac{1}{|\alpha|} X\left(\frac{\omega}{\alpha}\right).$$

39) Calculate the average value and the power of the signal

$$x(t) = Vu(\cos t).$$

40) For a given real signal $x(t)$, prove the following Parseval identity

$$E = \int_{-\infty}^{\infty} x^2(t)dt = \frac{1}{2\pi}\int_{-\infty}^{\infty} |X(\omega)|^2 d\omega.$$

Consider the signals $x(t)$ and $y(t) = x(t - \tau)$ and show, by using Parseval's identity, that both signals have the same energy E.

41) A signal $y(t)$ is given by the following expression

$$y(t) = \frac{1}{\pi} \int_{-\infty}^{\infty} \frac{x(\tau)}{t - \tau} d\tau,$$

where the signal $x(t)$ has a Fourier transform $X(\omega)$. By using properties of the Fourier transform determine the Fourier transform of $y(t)$.

42) Given that the Fourier transform of the signal $f(t) = \cos(\omega_o t)$ is $F(\omega) = \pi[\delta(w + \omega_o) + \delta(w - \omega_o)]$, determine the Fourier transform of the signal $g(t) = \sin(\omega_o t - \phi)$, where ϕ is a phase constant. Sketch the magnitude and phase graphs of the Fourier transform of this signal.

43) Calculate the Fourier transform of the radio frequency pulse

$$f(t) = \cos(\omega_o t)[u(t + T) - u(t - T)],$$

considering that $\omega_o \gg \frac{2\pi}{T}$. Sketch the magnitude and phase graphs of the Fourier transform of this signal.

44) Calculate the Fourier transform of the signal $f(t) = \delta(t) - \alpha e^{-\alpha t} u(t)$ and show, by using the property of the derivative in the time domain, that this signal is the derivative of the signal $g(t) = e^{-\alpha t} u(t)$. Sketch the respective time and frequency domain graphs of the signals, specifying the magnitude and the phase spectra of each signal.

45) By making use of properties of the Fourier transform show that the derivative of the signal $h(t) = f(t) * g(t)$ can be expressed as

$$h'(t) = f'(t) * g(t), \quad \text{or} \quad h'(t) = f(t) * g'(t).$$

46) Determine the Nyquist frequency for which the following signal can be recovered without distortion. Sketch the signal spectrum and give a graphical description of the procedure.

$$f(t) = \frac{\sin \alpha t \cdot \sin \beta t}{t^2}, \quad \alpha > \beta.$$

47) Using the Fourier transform show that the unit impulse function can be written as

$$\frac{1}{\pi} \int_{0}^{\infty} \cos(\omega t) d\omega.$$

48) Using properties of the Fourier transform determine the Fourier transform of the function $|t|$. Plot the corresponding magnitude and phase spectrum of that transform.

49) Show that $u(t) * u(t) = r(t)$, where $u(t)$ represents the unit step function and $r(t)$ denotes the ramp with slope 1.

50) Determine the Fourier transform for each one of the following functions: $u(t - T)$, t, $te^{-at}u(t)$, $\frac{1}{t}$, $\frac{1}{t^2}$. Plot the corresponding time domain diagrams and the respective magnitude and phase spectrum of the associated transforms.

51) Calculate the Fourier transform $P(\omega)$ of the signal $p(t) = v^2(t)$ representing the instantaneous power in a 1 Ω resistor, as a function of the Fourier transform $V(\omega)$ of $v(t)$. Using the expression obtained for $P(\omega)$ plot the instantaneous power spectrum for a sinusoidal input signal $v(t) = A\cos(\omega_o t)$.

52) Find the complex Fourier series for the signal

$$f(t) = \cos(\omega_o t) + \sin^2(\omega_o t).$$

53) Find the Fourier transform of the signal of the previous question and plot the corresponding magnitude spectrum.

54) Show that if $x(t)$ is a bandlimited signal, i.e., $X(\omega) = 0$ for $|\omega| > \omega_M$, then

$$x(t) * \frac{\sin(at)}{\pi t} = x(t), \text{ if } a > \omega_M.$$

Plot the corresponding graphs to illustrate the proof.

55) Prove the following Parseval equation,

$$\int_{-\infty}^{\infty} f(x)G(x)dx = \int_{-\infty}^{\infty} F(x)g(x)dx.$$

56) Find the Fourier transform of the current through a diode, represented by the expression $i(t) = I_o[e^{\alpha v(t)} - 1]$, given the voltage $v(t)$ applied to the diode and its Fourier transform $V(\omega)$, where α is a diode parameter and I_o is the reverse current. Plot the magnitude spectrum of the Fourier transform.

Chapter 2

PROBABILITY THEORY AND RANDOM PROCESSES

2.1 Set Theory, Functions and Measure

The theory of sets, in its more general form, began with Georg Cantor (1845-1918) in the nineteenth century. Cantor established the basis for this theory and demonstrated some of its most important results, including the concept of set cardinality. Cantor was born in St. Petersburg, Russia, but lived most of his life in Germany (Boyer, 1974). The ideas relative to the notions of universal set, empty set, set partition, discrete systems, continuous systems and infinity are, in reality, as old as philosophy itself.

In the time of Zenon, one of the most famous pre-Socrates philosophers, the notion of infinity was already discussed. Zenon, considered the creator of the dialectic, was born in Elea, Italy, around 504 B.C. and was the defendant of Parmenides, his master, against criticism from the followers of Pythagoras. Pythagoras was born in Samos around 580 B.C. and created a philosophic current based on the quantification of the universe. For the Pythagoreans, unity itself is the result of a being and of a not being. It can be noticed that the concept of emptiness is expressed in the above sentence as well as the concept of a universal set. The Pythagoreans established an association between the number one and the point, between the number two and the line, between the number three and the surface, between the number four and the volume (de Souza, 1996). In spite of dominating the notion of emptiness the Greeks still did not have the concept of zero.

Zenon, by his turn, defended the idea of a unique being continuous and indivisible, of Parmenides, against the multiple being, discontinuous and divisible of Pythagoras. Aristotle presents various of Zenon's arguments relative to movement, with the objective of establishing the concept of a continuum.

Aristotle was born in Estagira, Macedonia, in the year 384 B.C. The first of Aristotle's arguments suggests the idea of an infinite set, to be discussed later (Durant, 1996). This argument is as follows: If a mobile object has to cover a given distance, in a continuous space, it must then cover first the first half of the distance before covering the whole distance.

Reasoning in this manner Zenon's argument implies that infinite segments of distance can not be successively covered in a finite time. The expositive logics in Aristotle's counter argument is impressive(de Souza, 1996):

> In effect, length and time and in general all contents are called infinite in two senses, whether meaning division or whether with respect to extremes. No doubt, infinities in quantity can not be touched in a finite time; but infinities in division, yes, since time itself is also infinite in this manner. As a consequence, it is in an infinite time and not in a finite time that one can cross infinite and, if infinities are touched, they are touched by infinities and not by finite.

Despite the reflections of pre-Socratic philosophers and of others that followed, no one had yet managed to characterize infinite until 1872. In that year J. W. R. Dedekind (1831-1916) pointed to the universal property of infinite sets, which has found applications as far as in the study of fractals (Boyer, 1974):

> A system S is called infinite when it is similar to a part of itself. On the contrary, S is said to be finite.

Cantor also recognized the fundamental property of sets but, differing from Dedekind, he noticed that not all infinite sets are equal. This notion originated the cardinal numbers, which will be covered later, in order to establish a hierarchy of infinite sets in accordance with their respective powers. The results of Cantor led him to establish set theory as a fully developed subject. As a consequence of his results on transfinite arithmetics, too advanced for his time, Cantor suffered attacks of mathematicians like Leopold Kronecker (1823-1891), who disregarded him for a position at the University of Berlin.

Cantor spent most of his carrier life in the smaller University of Halle, in a city of medieval aspect with the same name in Germany, famous for its mines of rock salt, and died there in an institution for persons with mental health problems . However, his theory received from David Hilbert (1862-1943), one of the greatest mathematicians of the twentieth century, the following citation (Boyer, 1974):

> The new transfinite arithmetics is the most extraordinary product of human thought, and one of the most beautiful achievements of human activity in the domain of the purely intelligible.

Set Theory

The notion of a set is of fundamental importance and is axiomatic – in a manner that a set does not admit a problem free definition, i.e., a definition which will not resort to the original notion of a set. The mathematical concept of a set can be used as fundamental for all known mathematics.

Set theory will be developed based on a set of axioms, called fundamental axioms: Axiom of Extension, Axiom of Specification, Peano's Axioms, Axiom of Choice, besides Zorn's Lemma and Schröder - Bernstein's Theorem (Halmos, 1960).

The objective of this section is to develop the theory of sets in an informal manner, just quoting these fundamental axioms, since this theory is used as a basis for establishing a probability measure. Some examples of common sets are given next.

- The set of faces of a coin: $A = \{C_H, C_T\}$;
- The binary set: $B = \{0, 1\}$;
- The set of natural numbers: $N = \{1, 2, 3, \dots\}$;
- The set of integer numbers: $Z = \{\dots, -1, -2, 0, 1, 2, 3, \dots\}$;

The most important relations in set theory are the *belonging* relation, denoted as $a \in A$, where a is an element of the set A, and the *inclusion* relation, $A \subset B$, which is read "A is a subset of the set B", or B is a superset of the set A.

Sets may be specified by means of propositions as for example "The set of students that return their homework on time", or more formally $A = \{a \,|\, a \textit{ return their homework on time}\}$. This is, in a few cases, a way of denoting the empty set! Alias, the empty set can be written formally as $\emptyset = \{a \,|\, a \neq a\}$, i.e., the set the elements of which are not equal to themselves.

The notion of a universal set is of fundamental interest. A universal set is understood as that set which contains all other sets of interest. An example of a universal set is provided by the sample space in probability theory, usually denoted as S or Ω. The empty set is that set which contains no element and which is usually denoted as $\emptyset$ or $\{\ \}$. It is implicit that the empty set is contained in any set, i.e., that $\emptyset \subset A$, for any given set A. However, the empty set is not in general an element of any other set.

A practical way of representing sets is by means of the Venn diagram, as illustrated in Figure 2.1.

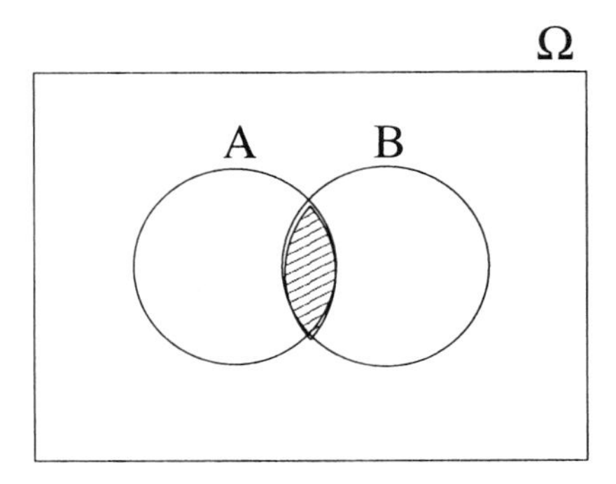

Figure 2.1. Venn diagram.

Two sets are said to be *disjoint* if they have no element in common, as illustrated in Figure 2.2. Thus, for example, the set of even natural numbers and the set of odd natural numbers are disjoint.

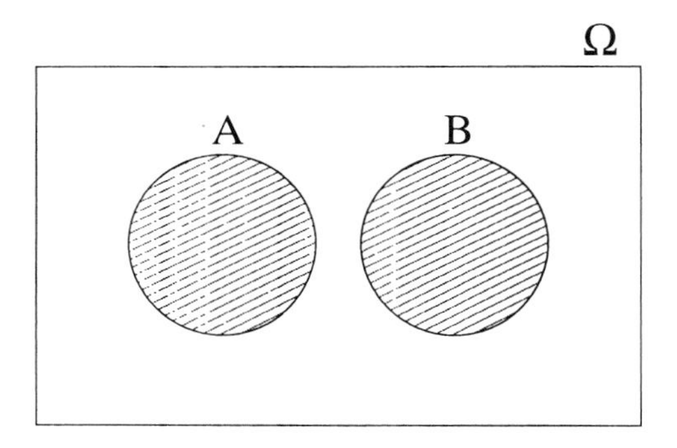

Figure 2.2. Disjoint sets.

Operations on Sets

- The operation $\overline{A}$ represents the *complement* of A with respect to the sample space Ω;

- The subtraction of sets, denoted $C = A - B$, gives as a result the set the elements of which belong to A and do not belong to B.
  Note: If B is completely contained in A : $A - B = A \cap \overline{B}$;

- The set of elements belonging to A and to B, but not belonging to $(A \cap B)$ é is specified by $A \Delta B = A \cup B - A \cap B$.

The generalization of these concepts to families of sets, as for example $\cup_{i=1}^{N} A_i$ and $\cap_{i=1}^{N} A_i$, is immediate. The following properties are usually employed as axioms in developing the theory of sets (Lipschutz, 1968).

- **Idempotent**

  $A \cup A = A, \qquad A \cap A = A$

- **Associative**

  $(A \cup B) \cup C = A \cup (B \cup C), \qquad (A \cap B) \cap C = A \cap (B \cap C)$

- **Commutative**

  $A \cup B = B \cup A, \qquad A \cap B = B \cap A$

- **Distributive**

  $A \cup (B \cap C) = (A \cup B) \cap (A \cup C)\,,$

  $A \cap (B \cup C) = (A \cap B) \cup (A \cap C)$

- **Identity**

  $A \cup \emptyset = A, \qquad A \cap U = A$

  $A \cup U = U, \qquad A \cap \emptyset = \emptyset$

- **Complementary**

  $A \cup \overline{A} = U, \qquad A \cap \overline{A} = \emptyset \qquad \overline{(\overline{A})} = A$

  $U = \emptyset, \qquad \overline{\emptyset} = U$

- **de Morgan laws**

  $\overline{A \cup B} = \overline{A} \cap \overline{B}, \qquad \overline{A \cap B} = \overline{A} \cup \overline{B}$

Families of Sets

Among the most interesting families of sets it is worth mentioning an increasing sequence of sets, such that $\lim_{i\to\infty} \cup A_i = A$. This sequence is used in proofs of limits over sets.

A decreasing sequence of sets is defined in a similar manner with $\lim_{i\to\infty} \cap A_i = A$.

Indexing

The Cartesian product is a way of expressing the idea of indexing of sets. The indexing of sets expands the possibilities for the use of sets, allowing to produce eventually entities known as vectors and signals.

Application 1: Consider $A_i = \{0, 1\}$. Starting from this set it is possible to construct an indexed sequence of sets by defining its indexing:

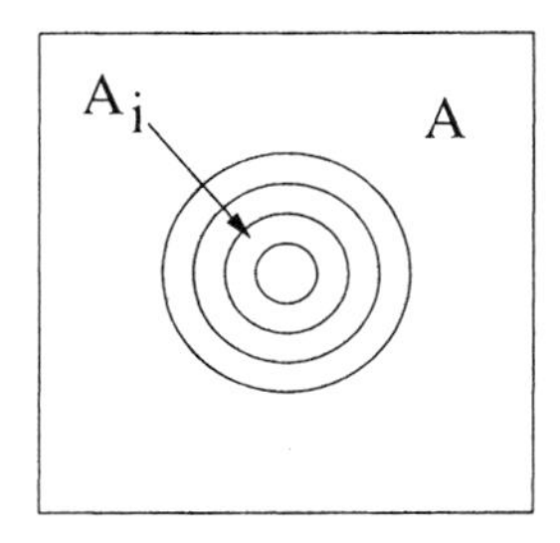

Figure 2.3. Increasing sequence of sets.

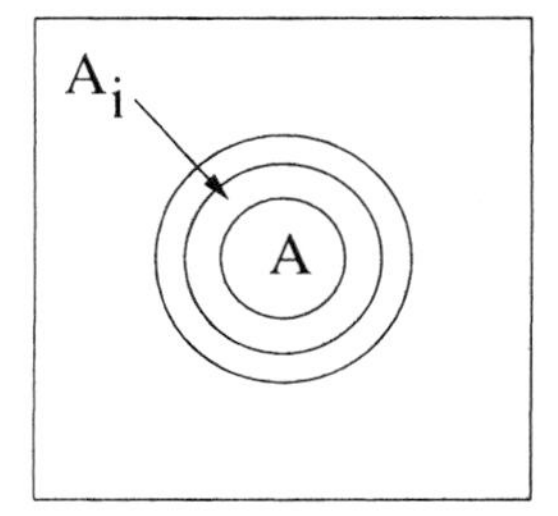

Figure 2.4. Decreasing sequence of sets.

$\{A_{i\epsilon I}\}$, $I = \{0, \cdots, 7\}$. This family of indexed sets A_i constitutes a finite discrete sequence, i.e., a vector.

Application 2: Consider again $A_i = \{0, 1\}$, but now let $I = Z$ (the set of positive and negative integers plus zero). It follows that $\{A_{i\epsilon Z}\}$, which represents an infinite series of 0's and 1's, i.e., it represents a binary digital signal. For example, $\cdots$ 0011111000 $\cdots$.

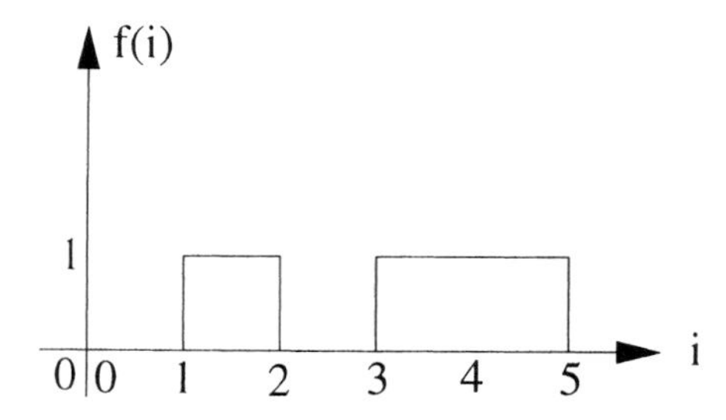

Figure 2.5. Signal discrete in time and in amplitude.

Application 3: Still letting $A_i = \{0, 1\}$, but considering the indexing over the set of real numbers, $\{A_{i\epsilon I}\}$, where $I = R$, a signal is formed which is discrete in amplitude but it is continuous in time, like the telegraph signal in Figure 2.6.

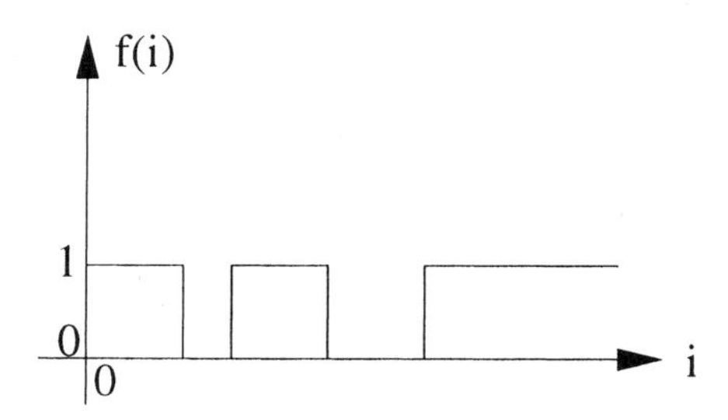

Figure 2.6. Telegraphic signal, discrete in amplitude and continuous in time.

Application 4: Finally, consider $A = R$ and $I = R$. In this manner the result is an analog signal, i.e., continuous in time and in amplitude, as shown in Figure 2.7.

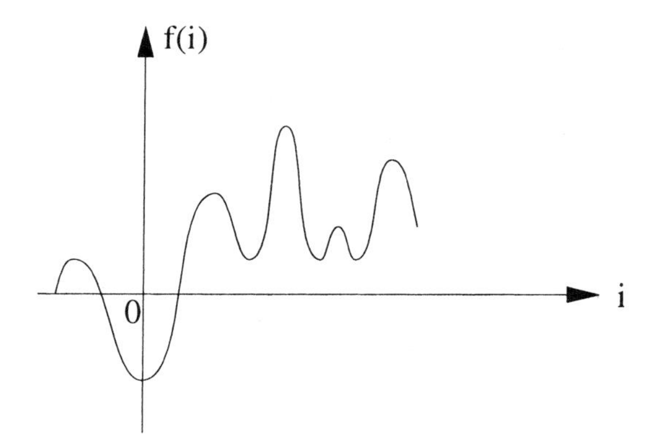

Figure 2.7. Analog signal.

Algebra of Sets

In order to construct an algebra of sets or, equivalently, to construct a field over which operations involving sets make sense, a few properties have to be obeyed.

1) If $A \in \mathcal{F}$ then $\overline{A} \in \mathcal{F}$. A is the set containing desired results, or over which one wants to operate;

2) If $A \in \mathcal{F}$ and $B \in \mathcal{F}$ then $A \cup B \in \mathcal{F}$.

The above properties guarantee the closure of the algebra with respect to finite operations over sets. It is noticed that the universal set Ω always belongs to the algebra, i.e., $\Omega \in \mathcal{F}$, because $\Omega = A \cup \overline{A}$. The empty set also belongs to the algebra, i.e., $\emptyset \in \mathcal{F}$, since $\emptyset = \overline{\Omega}$, follows by property 1.

Example: The family $\{\emptyset, \Omega\}$ complies with the above properties and therefore represents an algebra. In this case $\emptyset = \{\}$ and $\overline{\emptyset} = \Omega$. The union is also represented, as can be easily checked.

Example: Given the sets $\{C_H\}$ and $\{C_T\}$, representing the faces of a coin, respectively, if $\{C_H\} \in \mathcal{F}$ then $\overline{\{C_H\}} = \{C_T\} \in \mathcal{F}$. It follows that $\{C_H, C_T\} \in \mathcal{F} \Rightarrow \Omega \in \mathcal{F} \Rightarrow \emptyset \in \mathcal{F}$.

The previous example can be translated by the following expression. If there a measure for heads then there must be also a measure for tails, in order for the algebra to be properly defined. Whenever a probability is assigned to an event then a probability must also be assigned to the complementary event.

The cardinality of a finite set is defined as the number of elements belonging to this set. Sets with an infinite number of elements are said to have the same cardinality if they are equivalent, i.e., $A \sim B$ if $\sharp A = \sharp B$. Some examples of sets and their respective cardinals are presented next.

- $I = \{1, \cdots, k\} \Rightarrow C_I = k$;
- $N = \{0, 1, \cdots\} \Rightarrow C_N$ or $\aleph_0$;
- $Z = \{\cdots, -2, -1, 0, 1, 2, \cdots\} \Rightarrow C_Z$;
- $Q = \{\cdots, -1/3, 0, 1/3, 1/2, \cdots\} \Rightarrow C_Q$;
- $R = (-\infty, \infty) \Rightarrow C_R$ or $\aleph$;

For the above examples the following relations are verified: $C_R > C_Q = C_Z = C_N > C_I$. The notation $\aleph_0$, for the cardinality of the set of natural numbers was employed by Cantor.

The cardinality of the power set, i.e., of the family of sets consisting of all subsets of a given set I, $\mathcal{F} = 2^I$, is 2^{C_I}.

Borel Algebra

The Borel algebra $\mathcal{B}$, or σ-algebra, is an extension of the algebra so far discussed to operate with limits at infinity. The following properties are required from a σ-algebra.

1 $A \in \mathcal{B} \Rightarrow \overline{A} \in \mathcal{B}$

2 $A_i \in \mathcal{B} \Rightarrow \bigcup_{i=1}^{\infty} A_i \in \mathcal{B}$

The above properties guarantee the closure of the σ-algebra with respect to enumerable operations over sets. These properties allow the definition of limits in the Borel field.

Examples: Considering the above properties it can be verified that $A_1 \cap A_2 \cap A_3 \cdots \in \mathcal{B}$. In effect, it is sufficient to notice that

$$A \in \mathcal{B} \text{ and } \mathcal{B} \in \mathcal{B} \Rightarrow \mathcal{A} \cup \mathcal{B} \in \mathcal{B},$$

and

$$\overline{A} \in \mathcal{B} \text{ and } \overline{\mathcal{B}} \in \mathcal{B} \Rightarrow \overline{\mathcal{A}} \cup \overline{\mathcal{B}} \in \mathcal{B},$$

and finally

$$\overline{\overline{A} \cup \overline{B}} \in \mathcal{B} \Rightarrow \mathcal{A} \cap \mathcal{B} \in \mathcal{B}.$$

In summary, any combination of unions and intersections of sets belongs to the Borel algebra. In other words, operations of union or intersection of sets, or a combination of these operations, produce a set that belongs to the σ-algebra.

2.2 Probability Theory

This section summarizes the more basic definitions related to Probability Theory, Random Variables and Stochastic Processes, the main results and conclusions of which will be used in subsequent chapters.

Probability Theory began in France with studies about games of chance. Antoine Gombaud (1607-1684), known as *Chevalier de Méré*, was very keen on card games and would discuss with Blaise Pascal (1623-1662) about the probabilities of success in this game. Pascal, also interested on the subject, began a correspondence with Pierre de Fermat (1601-1665) in 1654, which originated the theory of finite probability (Zumpano and de Lima, 2004).

However, the first known work about probability is *De Ludo Aleae* (About Games of Chance), by the Italian medical doctor and mathematician Girolamo Cardano (1501-1576), published in 1663, almost 90 years after his death. This book was a handbook for players, containing some discussion about probability.

The first published treatise about the Theory of Probability, dated 1657, was written by the Dutch scientist Christian Huygens (1629-1695), a folder titled *De Ratiociniis in Ludo Aleae* (About Reasoning in Games of Chance).

Another Italian, the physicist and astronomer Galileo Galilei (1564-1642) was also concerned with random events. In a fragment probably written between 1613 and 1623, entitled *Sopra le Scorpete dei Dadi* (About Dice Games), Galileu answers a question asked, it is believed, by the Gran Duke of Toscana: When three dice are thrown, although both the number 9 as well as the number 10 may result from six distinct manners, in practice, the chances of getting a 9 are lower than those of obtaining a 10. How can that be explained?

The six distinct manners by which these numbers (9 and 10) can be obtained are: (1 3 6), (1 4 5), (2 3 5), (2 4 4), (2 6 2) and (3 3 4) for the number 10 and (1 2 6), (1 3 5), (1 4 4), (2 2 5), (2 3 4) and (3 3 3) for the number 9. Galileu concluded that, for this game, the permutations of the triplets must also be considered since (1 3 6) and (3 1 6) are distinct possibilities. He then calculated that in order to obtain the number 9 there are in fact 25 possibilities while there are 27 possibilities for the number 10. Therefore, combinations leading to the number 10 are more frequent.

Abraham de Moivre (1667-1754) was another important mathematician for the development of Probability Theory. He wrote a book of great influence at the time, called *Doctrine of Chances*. The law of large numbers was discussed by Jacques Bernoulli (1654-1705), Swiss mathematician, in his work Ars Conjectandi (The Art of Conjecturing).

The study of probability was deepened in the 18th and 19th centuries, being worth of mentioning the works of French mathematicians Pierre-Simon de Laplace (1749-1827) and Siméon Poisson (1781-1840), as well as the German mathematician Karl Friedrich Gauss (1777-1855).

Axiomatic Approach to Probability

Probability Theory is usually presented in one of the following manners. The classical approach, the relative frequency approach and the axiomatic approach. The classical approach is based on the symmetry of an experiment, but employs the concept of probability in a cyclic manner, because it is defined only for equiprobable events. The relative frequency approach to probability is more recent and relies on experiments.

Considering the difficulties found in the two previous approaches to probability, respectively the cyclic definition in the first case and the problem of convergence in a series of experiments for the second case, henceforth only the axiomatic approach will be followed in this text. Those readers interested in the classical or in the relative frequency approach are referred to the literature (Papoulis, 1981).

The axioms of probability were established by Andrei N. Kolmogorov (1903-1987), allowing the development of the complete theory, are just three statements as follows (Papoulis, 1983b):

Axiom 1) $P(S) = 1$, where S denotes the sample space or universal set and $P(\cdot)$ denotes the associated probability;
Axiom 2) $P(A) \geq 0$, where A denotes an event belonging to the sample space;
Axiom 3) $P(A \cup B) = P(A) + P(B)$, where A and B are mutually exclusive events and $A \cup B$ denotes the union of events A and B.

Using his axiomatic approach to Probability Theory, Kolmogorov established a firm mathematical basis on which other theories rely as, for example, the Theory of Stochastic Processes, Communications Theory and Information Theory.

Kolmogorov's fundamental work was published in 1933, in Russian, and soon afterwards was published in German with the title *Grundbegriffe der Wahrscheinlichkeits Rechnung* (Fundamentals of Probability Theory) (James, 1981). In this work Kolmogorov managed to combine Advanced Set Theory, of Cantor, with Measure Theory, of Lebesgue, in order to produce what to this date is the modern approach to Probability Theory.

By applying the above axioms it is possible to deduce all results relative to Probability Theory. For example, the probability of the empty set, $\emptyset = \{\}$, is easily calculated as follows. First it is noticed that

$$\emptyset \cup \Omega = \Omega,$$

since the sets $\emptyset$ and Ω are disjoint. Thus it follows that

$$P(\emptyset \cup \Omega) = P(\Omega) = P(\emptyset) + P(\Omega) = 1 \Rightarrow P(\emptyset) = 0.$$

In the case of sets A and B which are not disjoint it follows that

$$P(A \cup B) = P(A) + P(B) - P(A \cap B).$$

Bayes' Rule

Bayes' rule concerns the computation of conditional probabilities and can be expressed by the following rule

$$P(A|B) = \frac{P(A \cap B)}{P(B)},$$

assuming $P(B) \neq 0$. An equivalent manner of expressing the same result is the following.

$$P(A \cap B) = P(A|B) \cdot P(B) \quad , \quad P(B) \neq 0.$$

Some important properties of sets are presented next, where A and B denote events from a given sample space.

- If A is independent of B, then $P(A|B) = P(A)$. It then follows that $P(B|A) = P(B)$ and that B is independent of A.
- If $B \subset A$, then: $P(A|B) = 1$.
- If $A \subset B$, then: $P(A|B) = \frac{P(A)}{P(B)} \geq P(A)$.
- If A and B are independent events then $P(A \cap B) = P(A) \cdot P(B)$.
- If $P(A) = 0$ or $P(A) = 1$, then event A is independent of itself.
- If $P(B) = 0$, then $P(A|B)$ can assume any arbitrary value. Usually in this case one assumes $P(A|B) = P(A)$.
- If events A and B are disjoint, and non-empty, then they are dependent.

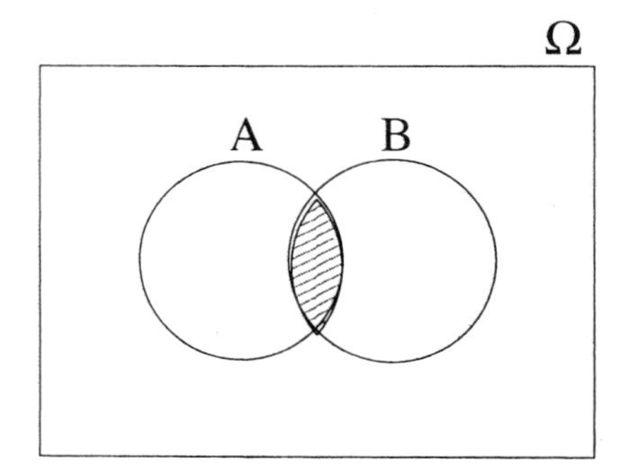

Figure 2.8. Venn diagram.

A partition is a possible splitting of the sample space into a family of subsets, in a manner that the subsets in this family are disjoint and their union coincides with the sample space. It follows that any set in the sample space can be expressed by using a partition of that sample space, and thus be written as a union of disjoint events.

The following property can be illustrated by means of a Venn diagram, as illustrated in Figure 2.9.

$$B = B \cap \Omega = B \cap \cup_{i=1}^{M} A_i = \cup_{i=1}^{N} B \cap A_i.$$

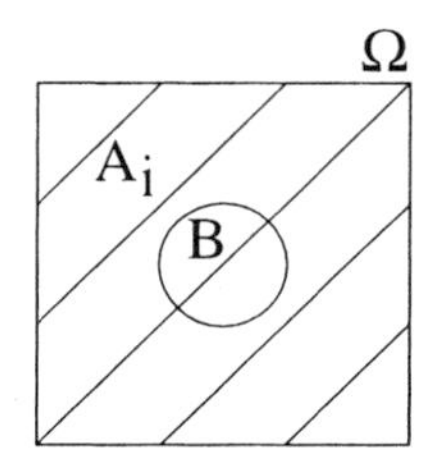

Figure 2.9. Partition of a set.

It now follows that

$$\begin{aligned} P(B) &= P(\cup_{i=1}^{N} B \cap A_i) = \sum_{i=1}^{N} P(B \cap A_i), \\ P(A_i|B) &= \frac{P(A_i \cap B)}{P(B)} = \frac{P(B|A_i) \cdot P(A_i)}{\sum_{i=1}^{N} P(B \cap A_i)} = \frac{P(B|A_i) \cdot P(A_i)}{\sum_{i=1}^{N} P(B|A_i) \cdot P(A_i)}. \end{aligned}$$

2.3 Random Variables

A random variable (r.v.) X represents a mapping of the sample space on the line (the set of real numbers). A random variable is usually characterized by a cumulative probability function (CPF) $P_X(x)$, or by a probability density function (pdf) $p_X(x)$.

Example: A random variable with a uniform probability density function $p_X(x)$ is described by the equation $p_X(x) = u(x) - u(x-1)$. It follows by axiom 1 that

$$\int_{-\infty}^{+\infty} p_X(x)dx = 1. \tag{2.1}$$

In general, for a given probability distribution, the probability of X belonging to the interval $(a, b]$ is given by

$$P(a < x \leq b) = \int_a^b p_X(x)dx. \tag{2.2}$$

The cumulative probability function $P_X(x)$, of a random variable X, is defined as the integral of $p_X(x)$, i.e.,

$$P_X(x) = \int_{-\infty}^{x} p_X(t)dt. \tag{2.3}$$

2.3.1 Average value of a random variable

Let $f(X)$ denote a function of a random variable X. The average value (or expected value) of $f(X)$ with respect to X is defined as

$$E[f(X)] = \int_{-\infty}^{+\infty} f(x)p_X(x)dx. \tag{2.4}$$

The following properties of the expected value follow from (2.4).

$$E[\alpha X] = \alpha E[X], \tag{2.5}$$

$$E[X+Y] = E[X] + E[Y] \tag{2.6}$$

and if X and Y are independent random variables then

$$E[XY] = E[X]E[Y]. \tag{2.7}$$

2.3.2 Moments of a random variable

The ith moment of a random variable X is defined as

$$m_i = E[X^i] = \int_{-\infty}^{+\infty} x^i p_X(x)dx. \tag{2.8}$$

Various moments of X have special importance and physical interpretation.

- $m_1 = E[X]$, arithmetic mean, average value, average voltage, statistical mean;
- $m_2 = E[X^2]$, quadratic mean, total power;
- $m_3 = E[X^3]$, measure of asymmetry of the probability density function;
- $m_4 = E[X^4]$, measure of flatness of the probability density function.

2.3.3 The variance of a random variable

The variance of a random variable X is an important quantity in communication theory (meaning AC power), defined as follows.

$$V[X] = \sigma_X^2 = E[(X - m_1)^2] = m_2 - m_1^2. \tag{2.9}$$

The *standard deviation* σ_X is defined as the square root of the variance of X.

2.3.4 The characteristic function of a random variable

The characteristic function $P_X(w)$, or moment generating function, of a random variable X is usually defined from the Fourier transform of the probability density function (pdf) of X, which is equivalent to making $f(x) = e^{-j\omega x}$ in (2.4), i.e.,

$$P_X(w) = E[e^{-j\omega x}] = \int_{-\infty}^{+\infty} e^{-j\omega x} p_X(x)dx, \text{where} \quad j = \sqrt{-1}. \tag{2.10}$$

The moments of a random variable X can also be obtained directly from then characteristic function as follows.

$$m_i = \frac{1}{(-j)^i} \frac{\partial^i P_X(w)}{\partial w^i} |_{w=0} \; . \tag{2.11}$$

Given that X is a random variable, it follows that $Y = f(X)$ is also a random variable, obtained by the application of the transformation $f(\cdot)$. The probability density function of Y is related to that of X by the formula (Blake, 1987)

$$p_Y(y) = \frac{p_X(x)}{|dy/dx|} |_{x=f^{-1}(y)} \; , \tag{2.12}$$

where $f^{-1}(\cdot)$ denotes the inverse function of $f(\cdot)$. This formula assumes the existence of the inverse function of $f(\cdot)$ as well as its derivative in all points.

2.3.4.1 Some important random variables

1) **Gaussian random variable**

The random variable X with pdf

$$p_X(x) = \frac{1}{\sigma_X \sqrt{2\pi}} e^{-\frac{(x-m_X)^2}{2\sigma_X^2}} \tag{2.13}$$

is called a Gaussian (or Normal) random variable. The Gaussian random variable plays an extremely important role in engineering, considering that many well known processes can be described or approximated by this pdf. The noise present in either analog or digital communications systems usually can be considered Gaussian as a consequence of the influence of many factors (Leon-Garcia, 1989).

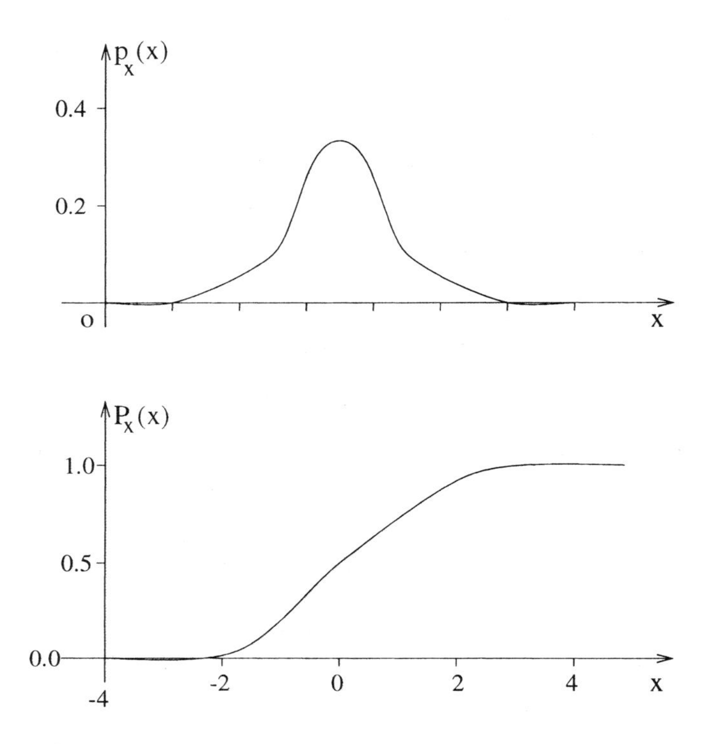

Figure 2.10. Gaussian probability density and corresponding cumulative probability function.

In (2.13) m_X represents the average value and σ_X^2 represents the variance of X. Figure 2.10 illustrates the Gaussian pdf and its corresponding cumulative probability function.

2) **Rayleigh random variable**
An often used model to represent the behavior of the amplitudes of signals subjected to fading employs the following pdf (Kennedy, 1969), (Proakis, 1990)

$$p_X(x) = \frac{x}{\sigma^2} e^{-\frac{x^2}{2\sigma^2}} u(x) \tag{2.14}$$

known as the Rayleigh pdf, with average $E[X] = \sigma\sqrt{\pi/2}$ and variance $V[X] = (2 - \pi)\frac{\sigma^2}{2}$.

The Rayleigh pdf represents the effect of multiple signals, reflected or refracted, which are captured by a receiver, in a situation where there is no main signal component or main direction of propagation

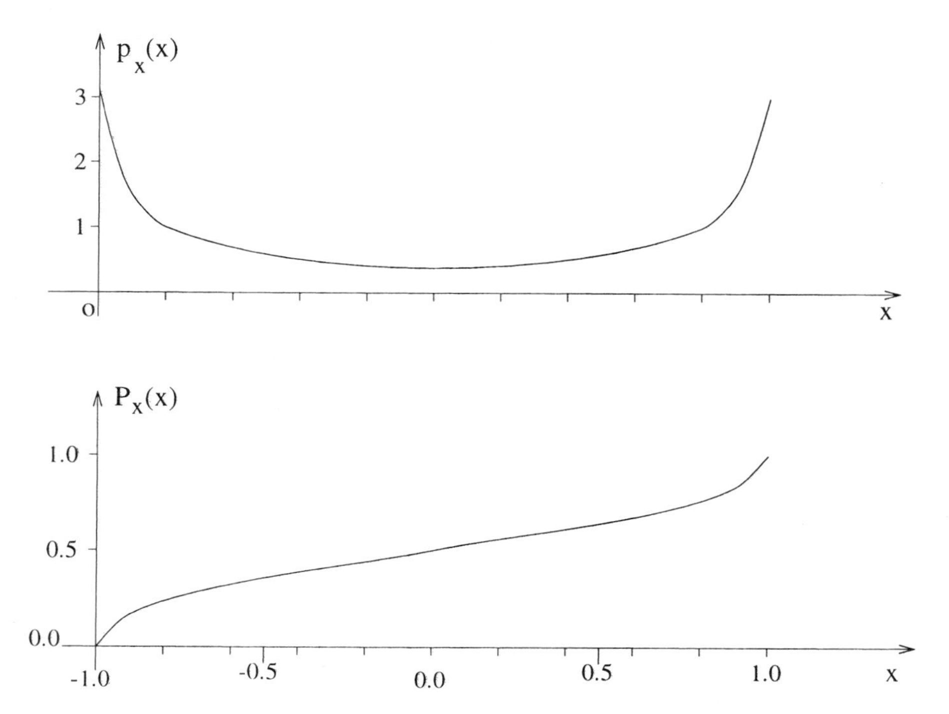

Figure 2.11. Probability density function and cumulative probability function of a sinusoidal random variable.

(Lecours et al., 1988). In this situation the phase distribution of the received signal can be considered uniform in the interval $(0, 2\pi)$. It is noticed that it is possible to closely approximate a Rayleigh pdf by considering only six waveforms with independently distributed phases (Schwartz et al., 1966).

3) **Sinusoidal random variable**
A sinusoidal tone X has the following pdf

$$p_X(x) = \frac{1}{\pi\sqrt{V^2 - x^2}}, \quad |x| < V. \tag{2.15}$$

The pdf and the CPF of X are illustrated in Figure 2.11.

Joint Random Variables

Considering that X and Y represent a pair of real random variables, with joint pdf $p_{XY}(x, y)$, as illustrated in Figure 2.12, then the probability of x and y being simultaneously in the region defined by the polygon [abcd] is given by the expression

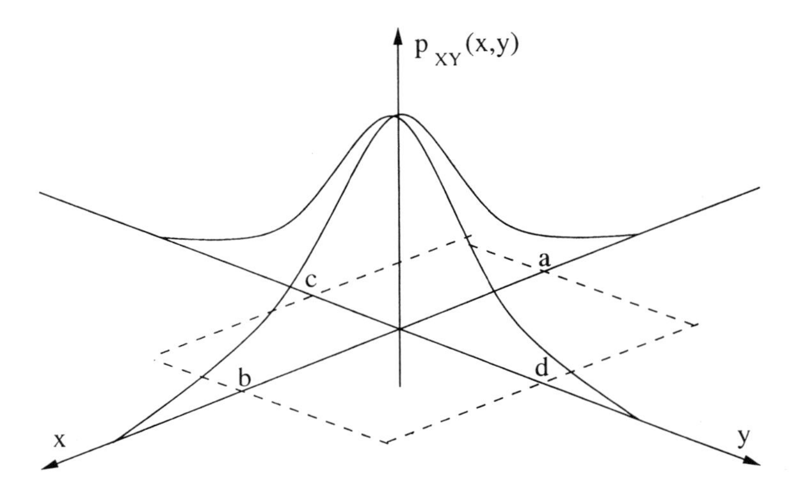

Figure 2.12. Joint probability density function.

$$\text{Prob}(a < x < b, c < y < d) = \int_a^b \int_c^d p_{XY}(x, y) dx dy. \tag{2.16}$$

The individual pdf's of X and Y, also called marginal pdf's, result from the integration of the joint pdf as follows.

$$p_X(x) = \int_{-\infty}^{+\infty} p_{XY}(x, y) dy, \tag{2.17}$$

and

$$p_Y(y) = \int_{-\infty}^{+\infty} p_{XY}(x, y) dx. \tag{2.18}$$

The joint average $E[f(X, Y)]$ is calculated as

$$E[f(X, Y)] = \int_{-\infty}^{+\infty} \int_{-\infty}^{+\infty} f(x, y) p_{XY}(x, y) dx dy, \tag{2.19}$$

for an arbitrary function $f(X, Y)$ of X and Y.

The joint moments m_{ik}, of order ik, are calculated as

$$m_{ik} = E[X^i, Y^k] = \int_{-\infty}^{+\infty} \int_{-\infty}^{+\infty} x^i y^k p_{XY}(xy) dx dy. \tag{2.20}$$

The two-dimensional characteristic function is defined as the two-dimensional Fourier transform of the joint probability density $p_{XY}(x,y)$

$$P_{XY}(\omega,\nu) = E[e^{-j\omega X - j\nu Y}]. \tag{2.21}$$

When the sum $Z = X + Y$ of two statistically independent r.v.'s is considered, it is noticed that the characteristic function of Z turns out to be

$$P_Z(\omega) = E[e^{-j\omega Z}] = E[e^{-j\omega(X+Y)}] = P_X(\omega) \cdot P_Y(\omega). \tag{2.22}$$

As far as the pdf of Z is concerned, it can be said that

$$p_Z(z) = \int_{-\infty}^{\infty} p_X(\rho) p_Y(z-\rho) d\rho, \tag{2.23}$$

or

$$p_Z(z) = \int_{-\infty}^{\infty} p_X(z-\rho) p_Y(\rho) d\rho. \tag{2.24}$$

Equivalently, the sum of two statistically independent r.v.'s has a pdf given by the convolution of the respective pdf's of the r.v.'s involved in the sum.

The r.v.'s X and Y are called uncorrelated if $E[XY] = E[X]E[Y]$. The criterion of statistical independence of random variables, which is stronger than that for the r.v.'s being uncorrelated, is satisfied if $p_{XY}(x,y) = p_X(x).p_Y(y)$.

2.4 Stochastic Processes

A random process, or a stochastic process, is an extension of the concept of a random variable, involving a sample space, a set of signals and the associated probability density functions. Figure 2.13 illustrates a random signal and its associated probability density function.

A random process (or stochastic process) $X(t)$ defines a random variable for each point on the time axis. A stochastic process is said to be stationary if the probability densities associated with the process are time independent.

The Autocorrelation Function

An important joint moment of the random process $X(t)$ is the autocorrelation function

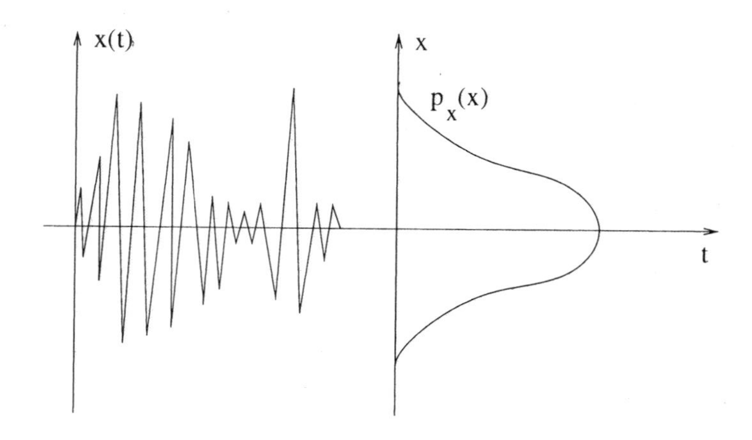

Figure 2.13. Example of a random process and its corresponding probability density function.

$$R_X(\xi, \eta) = E[X(\xi)X(\eta)], \tag{2.25}$$

where

$$E[X(\xi)X(\eta)] = \int_{-\infty}^{+\infty} \int_{-\infty}^{+\infty} x(\xi)x(\eta)p_{X(\xi)X(\eta)}(x(\xi)x(\eta))dx(\xi)dy(\eta) \tag{2.26}$$

denotes the joint moment of the r.v. $X(t)$ at $t = \xi$ and at $t = \eta$.

The random process is called wide sense stationary if its autocorrelation depends only on the interval of time separating $X(\xi)$ and $X(\eta)$, i.e., depends only on $\tau = \xi - \eta$. Equation 2.25 in this case can be written as

$$R_X(\tau) = E[X(t)X(t + \tau)]. \tag{2.27}$$

Stationarity

In general, the statistical mean of a time signal is a function of time. Thus, the mean value

$$E[X(t)] = m_X(t),$$

the power

$$E[X^2(t)] = P_X(t)$$

and the autocorrelation

$$R_X(\tau, t) = E[X(t)X(t + \tau)],$$

are, in general, time dependent. However, there exists a set of time signals the mean value of which are time independent. These signals are called stationary signals, as illustrated in the following example.

Example: Calculate the power (RMS square value) of the random signal $X(t) = V\cos(\omega t + \phi)$, where V is a constant and where the phase ϕ is a r.v. with a uniform probability distribution over the interval $[0, 2\pi]$. Applying the definition of power it follows that

$$E[X^2(t)] = \int_{-\infty}^{\infty} X^2(t) p_X(x)\, dx = \frac{1}{2\pi}\int_0^{2\pi} V^2 \cos^2(\omega t + \phi)\, d\phi.$$

Recalling that $\cos^2\theta = \frac{1}{2} + \frac{1}{2}\cos 2\theta$, it follows that

$$E[X^2(t)] = \frac{V^2}{4\pi}\int_0^{2\pi} (1 + \cos(2\omega t + 2\phi))\, d\phi = \frac{V^2}{4\pi}\phi\,\Big|_o^{2\pi} = \frac{V^2}{2}.$$

Since the mean value m_X of $X(t)$ is zero, i.e.,

$$m_X = E[X(t)] = E[V\cos(\omega t + \phi)] = 0,$$

the variance, or AC power, becomes

$$V[X] = E[(X - m_X)^2] = E[X^2 - 2Xm_X + m_X^2] = E[X^2] = \frac{V^2}{2}.$$

Therefore,

$$\sigma_X = \frac{V}{\sqrt{2}} = \frac{V\sqrt{2}}{2}.$$

Example: Consider the digital signal shown in Figure 2.14, with equiprobable amplitudes A and $-A$.

The probability density function, as shown in Figure 2.15, is given by

$$p_X(x) = \frac{1}{2}\left[\delta(x + A) + \delta(x - A)\right].$$

By applying the definition of the mean to $E[X(t)]$ it follows that

$$E[X(t)] = 0.$$

The power also follows as

$$E[X^2(t)] = \frac{1}{2}A^2 + \frac{1}{2}(-A)^2 = A^2.$$

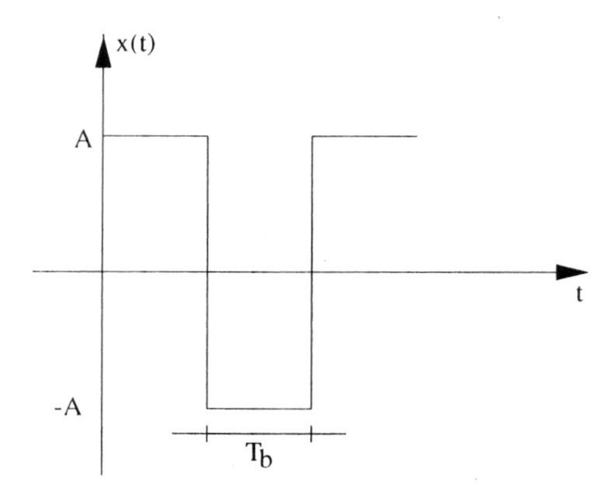

Figure 2.14. Digital signal.

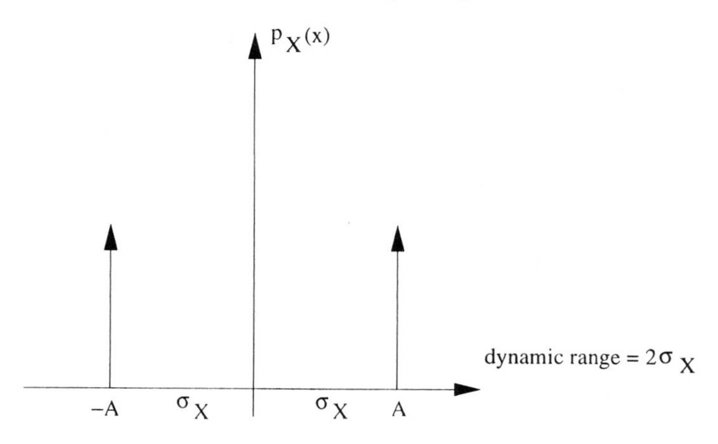

Figure 2.15. Probability density function for the digital signal of Figure 2.14.

Finally, the variance and the standard deviation are calculated as

$$\sigma_X^2 = E[X^2(t)] = A^2 \Rightarrow \sigma_X = A.$$

Application: The dynamic range of a signal, from a probabilistic point of view, is illustrated in Figure 2.16. As can be seen, the dynamic range depends on the standard deviation, or RMS voltage, being usually specified for $2\sigma_X$ or $4\sigma_X$. For a signal with a Gaussian probability distribution of amplitudes, this corresponds to a range encompassing, respectively, 97% and 99, 7% of all signal amplitudes.

However, since the signal is time varying, its statistical mean can also change with time, as illustrated in Figure 2.17. In the example considered, the variance is initially diminishing with time and later it is growing with time. In this case, an adjustment in the signal variance,

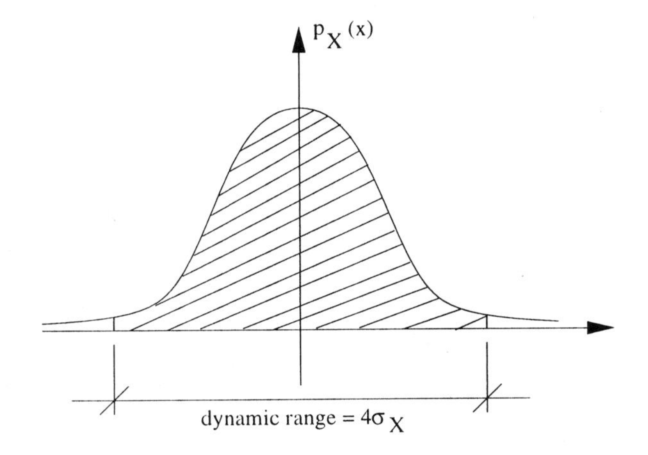

Figure 2.16. Dynamic range of a signal.

by means of an automatic gain control mechanism, can remove the pdf dependency on time.

A signal is stationary whenever its pdf is time independent, i.e., whenever $p_X(x,t) = p_X(x)$, as illustrated in Figure 2.18.

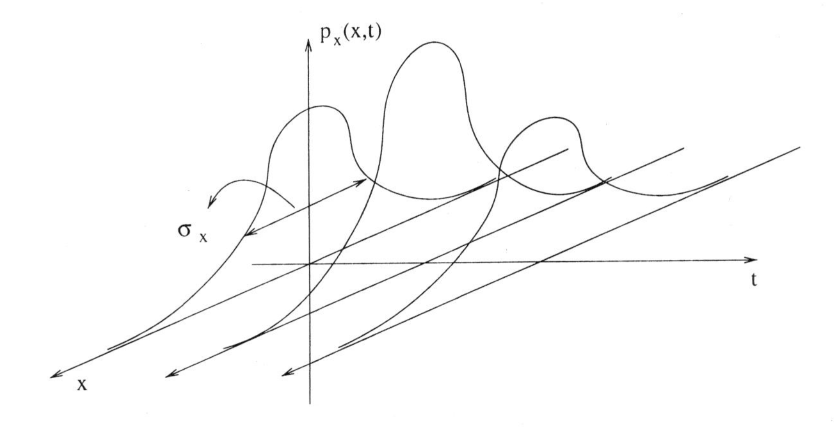

Figure 2.17. Time varying probability density function.

Stationarity may occur in various instances:

1) Stationary mean $\Rightarrow$ $m_X(t) = m_X$;

2) Stationary power $\Rightarrow$ $P_X(t) = P_X$;

3) First order stationarity implies that the first order moment is also time independent;

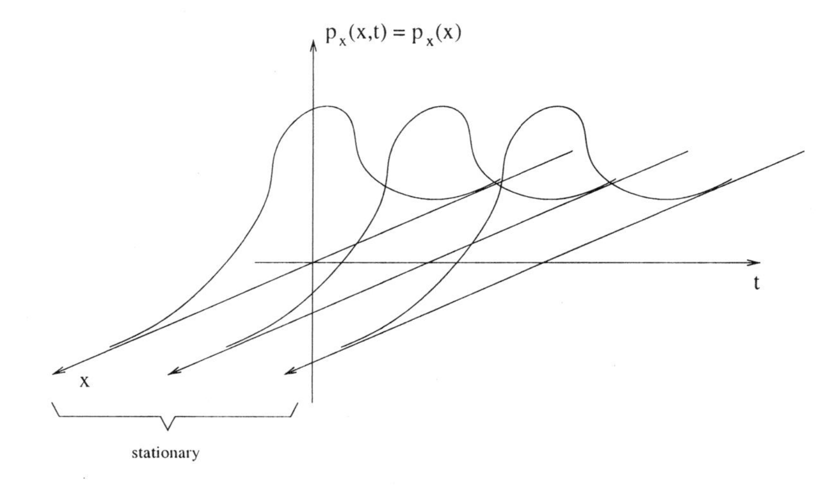

Figure 2.18. Time invariant probability density function.

4) Second order stationarity implies that the second order moments are also time independent;

5) Narrow sense stationarity implies that the signal is stationary for all orders, i.e., $p_{X_1 \cdots X_M}(x_1, \cdots, x_M; t) = p_{X_1 \cdots X_M}(x_1, \cdots, x_M)$

Wide Sense Stationarity

The following conditions are necessary to guarantee that a stochastic process is wide sense stationary.

1) The autocorrelation is time independent;

2) The mean and the power are constant;

3) $R_X(t_1, t_2) = R_X(t_2 - t_1) = R_X(\tau)$. The autocorrelation depends on the time interval and not on the origin of the time interval.

Stationarity Degrees

A pictorial classification of degrees of stationarity are illustrated in Figure 2.19. The universal set includes all signals in general. As subsets of the universal set come the signals with a stationary mean, wide sense stationary signals and narrow sense stationary signals.

Ergodic Signals

Ergodicity is another characteristic of random signals. A given expected value of a function of the pdf is ergodic if the time expected value coincides with the statistical expected value. Thus ergodicity can occur

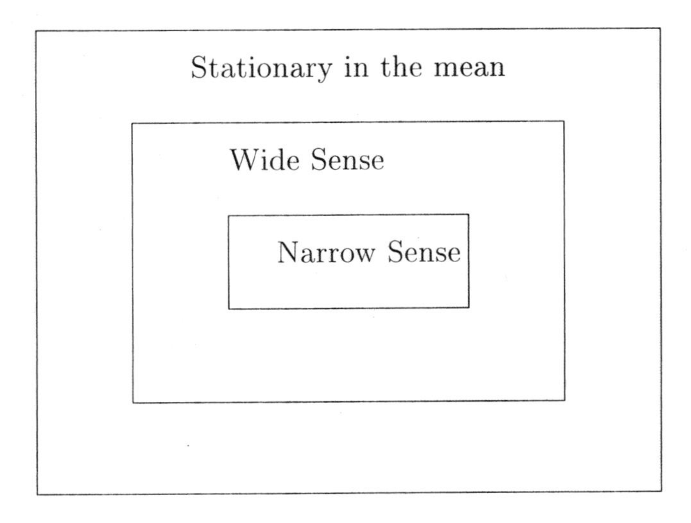

Figure 2.19. Stationarity Degrees.

on the mean, on the power, on the autocorrelation, or with respect to other quantities. Ergodicity of the mean implies that the time average is equivalent to the statistical signal average. Therefore,

- Ergodicity of the mean: $\overline{X(t)} \sim E[X(t)]$;
- Ergodicity of the power: $\overline{X^2(t)} \sim \overline{R_X}(\tau) \sim R_X(\tau)$;
- Ergodicity of the autocorrelation: $\overline{R_X}(\tau) \sim R_X(\tau)$.

A strictly stationary stochastic process has time independent joint pdf's of all orders. A stationary process of second order is that process for which all means are constant and the autocorrelation depends only on the measurement time interval.

Summarizing, a stochastic process is ergodic whenever its statistical means, which are functions of time, can be approximated by their corresponding time averages, which are random processes, with a standard deviation which is close to zero. The ergodicity may appear only on the mean value of the process, in which case the process is said to be ergodic on the mean.

Properties of the Autocorrelation

The autocorrelation function has some important properties as follows.

1) $R_X(0) = E[X^2(t)] = P_X$, (Total power);

2) $R_X(\infty) = \lim_{\tau\to\infty} R_X(\tau) = \lim_{\tau\to\infty} E[X(t+\tau)X(t)] = E^2[X(t)]$, (Average power or DC level);

3) Autocovariance: $C_X(\tau) = R_X(\tau) - E^2[X(t)]$;

4) Variance: $V[X(t)] = E[(X(t) - E[X(t)])^2] = E[X^2(t)] - E^2[X(t)]$ or $P_{AC}(0) = R_X(0) - R_X(\infty)$;

5) $R_X(0) \geq |R_X(\tau)|$, (Maximum at the origin);
This property is demonstrated by considering the following tautology

$$E[(X(t) - X(t+\tau))^2] \geq 0.$$

Thus,

$$E[X^2(t) - 2X(t)X(t+\tau)] + E[X^2(t+\tau)] \geq 0,$$

i.e.,

$$2R_X(0) - 2RX(\tau) \geq 0 \Rightarrow R_X(0) \geq R_X(\tau).$$

6) Symmetry: $R_X(\tau) = R_X(-\tau)$;
In order to prove this property it is sufficient to use the definition $R_X(-\tau) = E[X(t)X(t-\tau)]$
Letting $t - \tau = \sigma \Rightarrow t = \sigma + \tau$

$$R_X(-\tau) = E[X(\sigma+\tau)\cdot X(\sigma)] = R_X(\tau).$$

7) $E[X(t)] = \sqrt{R_X(\infty)}$, (Signal mean value).

Application: The relationship between the autocorrelation and various other power measures is illustrated in Figure 2.20.

Example A digital signal $X(t)$, with equiprobable amplitude levels A and $-A$, has the following autocorrelation function.

$$R_X(\tau) = A^2[1 - \frac{|\tau|}{T_b}][u(\tau+T_b) - u(\tau - T_b)],$$

where T_b is the bit duration.

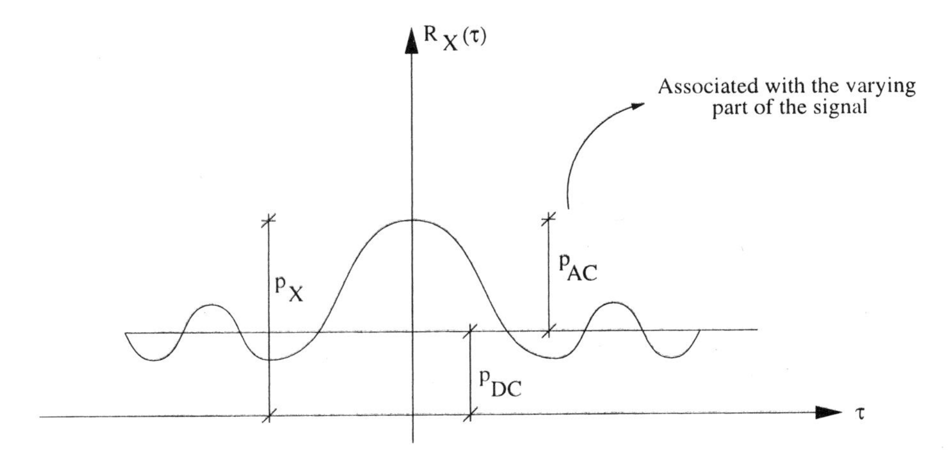

Figure 2.20. Relationship between the autocorrelation and various power measures.

The Power Spectral Density

By using the autocorrelation function it is possible to define the following Fourier transform pair, known as the Wiener-Khintchin theorem.

$$S_X(w) = \int_{-\infty}^{+\infty} R_X(\tau) e^{-jw\tau} d\tau \tag{2.28}$$

$$R_X(\tau) = \frac{1}{2\pi} \int_{-\infty}^{+\infty} S_X(w) e^{jw\tau} dw. \tag{2.29}$$

The function $S_X(w)$ is called the *power spectral density* (PSD) of the random process.

The Wiener-Khintchin theorem relates the autocorrelation function with the power spectral density, i.e., it plays the role of a bridge between the time domain and the frequency domain for random signals. This theorem will be proved in the sequel. Figure 2.21 shows a random signal truncated in an interval T.

The Fourier transform for the signal $x(t)$ given in Figure 2.21 is given by $\mathcal{F}[x_T(t)] = X_T(\omega)$. The time power spectral density of $x(t)$ is calculated as follows.

$$\lim_{T\to\infty} \frac{1}{T} |X_T(\omega)|^2 = \overline{S_X}(\omega)$$

The result obtained is obviously a random quantity, and it is possible to compute its statistical mean to obtain the power spectral density

$$S_X(\omega) = E[\overline{S_X}(\omega)]$$

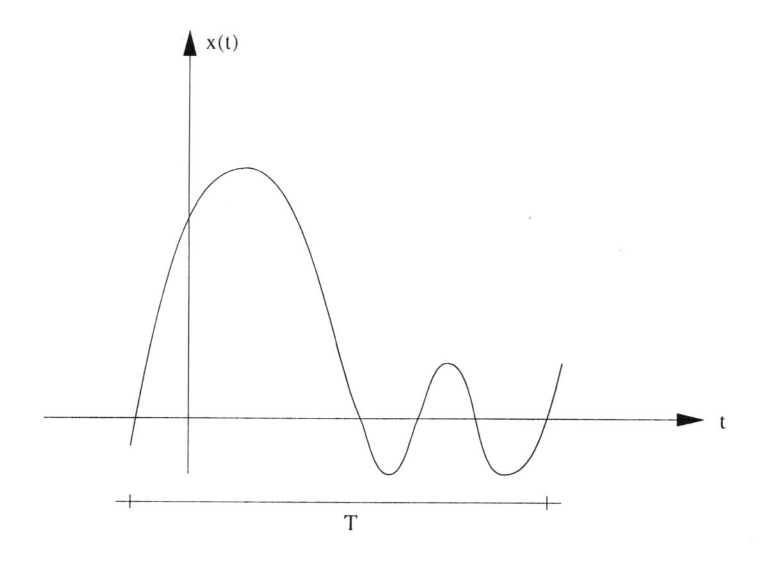

Figure 2.21. Random signal truncated in an interval T.

Using the fact that

$$|X_T(\omega)|^2 = X_T(\omega) \cdot X_T^*(\omega) = X_T(\omega) \cdot X_T(-\omega)$$

for X(t) real, where $X_T(\omega) = \int_{-\infty}^{\infty} X_T(t)e^{-j\omega t}\, dt$, it follows that

$$S_X(\omega) = \lim_{T\to\infty} \frac{1}{T} E[|X_T(\omega)|^2] = \lim_{T\to\infty} \frac{1}{T} E[X_T(\omega) \cdot X_T(-\omega)] =$$

$$= \lim_{T\to\infty} E\left[\int_{-\infty}^{\infty} X_T(t)e^{-j\omega t}\, dt \cdot \int_{-\infty}^{\infty} X_T(\tau)e^{j\omega\tau}\, d\tau\right] =$$

$$= \lim_{T\to\infty} \frac{1}{T} \int_{-\infty}^{\infty}\int_{-\infty}^{\infty} E[X_T(t)X_T(\tau)]e^{-j(t-\tau)\omega}\, dt\, d\tau =$$

$$= \lim_{T\to\infty} \frac{1}{T} \int_{-\infty}^{\infty}\int_{-\infty}^{\infty} R_{X_T}(t-\tau)e^{-j(t-\tau)\omega}\, dt\, d\tau.$$

Letting $t - \tau = \sigma$, it follows that $t = \sigma + \tau$ and $dt = d\sigma$. Thus,

$$S_X(\omega) = \lim_{T\to\infty} \frac{1}{T} \int_{-\infty}^{\infty}\int_{-\infty}^{\infty} R_{X_T}(\sigma)e^{-j\sigma\omega}\, dt\, d\sigma\ .$$

This result implies that $S_X(\omega) = \mathcal{F}[R_X(\tau)]$, i.e., implies that

$$S_X(\omega) = \int_{-\infty}^{\infty} R_X(\tau)e^{-j\omega\tau}\, d\tau.$$

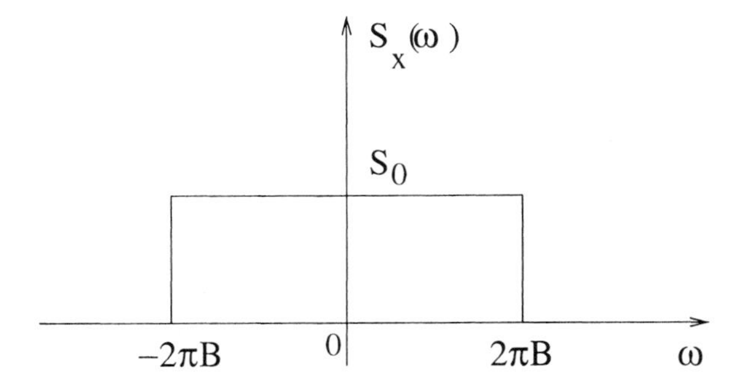

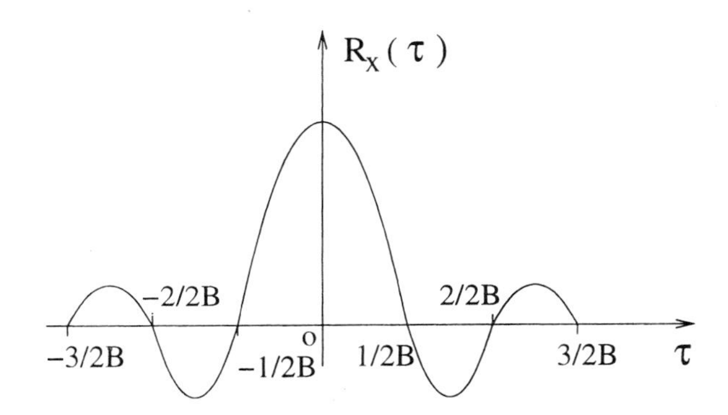

Figure 2.22. Power spectral density function and the autocorrelation function for a bandlimited signal.

The function $S_X(\omega)$ represents the power spectral density, which measures power per unit frequency. The corresponding inverse transform is the autocorrelation function $R_X(\tau) = \mathcal{F}^{-1}[S_X(\omega)]$, or

$$R_X(\tau) = \frac{1}{2\pi} \int_{-\infty}^{\infty} S_X(\omega) e^{j\omega\tau} \, d\omega.$$

Figure 2.22 illustrates the power spectral density function and the autocorrelation function for a bandlimited signal. Figure 2.23 illustrates the power spectral density and the correspondent autocorrelation function for white noise. It is noticed that $S_X(\omega) = S_0$, which indicates a uniform distribution for the power density along the spectrum, and $R_X(\tau) = S_0\delta(\tau)$, which shows white noise as the most uncorrelated, or random, of all signals. Correlation is non-zero for this signal only at $\tau = 0$. On the other hand, Figure 2.24 illustrates the power spectral

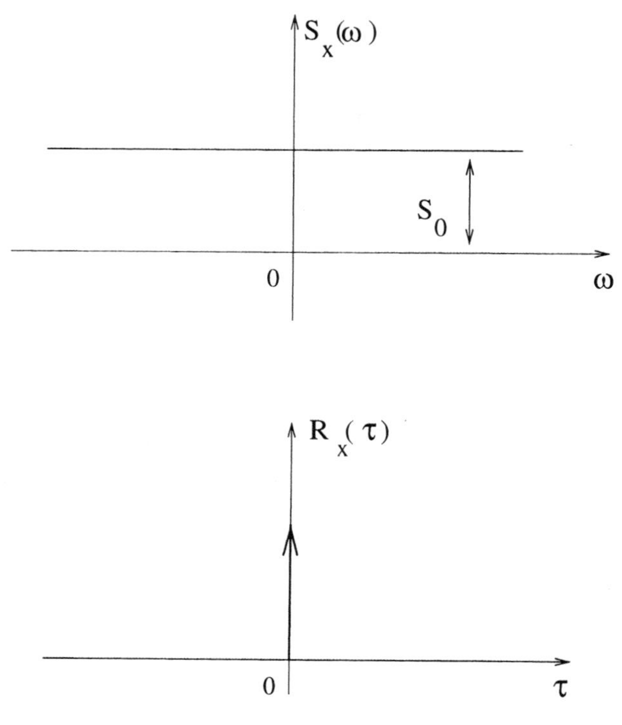

Figure 2.23. Power spectral density function and the autocorrelation function for white noise.

density for a constant signal, with autocorrelation $R_S(\tau) = R_0$. This is, no doubt, the most predictable among all signals.

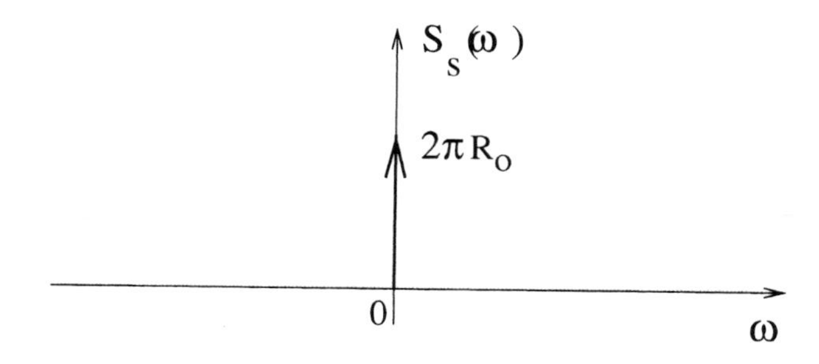

Figure 2.24. Power spectral density for a constant signal.

Properties of the Power Spectral Density

In the sequel a few properties of the power spectral density function are listed.

- The area under the curve of the power spectral density is equal to the total power of the random process, i.e.

$$P_X = \frac{1}{2\pi}\int_{-\infty}^{+\infty} S_X(\omega)d\omega. \tag{2.30}$$

This fact can be verified directly as

$$\begin{aligned} P_X &= R_X(0) = \frac{1}{2\pi}\int_{-\infty}^{\infty} S_X(\omega)e^{j\omega 0}\,d\omega \\ &= \frac{1}{2\pi}\int_{-\infty}^{\infty} S_X(\omega)\,d\omega = \int_{-\infty}^{\infty} S_X(f)\,df, \end{aligned}$$

where $\omega = 2\pi f$.

- $S_X(0) = \int_{-\infty}^{\infty} R_X(\tau)\,d\tau$;

Application: Figure 2.25 illustrates the fact that the area under the curve of the autocorrelation function is the value of the PSD at the origin.

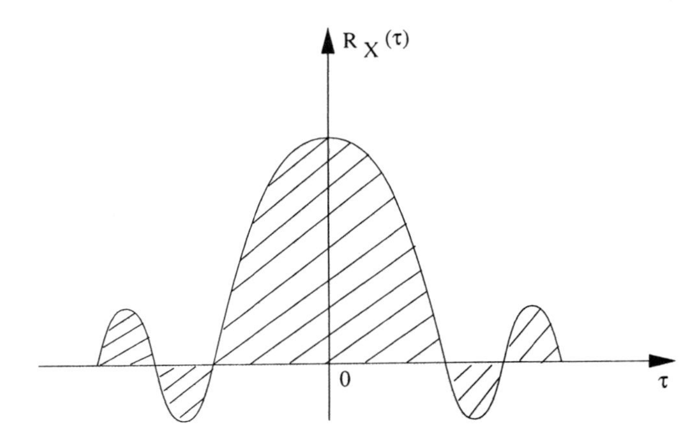

Figure 2.25. Area under the curve of the autocorrelation function.

- If $R_X(\tau)$ is real and even then

$$\begin{aligned} S_X(\omega) &= \int_{-\infty}^{\infty} R_X(\tau)[\cos\omega\tau - j\sin\omega\tau]\,d\tau, \\ &= \int_{-\infty}^{\infty} R_X(\tau)\cos\omega\tau\,d\tau, \end{aligned} \tag{2.31}$$

i.e., $S_X(\omega)$ is real and even.

- $S_X(\omega) \geq 0$, since the density reflects a power measure.
- The following identities hold.

$$\int_{-\infty}^{+\infty} R_X(\tau) R_Y(\tau) d\tau = \frac{1}{2\pi} \int_{-\infty}^{+\infty} S_X(w) S_Y(w) dw \tag{2.32}$$

$$\int_{-\infty}^{+\infty} R_X^2(\tau) d\tau = \frac{1}{2\pi} \int_{-\infty}^{+\infty} S_X^2(w) dw. \tag{2.33}$$

Finally, the cross-correlation between two random processes $X(t)$ and $Y(t)$ is defined as

$$R_{XY}(\tau) = E[X(t)Y(t+\tau)], \tag{2.34}$$

which leads to the definition of the cross-power spectral density $S_{XY}(\omega)$.

$$S_{XY}(\omega) = \int_{-\infty}^{+\infty} R_{XY}(\tau) e^{-j\omega\tau} d\tau. \tag{2.35}$$

Example: By knowing that $\hat{m}(t) = \frac{1}{\pi t} * m(t)$ is the Hilbert transform of $m(t)$ and using properties of the autocorrelation and of the cross-correlation, it can be shown that

$$E[m(t)^2] = E[\hat{m}(t)^2]$$

and that

$$E[m(t)\hat{m}(t)] = 0.$$

If two stationary processes, $X(t)$ and $Y(t)$, are added to form a new process $Z(t) = X(t) + Y(t)$, then the autocorrelation function of the new process is given by

$$\begin{aligned} R_Z(\tau) &= E[Z(t) \cdot Z(t+\tau)] \\ &= E[(x(t) + y(t))(x(t+\tau) + y(t+\tau))], \end{aligned}$$

which implies

$$R_Z(\tau) = E[x(t)x(t+\tau) + y(t)y(t+\tau) + x(t)y(t+\tau) + x(t+\tau)y(t)].$$

By applying properties of the expected value to the above expression it follows that

$$R_Z(\tau) = R_X(\tau) + R_Y(\tau) + R_{XY}(\tau) + R_{YX}(\tau). \quad (2.36)$$

If the processes $X(t)$ and $Y(t)$ are uncorrelated, then $R_{XY}(\tau) = R_{YX}(\tau) = 0$. Thus, $R_Z(\tau)$ can be written as

$$R_Z(\tau) = R_X(\tau) + R_Y(\tau), \quad (2.37)$$

and the associated power can be written as

$$P_Z = R_Z(0) = P_X + P_Y.$$

The corresponding power spectral density is given by

$$S_Z(\omega) = S_X(\omega) + S_Y(\omega).$$

2.5 Linear Systems

Linear systems when examined under the light of the theory of stochastic processes provide more general and more interesting solutions then those resulting from classical analysis. This section deals with the response of linear systems to a random input $X(t)$.

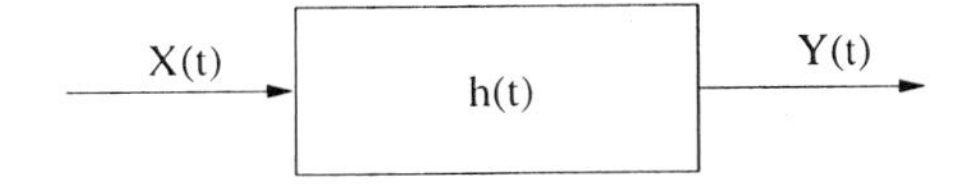

Figure 2.26. Linear system fed with a random input signal.

For a linear system, as illustrated in Figure 2.26, the Fourier transform of its impulse response $h(t)$ is given by

$$H(\omega) = \int_{-\infty}^{\infty} h(t)e^{-j\omega t}\, dt. \quad (2.38)$$

The linear system response $Y(t)$ is obtained by means of the convolution of the input signal with the impulse response as follows.

$$\begin{aligned} Y(t) &= X(t) * h(t) \;\Rightarrow\; Y(t) = \int_{-\infty}^{\infty} X(t-\alpha)\, h(\alpha)\, d\alpha \\ &= \int_{-\infty}^{\infty} X(\alpha)\, h(t-\alpha)\, d\alpha. \end{aligned}$$

Expected Value of Output Signal

The mean value of the random signal at the output of a linear system is calculated as follows.

$$E[Y(t)] = E\left[\int_{-\infty}^{\infty} X(t-\alpha)\,h(\alpha)\,d\alpha\right] = \int_{-\infty}^{\infty} E[X(t-\alpha)]\,h(\alpha)\,d\alpha$$

Considering the random signal $X(t)$ to be narrow-sense stationary, it follows that $E[X(t-\alpha)] = E[X(t)] = m_X$, and thus

$$E[Y(t)] = m_X \int_{-\infty}^{\infty} h(\alpha)\,d\alpha = m_X H(0),$$

where $H(0) = \int_{-\infty}^{\infty} h(\alpha)\,d\alpha$ follows from (2.38) computed at $\omega = 0$. Therefore, the mean value of the output signal depends only on the mean value of the input signal and on the value assumed by the transfer function at $\omega = 0$.

The Response of Linear Systems to Random Signals

The computation of the autocorrelation of the output signal, given the autocorrelation of the input signal to a linear system can be performed as follows.

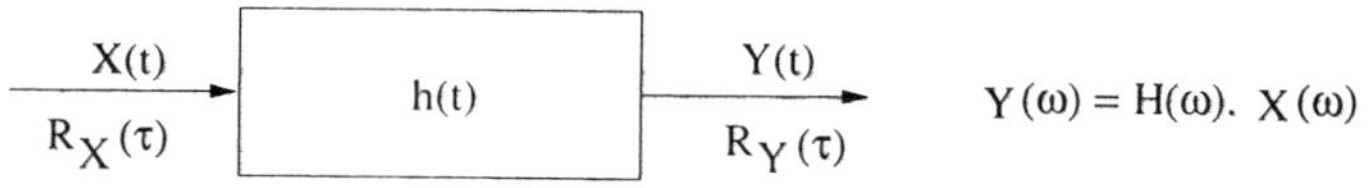

Figure 2.27. Linear system and input-output relationships.

The relationship between the input and the output of a linear system was shown earlier to be given by

$$Y(t) = \int_{-\infty}^{\infty} X(\rho)\,h(t-\rho)\,d\rho = \int_{-\infty}^{\infty} X(t-\rho)\,h(\rho)\,d\rho = X(t) * h(t).$$

The output autocorrelation function can be calculated directly from its definition as

$$\begin{aligned} R_Y(\tau) &= E[Y(t)Y(t+\tau)] \\ &= E\left[\int_{-\infty}^{\infty} X(t-\rho)\, h(\rho\, d\rho \cdot \int_{-\infty}^{\infty} X(t+\tau-\sigma)\, h(\sigma)\, d\sigma\right] \\ &= \int_{-\infty}^{\infty}\int_{-\infty}^{\infty} E[X(t-\rho)X(t+\tau-\sigma)] \cdot h(\rho) \cdot h(\sigma)\, d\rho\, d\sigma \\ &= \int_{=\infty}^{\infty}\int_{-\infty}^{\infty} R_{XX}(\tau+\rho-\sigma)\, h(\rho)\, h(\sigma)\, d\rho\, d\sigma. \end{aligned}$$

Example: Suppose that white noise with autocorrelation $R_X(\tau) = \delta(\tau)$ is the input signal to a linear system. The corresponding autocorrelation function of the output signal is given by

$$\begin{aligned} R_Y(\tau) &= \int_{-\infty}^{\infty}\int_{-\infty}^{\infty} \delta(\tau+\rho-\sigma)\, h(\rho)\, h(\sigma)\, d\rho\, d\sigma \\ &= \int_{-\infty}^{\infty} h(\sigma-\tau) \cdot h(\sigma)\, d\sigma \\ &= h(-\tau) * h(\tau). \end{aligned}$$

The Fourier transform of $R_Y(\tau)$ leads to the following result

$$R_Y(\tau) = h(-t) * h(t) \iff S_Y(\omega) = H(-\omega) \cdot H(\omega),$$

and for $h(\tau)$ a real function of τ it follows that $H(-\omega) = H^*(\omega)$, and consequently

$$S_Y(\omega) = H(-\omega) \cdot H(\omega) = H^*(\omega) \cdot H(\omega) = |H(\omega)|^2.$$

Summarizing, the output power spectral density is $S_Y(\omega) = |H(\omega)|^2$ when white noise is the input to a linear system.

In general the output power spectral density can be computed by applying the Wiener-Khintchin theorem $S_Y(\omega) = \mathcal{F}[R_Y(\tau)]$,

$$S_Y(\omega) = \int_{-\infty}^{\infty}\int_{-\infty}^{\infty}\int_{-\infty}^{\infty} R_X(\tau+\rho-\sigma)\, h(\rho)\, h(\sigma) \cdot e^{-j\omega\tau}\, d\rho\, d\sigma\, d\tau.$$

Integrating on the variable τ, it follows that

$$S_Y(\omega) = \int_{-\infty}^{\infty}\int_{-\infty}^{\infty} S_X(\omega) e^{j\omega(\rho-\sigma)}\, h(\rho)\, h(\sigma)\, d\rho\, d\sigma.$$

Finally, removing $S_X(\omega)$ from the double integral and then separating the two variables in this double integral it follows that

$$\begin{aligned} S_Y(\omega) &= S_X(\omega) \int_{-\infty}^{\infty} h(\rho) e^{j\omega\rho}\, d\rho \int_{-\infty}^{\infty} h(\sigma) e^{-j\omega\sigma}\, d\sigma \\ &= S_X(\omega) \cdot H(-\omega) \cdot H(\omega). \end{aligned}$$

Therefore, $S_Y(\omega) = S_X(\omega) \cdot |H(\omega)|^2$ will result whenever the system impulse response is a real function.

Example: Consider again white noise with autocorrelation function $R_X(\tau) = \delta(\tau)$ applied to a linear system. The white noise power spectral density is calculated as follows

$$S_X(\omega) = \int_{-\infty}^{\infty} R_X(\tau) e^{-j\omega\tau}\, d\tau = \int_{-\infty}^{\infty} \delta(\tau) e^{-j\omega\tau}\, d\tau = 1,$$

from which it follows that

$$S_Y(\omega) = |H(\omega)|^2,$$

similar to the previous example.

Example: The linear system shown in Figure 2.28 is a differentiator, used in control systems or demodulator/detector for frequency modulated signals.

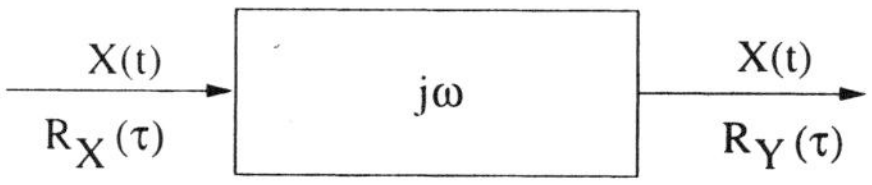

Figure 2.28. Differentiator circuit.

The output power spectral density for this circuit (or its frequency response) is equal to

$$S_Y(\omega) = |j\omega|^2 \cdot S_X(\omega) = \omega^2 S_X(\omega).$$

It is thus noticed that, for frequency modulated signals, the noise at the detector output follows a square law, i.e. , the output power spectral density grows with the square of the frequency. In this manner, in a frequency division multiplexing of frequency modulated channels, the noise will affect more intensely those channels occupying the higher frequency region of the spectrum.

Figure 2.29 shows, as an illustration of what has been discussed so far about square noise, the spectrum of a low-pass flat noise (obtained by passing white noise through an ideal low-pass filter). This filtered white noise is applied to the differentiator circuit of the example, which in turn produces at the output the square law noise shown in Figure 2.30.

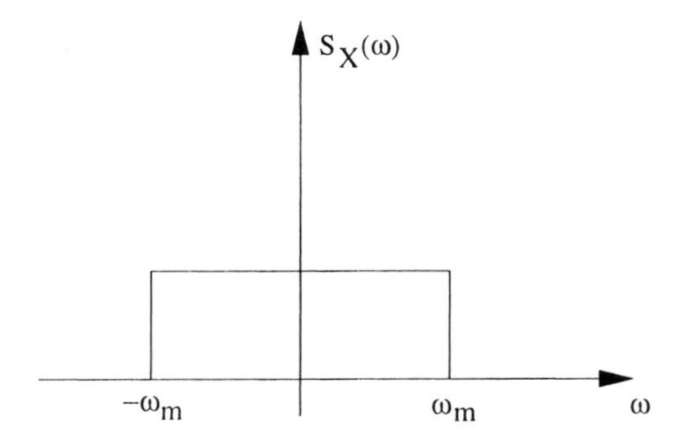

Figure 2.29. Low-pass noise.

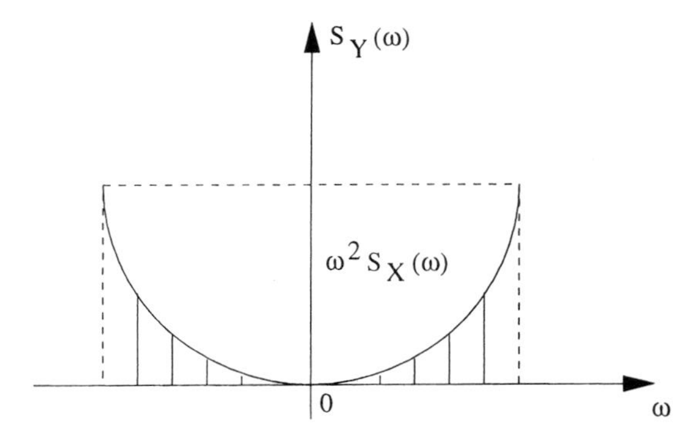

Figure 2.30. Quadratic noise.

Observation: Pre-emphasis circuits are used in FM modulators to compensate for the effect of square noise.

Other relationships among different correlation functions can be derived, as illustrated in Figure 2.31. The correlation measures between input and output (input-output cross-correlation) and correlation between output and input can also be calculated from the input signal autocorrelation.

The correlation between the input and the output can be calculated with the formula

$$R_{XY}(\tau) = E[X(t)Y(t+\tau)],$$

and in an analogous manner, the correlation between the output and the input can be calculated as

$$R_{YX}(\tau) = E[Y(t)X(t+\tau)].$$

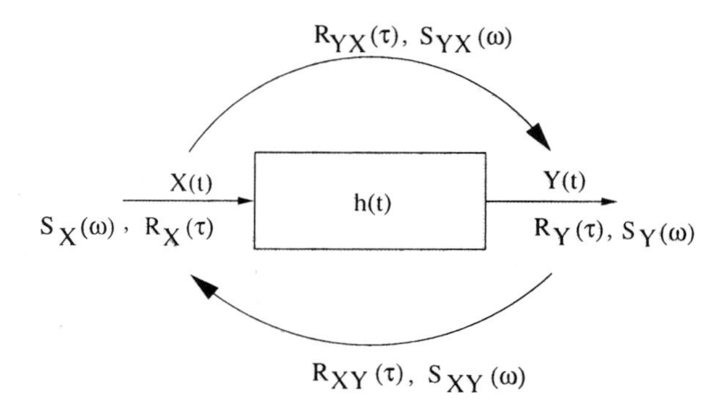

Figure 2.31. Linear system and its correlation measures.

For a linear system, the correlation between output and input is given by

$$R_{YX}(\tau) = E\left[\int_{-\infty}^{\infty} X(t-\rho)\,h(\rho)\,d\rho \cdot X(t+\tau)\right].$$

Exchanging the order of the expected value and integral computations, due to their linearity, it follows that

$$R_{YX}(\tau) = \int_{-\infty}^{\infty} E[X(t-\rho)X(t+\tau)]\,h(\rho)\,d\rho = \int_{-\infty}^{\infty} R_X(\tau+\rho)\,h(\rho)\,d\rho.$$

In a similar manner, the correlation between the input and the output is calculated as

$$\begin{aligned} R_{XY}(\tau) &= E\left[X(t)\cdot\int_{-\infty}^{\infty} X(t+\tau-\rho)\,h(\rho)\,d\rho\right] \\ &= \int_{-\infty}^{\infty} E[X(t)X(t+\tau-\rho)]\,h(\rho)\,d\rho, \end{aligned}$$

and finally,

$$R_{XY}(\tau) = \int_{-\infty}^{\infty} R_X(\tau-\rho)\,h(\rho)\,d\rho.$$

Therefore

$$R_{XY}(\tau) = R_X(\tau) * h(\tau)$$

and

$$R_{YX}(\tau) = R_X(\tau) * h(-\tau).$$

The resulting cross-power spectral densities, respectively between input-output and between output-input are given by

$$S_{XY}(\tau) = S_X(\omega) \cdot H(\omega),$$

$$S_{YX}(\tau) = S_X(\omega) \cdot H^*(\omega).$$

By assuming $S_Y(\omega) = |H(\omega)|^2 S_X(\omega)$, the following relationships are immediate.

$$\begin{aligned} S_Y(\omega) &= H^*(\omega) \cdot S_{XY}(\omega) \\ &= H(\omega) \cdot S_{YX}(\omega). \end{aligned}$$

Since it is usually easier to determine power spectral densities rather than autocorrelations, the determination of $S_Y(\omega)$ from $S_X(\omega)$ is of fundamental interest. Other power spectral densities are calculated afterwards and the correlations are then obtained by means of the Wiener-Khintchin theorem. This is pictorially indicated below.

$$\begin{array}{ccccccc} R_X(\tau) & \longleftrightarrow & S_X(\omega) & & & & \\ & & \downarrow & & & & \\ R_Y(\tau) & \longleftrightarrow & S_Y(\omega) & \longrightarrow & S_{XY}(\omega) & \longleftrightarrow & R_{XY}(\tau) \\ & & \downarrow & & & & \\ & & S_{YX}(\omega) & & & & \\ & & \updownarrow & & & & \\ & & R_{YX}(\tau) & & & & \end{array}$$

Phase Information

The autocorrelation is a special measure of average behavior of a signal. Consequently, it is not always possible to recover a signal from its autocorrelation. Since the power spectral density is a function of the autocorrelation, it also follows that signal recovery from its PSD is not always possible because phase information about the signal has been lost in the averaging operation involved. However, the cross-power spectral densities, relating input-output and output-input, preserve signal phase information and can be used to recover the phase function explicitly.

The transfer function of a linear system can be written as

$$H(\omega) = |H(\omega)|e^{j\theta(\omega)},$$

where the modulus $|H(\omega)|$ and the phase $\theta(\omega)$ are clearly separated. The complex conjugate of the transfer function is

$$H^*(\omega) = |H(\omega)|e^{-j\theta(\omega)}.$$

Since

$$S_Y(\omega) = H^*(\omega)S_{XY}(\omega) = H(\omega) \cdot S_{YX}(\omega),$$

it follows for a real $h(t)$ that

$$|H(\omega)|e^{-j\theta\omega} \cdot S_{XY}(\omega) = |H(\omega)| \cdot e^{j\theta\omega} \cdot S_{YX}(\omega)$$

and finally,

$$e^{2j\theta(\omega)} = \frac{S_{XY}(\omega)}{S_{YX}(\omega)}.$$

The function $\theta(\omega)$ can then be extracted thus giving

$$\theta(\omega) = \frac{1}{2j} \ln \frac{S_{XY}(\omega)}{S_{YX}(\omega)},$$

which is the desired signal phase information.

Example: The Hilbert transform provides a simple example of application of the preceding theory. The time domain representation (filter impulse response) of the Hilbert transform is shown in Figure 2.32. The frequency domain representation (transfer function) of the Hilbert transform is shown in Figures 2.33 and 2.34.

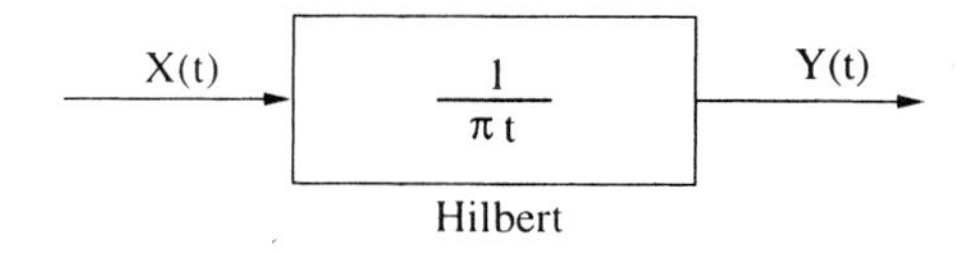

Figure 2.32. Hilbert filter fed with a random signal.

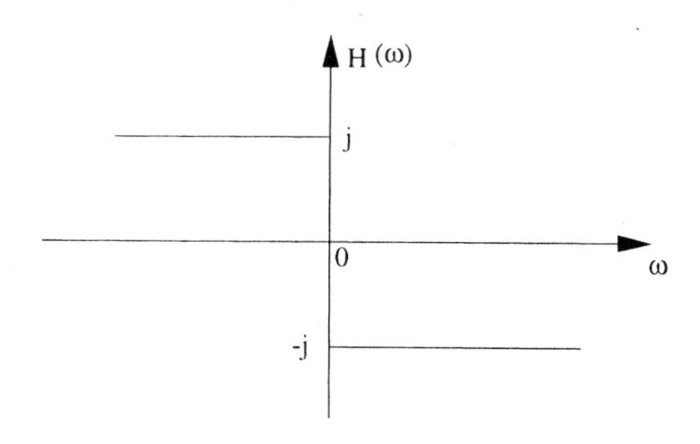

Figure 2.33. Frequency domain representation of the Hilbert transform.

Since for the Hilbert transform $H(\omega) = j\,[u(-\omega) - u(\omega)]$, it follows that $|H(\omega)|^2 = 1$, and thus from $S_Y(\omega) = |H(\omega)|^2 \cdot S_X(\omega)$ it follows that

$$S_Y(\omega) = S_X(\omega).$$

The fact that $S_Y(\omega) = S_X(\omega)$ comes as no surprise since the Hilbert transform acts only on the signal phase and the PSD does not contain phase information.

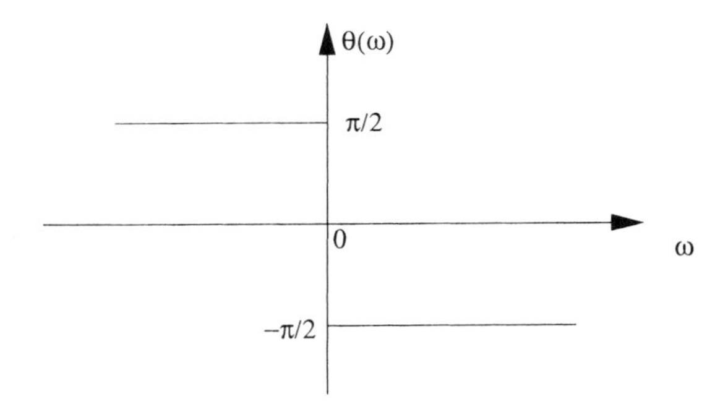

Figure 2.34. Frequency domain representation of the phase function of the Hilbert transform.

2.6 Mathematical Formulation for the Digital Signal

This section presents a mathematical formulation for the digital signal, including the computation of the autocorrelation function and the power spectrum density.

The digital signal, which can be produced by the digitization of the speech signal, or directly generated by a computer hooked to the Internet, or other equipment, can be mathematically expressed as

$$m(t) = \sum_{k=-\infty}^{\infty} m_k p(t - kT_b), \tag{2.39}$$

where m_k represents the k-th randomly generated symbol from the discrete alphabet, $p(t)$ is the pulse function that shapes the transmitted signal and T_b é is the bit interval.

2.6.1 Autocorrelation for the Digital Signal

The autocorrelation function for signal $m(t)$, which can be non-stationary, is given by the formula

$$R_M(\tau, t) = E[m(t)m(t+\tau)]. \tag{2.40}$$

Substituting $m(t)$ into Formula 2.40,

$$R_M(\tau, t) = E\left[\sum_{k=-\infty}^{\infty} \sum_{i=-\infty}^{i=\infty} m_k p(t - kT_b) m_j p(t + \tau - iT_b)\right]. \tag{2.41}$$

The expected value operator applies directly to the random signals, because of the linearity property, giving

$$R_M(\tau, t) = \sum_{k=-\infty}^{\infty} \sum_{i=-\infty}^{i=\infty} E\left[m_k m_j\right] p(t - kT_b) p(t + \tau - iT_b). \tag{2.42}$$

In order to eliminate the time dependency, the time average is taken in Equation 2.42, producing

$$R_M(\tau) = \frac{1}{T_b} \int_0^{T_b} R_M(\tau, t) dt, \tag{2.43}$$

or, equivalently

$$R_M(\tau) = \frac{1}{T_b} \int_0^{T_b} \sum_{k=-\infty}^{\infty} \sum_{i=-\infty}^{i=\infty} E\left[m_k m_j\right] p(t - kT_b) p(t + \tau - iT_b) dt. \tag{2.44}$$

Changing the integral and summation operations, it follows that

$$R_M(\tau) = \frac{1}{T_b} \sum_{k=-\infty}^{\infty} \sum_{i=-\infty}^{i=\infty} E\left[m_k m_j\right] \int_0^{T_b} p(t - kT_b) p(t + \tau - iT_b) dt. \tag{2.45}$$

Defining the discrete autocorrelation as

$$R(k - i) = E\left[m_k m_j\right], \tag{2.46}$$

the signal autocorrelation can be written as

$$R_M(\tau) = \frac{1}{T_b} \sum_{k=-\infty}^{\infty} \sum_{i=-\infty}^{i=\infty} R(k-i) \int_0^{T_b} p(t-kT_b)p(t+\tau-iT_b)dt. \tag{2.47}$$

For a rectangular pulse, with independent and equiprobable symbols, the autocorrelation function is given by

$$R_M(\tau) = A^2[1 - \frac{|\tau|}{T_b}][u(\tau+T_b) - u(\tau - T_b)], \tag{2.48}$$

where T_b is the bit interval, A represents the pulse amplitude.

Figure 2.35 shows that this function has a triangular shape. Its maximum occurs at the origin (signal power) and is equal to A^2. The autocorrelation decreases linearly with the time interval, and reaches zero at time T_b.

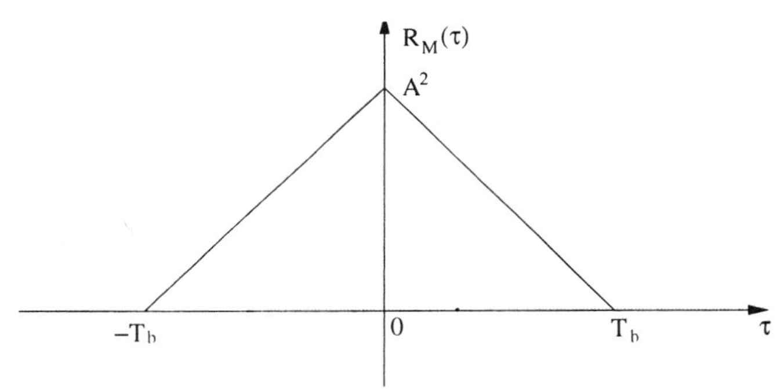

Figure 2.35. Autocorrelation for the digital signal.

2.6.2 Power Spectrum Density for the Digital Signal

The power spectrum density for the digital signal can be obtained by taking the Fourier transform of the autocorrelation function, Equation 2.47. This is the well-known Wiener-Khintchin theorem.

$$S_M(\omega) = \int_{-\infty}^{\infty} R_M(\tau)e^{-j\omega\tau}d\tau, \tag{2.49}$$

Therefore,

$$S_M(\omega) = \frac{1}{T_b} \sum_{k=-\infty}^{\infty} \sum_{i=-\infty}^{i=\infty} R(k-i) \cdot \int_{-\infty}^{\infty} \int_{0}^{T_b} p(t-kT_b)p(t+\tau-iT_b)e^{-j\omega\tau}dtd\tau. \quad (2.50)$$

Changing the order of integration, one can compute the Fourier integral of the shifted pulse can be written as

$$S_M(\omega) = \frac{1}{T_b} \sum_{k=-\infty}^{\infty} \sum_{i=-\infty}^{i=\infty} R(k-i) \int_{0}^{T_b} p(t-kT_b)P(\omega)e^{-j\omega(kT_b-t)}dt. \quad (2.51)$$

The term $P(\omega)e^{-j\omega kT_b}$ is independent of time and can be taken out of the integral, i.e.

$$S_M(\omega) = \frac{1}{T_b} \sum_{k=-\infty}^{\infty} \sum_{i=-\infty}^{i=\infty} R(k-i)P(\omega)e^{-j\omega kT_b} \int_{0}^{T_b} p(t-kT_b)e^{j\omega t}dt. \quad (2.52)$$

Computing the integral in 2.52 gives

$$S_M(\omega) = \frac{1}{T_b} \sum_{k=-\infty}^{\infty} \sum_{i=-\infty}^{i=\infty} R(k-i)P(\omega)P(-\omega)e^{-j\omega(k-j)T_b}. \quad (2.53)$$

As can be observed from the previous equation the shape of the spectrum for the random digital signal depends on the pulse shape, defined by $P(\omega)$, and also on the manner the symbols relate to each other, specified by the discrete autocorrelation function $R(k-i)$.

Therefore, the signal design involves pulse shaping as well as the control of the correlation between the transmitted symbols, that can be obtained by signal processing.

For a rectangular pulse, one can write $P(-\omega) = P^*(\omega)$, and the power spectrum density can be written as

$$S_M(\omega) = \frac{|P(\omega)|^2}{T_b} \sum_{k=-\infty}^{\infty} \sum_{i=-\infty}^{i=\infty} R(k-i)e^{-j\omega(k-j)T_b}, \quad (2.54)$$

which can be simplified to

$$S_M(\omega) = \frac{|P(\omega)|^2}{T_b} S(\omega), \tag{2.55}$$

letting $l = k - i$, the summations can be simplified and the power spectrum density for the discrete sequence of symbols is given by

$$S(\omega) = \sum_{l=-\infty}^{l=\infty} R(l) e^{-j\omega l T_b}. \tag{2.56}$$

For the example of Equation 2.48, the corresponding power spectrum density is

$$S_M(\omega) = A^2 T_b \frac{\sin^2(\omega T_b)}{(\omega T_b)^2}, \tag{2.57}$$

which is the sample function squared, and shows that the random digital signal has a continuous spectrum that occupies a large portion of the spectrum. The function is sketched in Figure 2.36. The first null is a usual measure of the bandwidth, and is given by $\omega_M = \pi/T_b$.

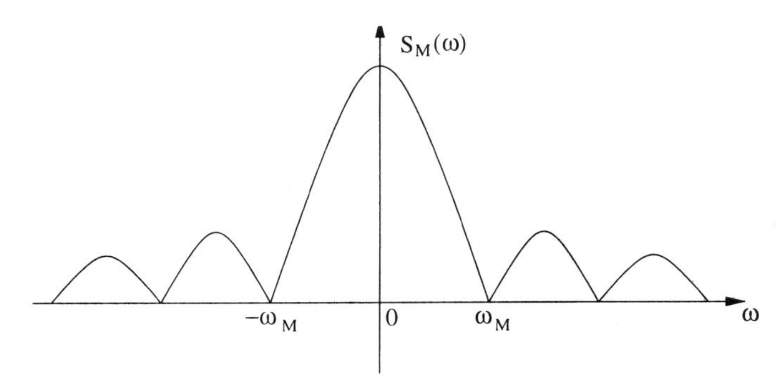

Figure 2.36. Power spectrum density for the random digital signal.

The signal bandwidth can be defined in several ways. The most common is the half-power bandwidth (ω_{3dB}). This bandwidth is computed by taking the maximum of the power spectrum density, dividing by two, and finding the frequency for which this value occurs.

The root mean square (RMS) bandwidth is computed using the frequency deviation around the carrier, if the signal is modulated, or around the origin, for a baseband signal. The frequency deviation, or RMS bandwidth, for a baseband signal is given by

$$\omega_{RMS} = \frac{\int_{-\infty}^{\infty} \omega^2 S_M(\omega) d\omega}{\int_{-\infty}^{\infty} S_M(\omega) d\omega}. \tag{2.58}$$

The previous formula is equivalent to

$$\omega_{RMS} = \frac{-R_M''(0)}{R_M(0)}. \tag{2.59}$$

The white noise bandwidth can be computed by equating the maximum of the signal power spectrum density to the noise power spectrum density $S_N = \max S_M(\omega)$. After that, the power for both signal and noise are equated and ω_N is obtained. The noise power is $P_N = 2\omega_N S_N$ and the signal power, P_M, is given by the formula

$$P_M = R_M(0) = \frac{1}{2\pi} \int_{-\infty}^{\infty} S_M(\omega) d\omega. \tag{2.60}$$

There are signals which exhibit a finite bandwidth, when the spectrum vanishes after a certain frequency. Finally, the percent bandwidth is computed by finding the frequency which includes 90% of the signal power.

2.7 Problems

1) Given two events A and B, under which conditions are the following relations true?

   a. $A \cap B = \Omega$

   b. $A \cup B = \Omega$

   c. $A \cap B = \bar{A}$

   d. $A \cup B = \emptyset$

   e. $A \cup B = A \cap B$

2) If A, B and C are arbitrary events in a sample space Ω, express $A \cup B \cup C$ as the union of three disjoint sets.

3) Show that $P\{A \cap B\} \leq P\{A\} \leq P\{A \cup B\} \leq P\{A\} + P\{B\}$ and specify the conditions for which equality holds.

4) If $P\{A\} = a$, $P\{B\} = b$ and $P\{A \cap B\} = ab$, find $P\{A \cap \bar{B}\}$ and $P\{\bar{A} \cap \bar{B}\}$.

5) Given that A, B and C are events in a given random experiment, show that the probability that exactly one of the events A, B or C

occurs is $P\{A\}+P\{B\}+P\{C\}-2P\{A\cap B\}-2P\{B\cap C\}-2P\{A\cap C\}+3P\{A\cap B\cap C\}$. Illustrate the solution with a Venn diagram.

6) Prove that a set with N elements has 2^N subsets.

7) Let A, B and C be arbitrary events in a sample space Ω, each one with a nonzero probability. Show that the sample space Ω, conditioned on A, provides a valid probability measure by proving the following.

   a. $P\{\Omega|A\}=1$
   b. $P\{B|A\}\leq P\{C|A\}$, if $B\subset C$
   c. $P\{B|A\}+P\{C|A\}=P\{B\cup C/A\}$ if $B\cap C=\emptyset$

   Show also that $P\{B|A\}+P\{\bar{B}|A\}=1$.

8) Show that if $A_1, A_2, \ldots, A_N$ are independent events then

$$P\{A_1\cup A_2\cdots\cup A_N\}=1-(1-P\{A_1\})(1-P\{A_2\})\ldots(1-P\{A_n\}).$$

9) Show that if A and B are independent events then A and $\bar{B}$ are also independent events.

10) Consider the events $A_1, A_2, \ldots, A_N$ belonging to the sample space Ω. If $\sum_{i=1}^{n} P\{A_i\}=1$, for which conditions $\bigcup_{i=1}^{n} A_i=\Omega$? If $A_1, A_2, \ldots, A_N$ are independent and $P\{A_i\}=\theta$, $i=1,\ldots,n$, find an expression for $P\{\bigcup_{i=1}^{n} A_i\}$.

11) The sample space Ω consists of the interval $[0,1]$. If sets represented by equal lengths in the interval are equally likely, find the conditions for which two events are statistically independent.

12) Using mathematical induction applied to Kolmogorov's axioms, for $A_1,\ A_2,\ldots,A_n$ mutually exclusive events, prove the following property:

$$P\left[\bigcup_{k=1}^{n} A_k\right]=\sum_{k=1}^{n} P\left[A_k\right],\ \text{for } n\geq 2.$$

13) A family of sets A_n, $n=1,2,\ldots$ is called limited above by A, by denoting $A_n\uparrow A$, if $A_n\subset A_{n+1}$ and $\bigcup_{n\geq 1} A_n=A$. Using finite additivity, show that if $A_n\uparrow A$, then $P(A_n)\uparrow P(A)$.

14) For the Venn diagram below, consider that the elements ω are represented by points and are equiprobable, i.e., $P(\omega)=\frac{1}{4}$, $\forall\omega\in\Omega$. Prove that the events A, B and C are not independent.

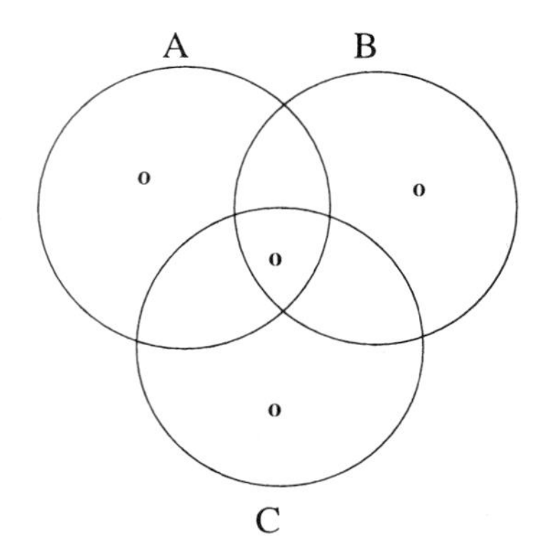

Figure 2.37. Venn diagram for problem 14.

15) For the r.v. X it is known that

$$P\{X > t\} = e^{-\mu t}(\mu t + 1),\ \mu > 0,\ t > 0.$$

Find $P_X(x)$, $p_X(x)$ e $P\{X > 1/\mu\}$.

16) The ratio between the deviations in the length and width of a substrate has the following pdf

$$p_X(x) = \frac{a}{1 + x^2},\ -\infty < x < \infty.$$

Calculate the value of a and find the CPF of X.

17) The r.v. X is uniformly distributed in the interval (a, b). Derive an expression for the nth moments $E[X^n]$ and $E[(X - m_X)^n]$, where $m_X = E[X]$.

18) For a Poisson r.v. X with parameter λ show that $P\{X \text{ even}\} = \frac{1}{2}(1 + e^{-2\lambda})$.

19) By considering a r.v. with mean m_X and variance σ_X^2, compute the moment expression $E[(X - C)^2]$, for an arbitrary constant C. For which value of C the moment $E[(X - C)^2]$ is minimized?

20) An exponential r.v. X, with parameter α, has $P\{X \geq x\} = e^{-\alpha x}$. Show that $P\{X > t + s | X > t\} = P\{X > s\}$.

21) A communication signal has a normal pdf with zero mean and variance σ^2. Design a compressor/expansor for the ideal quantizer for this distribution.

22) A voice signal with a Gamma bilateral amplitude probability distribution, given below, is fed through a compressor obeying the μ-Law of ITU-T. Compute the pdf of the signal at the compressor output and sketch an input versus output diagram for $\gamma = 1$ and $\alpha = 1/2, 1$ and 2.

$$p_X(x) = \frac{\gamma^\alpha}{\Gamma(\alpha)} |x|^{\alpha-1} e^{-|\gamma x|}, \; \alpha, \; \gamma > 0.$$

23) For a given exponential probability distribution with parameter α, compute the probability that the r.v. will take values exceeding $2/\alpha$. Estimate this probability using Tchebychev's inequality.

24) Calculate all the moments of a Gaussian distribution.

25) Show, for the Binomial distribution

$$p_X(x) = \sum_{k=0}^{\infty} \binom{N}{k} p^k (1-p)^{N-k} \delta(x-k)$$

that its characteristic function is given by

$$P_X(\omega) = \left[1 - p + pe^{j\omega}\right]^N.$$

26) The Erlang distribution has a characteristic function given by

$$P_X(\omega) = \left[\frac{a}{a + j\omega}\right]^N, \; a > 0, \; N = 1, \; 2, \; \ldots.$$

Show that $E[X] = N/a$, $E[X^2] = N(N+1)/a^2$ and $\sigma_X^2 = N/a^2$.

27) The Weibull distribution is given by

$$p_X(x) = abx^{b-1} e^{-ax^b} u(x).$$

Calculate the mean, the second moment and the variance for this distribution.

28) For the Poisson distribution

$$p_X(x) = e^{-b} \sum_{k=0}^{\infty} \frac{b^k}{k!} \delta(x-k).$$

calculate the cumulative probability function, $P_X(x)$, and show that

$$P_X(\omega) = e^{-b(1-e^{j\omega})}.$$

29) Calculate the statistical mean and the second moment of Maxwell's distribution

$$p_Y(y) = \sqrt{\frac{2}{\pi}}\frac{y^2}{\sigma^3}e^{-\frac{y^2}{2\sigma^2}}u(y),$$

exploiting the relationship between Gauss's distribution and its characteristic function:

$$p_X(x) = \frac{1}{\sqrt{2\pi}\sigma}e^{-\frac{x^2}{2\sigma^2}}$$

$$P_X(\omega) = e^{-\frac{\sigma^2\omega^2}{2}}.$$

30) Show that, for a non-negative r.v. X, it is possible to calculate its mean value through the formula

$$E[X] = \int_0^{\infty}(1 - P_X(x))dx.$$

Using this formula calculate the mean of the exponential distribution of the previous problem.

31) The dynamic range of a time discrete signal $X(n)$ is defined as $W = X_{\text{max}} - X_{\text{min}}$. Assuming that the samples of the signal $X(n)$, $n = 1, 2, \cdots, N$, are identically distributed, calculate the probability distribution of W.

32) For the following joint distribution

$$p_{XY}(x, y) = u(x)u(y)xe^{-x(y+1)}$$

calculate the marginal distributions, $p_X(x)$ and $p_Y(y)$, and show that the conditional probability density function is given by

$$p_{Y|X}(y|x) = u(x)u(y)xe^{-xy}.$$

33) The r.v.'s V and W are defined in terms of X and Y as $V = X + aY$ and $W = X - aY$, where a is a real number. Determine a as a function of the moments of X and Y, such that V and W are orthogonal.

34) Show that $E[E[Y|X]] = E[Y]$, where

$$E[E[Y|X]] = \int_{-\infty}^{\infty} E[Y|x] p_X(x) dx.$$

35) Find the distribution of the r.v. $Z = X/Y$, assuming that X and Y are statistically independent r.v.'s having an Exponential distribution with parameter equal to 1. Use the formula

$$p_Z(z) = \int_{-\infty}^{\infty} |y| p_{XY}(yz, y) dy.$$

36) A random variable Y is defined by the equation $Y = X + \beta$, where X is a r.v. with an arbitrary probability distribution and β is a constant. Determine the value of β which minimizes $E[X^2]$. Use this value and the expression for $E[(X \pm Y)^2]$ to determine a lower and an upper limit for $E[XY]$.

37) Two r.v.'s, X and Y, have the following characteristic functions, respectively,

$$P_X(\omega) = \frac{\alpha}{\alpha + j\omega} \quad \text{and} \quad P_Y(\omega) = \frac{\beta}{\beta + j\omega}.$$

Calculate the statistical mean of the sum $X + Y$.

38) Determine the probability density function of $Z = X/Y$. Given that X and Y are r.v.'s with zero mean and Gaussian distribution, show that Z is Cauchy distributed.

39) Determine the probability density function of $Z = XY$, given that X is a r.v. with zero mean and a Gaussian distribution, and Y is a r.v. with the following distribution

$$p_Y(y) = \frac{1}{2}[\delta(y+1) + \delta(y-1)].$$

40) Let X and Y be statistically independent r.v.'s having Exponential distribution with parameter $\alpha = 1$. Show that $Z = X + Y$ and $W = X/Y$ are statistically independent and find their respective probability density function and cumulative probability function.

41) Show that if X and Y are statistically independent r.v.'s then

$$P(X < Y) = \int_{\infty}^{\infty} (1 - P_Y(x)) p_X(x) dx.$$

Suggestion: sketch the region $\{X < Y\}$.

42) By using the Cauchy-Schwartz inequality show that

$$P_{XY}(x,y) \leq \sqrt{P_X(x)P_Y(y)}.$$

43) Given the joint distribution

$$p_{XY}(x,y) = kxye^{-x^2-y^2},$$

where X and Y are non-negative r.v.'s, determine k, $p_X(x)$, $p_Y(x)$, $p_X(x|y)$, $p_Y(y|x)$ and the first and second moments of this distribution.

44) Determine the relationship between the cross-correlation and the mean of two uncorrelated processes.

45) Design an equipment to measure the cross-correlation between two signals, employing delay units, multipliers and integrators.

46) Show that the RMS bandwidth of a signal $X(t)$ is given by

$$B^2_{RMS} = \frac{-1}{R_X(0)} \frac{d^2 R_X(\tau)}{d\tau^2}, \text{ for } \tau = 0.$$

47) For the complex random process

$$X(t) = \sum_{n=1}^{N} A_n e^{j\omega_o t + j\theta_n}$$

where A_n and θ_n are statistically independent r.v.'s, θ_n is uniformly distributed in the interval $[0, 2\pi]$ and $n = 1, 2, \ldots, N$, show that

$$R_X(\tau) = E[X^*(t)X(t+\tau)] = e^{j\omega_o \tau} \sum_{n=1}^{N} E[A_n^2].$$

48) The process $X(t)$ is stationary with autocorrelation $R_X(\tau)$. Let

$$Y = \int_a^{a+T} X(t)dt, T > 0, \ a \text{ real},$$

and then show that

$$E[|Y|^2] = \int_{-T}^{T} (T - |\tau|) R_X(\tau) d\tau.$$

49) The processes $X(t)$ and $Y(t)$ are jointly wide sense stationary. Let

$$Z(t) = X(t)\cos w_c t + Y(t)\sin w_c t.$$

Under which conditions, in terms of means and correlation functions of $X(t)$ and $Y(t)$, is $Z(t)$ wide sense stationary? Applying these conditions compute the power spectral density of the process $Z(t)$. What power spectral density would result if $X(t)$ and $Y(t)$ were uncorrelated?

50) For a given complex random process $Z(t) = X(t) + jY(t)$ show that

$$E[|Z(t)|^2] = R_X(0) + R_Y(0).$$

51) Considering that the geometric mean of two positive numbers can not exceed the correspondent arithmetic mean, and that $E[(Y(t+\tau) + \alpha X(t))^2] \geq 0$, show that

$$|R_{XY}(\tau)| \leq \frac{1}{2}[R_X(0) + R_Y(0)].$$

52) Let $X(t)$ be a wide sense stationary process, with mean $E[X(t)] \neq 0$. Show that

$$S_X(\omega) = 2\pi E^2[X(t)]\delta(\omega) + \int_{-\infty}^{\infty} C_X(\tau)e^{-j\omega\tau}d\tau,$$

where $C_X(\tau)$ is the auto-covariance function of $X(t)$.

53) Calculate the RMS bandwidth

$$B_{RMS}^2 = \frac{\int_{-\infty}^{\infty} \omega^2 S_X(\omega)d\omega}{\int_{-\infty}^{\infty} S_X(\omega)d\omega}$$

of a modulated signal having the following power spectral density

$$S_X(\omega) = \frac{\pi A^2}{2\Delta_{FM}}\left[p_X\left(\frac{\omega+\omega_c}{\Delta_{FM}}\right) + p_X\left(\frac{\omega-\omega_c}{\Delta_{FM}}\right)\right],$$

where $p_X(\cdot)$ denotes the probability density function of the signal $X(t)$.

54) The autocorrelation $R_X(\tau)$ can be seen as a measure of similarity between $X(t)$ e $X(t+\tau)$. In order to illustrate this point, consider the process $Y(t) = X(t) - \rho X(t+\tau)$ and determine the value for ρ which minimizes the mean square value $Y(t)$.

55) Calculate the cross-correlation between the processes $U(t) = X(t) + Y(t)$ and $V(t) = X(t) - Y(t)$, given that $X(t)$ and $Y(t)$ have zero mean and are statistically independent.

56) Determine the probability density function (pdf) for the stochastic process $X(t) = e^{At}$, where A is a uniformly distributed r.v. over the interval $[-1, 1]$. Analyze whether the process is stationary and compute its autocorrelation. Sketch a few typical representations of the process $X(t)$, for varying A, as well as the resulting pdf.

57) A time series $X(t)$ is used to predict $X(t + \tau)$. Calculate the correlation between the current value and the predicted value, given that:

    (a) The predictor uses only the current value of the series, $X(t)$.

    (b) The predictor uses the current value $X(t)$ and its derivative, $X'(t)$.

58) Find the correlation between the processes $V(t)$ and $W(t)$, where

$$V(t) = X\cos\omega_o t - Y\sin\omega_o t,$$

and

$$W(t) = Y\cos\omega_o t + X\sin\omega_o t,$$

where X and Y are statistically independent r.v.'s with zero mean and variance σ^2.

59) Calculate the power spectral density for a signal with autocorrelation

$$R(\tau) = Ae^{-\alpha|\tau|}(1 + \alpha|\tau| + \frac{1}{3}\alpha^2\tau^2).$$

60) Determine the power spectral density of a signal with autocorrelation given by the expression

$$R(\tau) = Ae^{-\alpha|\tau|}(1 + \alpha|\tau| - 2\alpha^2\tau^2 + \frac{1}{3}\alpha^3|\tau^3|).$$

61) Prove the following properties of narrow-band random processes.

    (a) $S_{XY}(\omega) = S_{YX}(-\omega)$;

    (b) $\text{Re}[S_{XY}(\omega)]$ and $\text{Re}[S_{YX}(\omega)]$ both even;

    (c) $\text{Im}[S_{XY}(\omega)]$ and $\text{Im}[S_{YX}(\omega)]$ both odd;

    (d) $S_{XY}(\omega) = S_{YX}(\omega) = 0$, if $X(t)$ and $Y(t)$ are orthogonal.

62) Show that, for uncorrelated $X(t)$ and $Y(t)$ in a narrow-band process,

$$S_{XY}(\omega) = S_{YX}(\omega) = 2\pi E[X(t)]E[Y(t)]\delta(\omega).$$

63) Prove that the RMS bandwidth is given by

$$B_{RMS}^2 = \frac{-1}{R_X(0)} \frac{d^2 R_X(\tau)}{d\tau^2}, \text{ for } \tau = 0.$$

64) For the stochastic processes

$$X(t) = Z(t)\cos(w_c t + \theta)$$

and

$$Y(t) = Z(t)\sin(w_c t + \theta),$$

where A, $\omega_c > 0$, and θ is a uniformly distributed r.v. over the interval $[0, 2\pi]$, and statistically independent of $Z(t)$, show that

$$S_{XY}(\omega) = \frac{\pi A}{2} E[Z(t)][\delta(w - w_c) + \delta(w + w_c)]$$

65) Calculate the bandwidth

$$B_N = \frac{1}{|H(0)|^2} \int_0^{\infty} |H(\omega)|^2 d\omega$$

of the noise for a system with the following transfer function

$$|H(\omega)|^2 = \frac{1}{1 + (\omega/W)^2}.$$

66) The control system given below has transfer function $H(\omega)$.

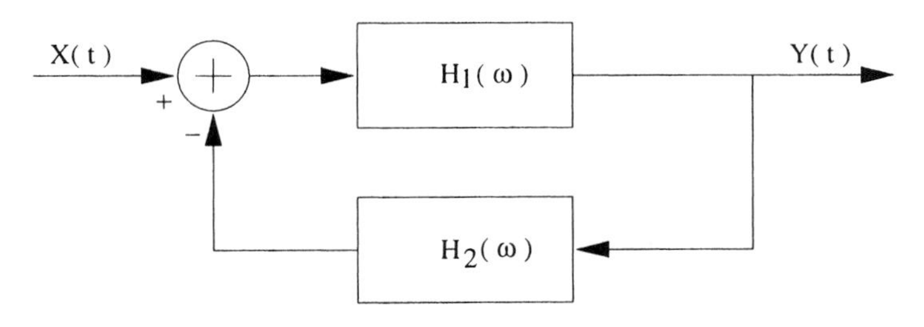

Figure 2.38. Linear control system.

It is known that the Wiener filter, which minimizes the mean square error, and is called optimum in this sense, has a transfer function given by

$$G(\omega) = \frac{S_X(\omega)}{S_X(\omega) + S_N(\omega)},$$

where $S_X(\omega)$ denotes the power spectral density of the desired signal and $S_N(\omega)$ denotes the noise power spectral density. Determine $H_1(\omega)$ and $H_2(\omega)$ such that this system operates as a Wiener filter.

67) Calculate the correlation between the input signal and the output signal for the filter with transfer function

$$H(\omega) = u(\omega + \omega_M) - u(\omega + \omega_M),$$

given that the input has autocorrelation $R_X(\tau) = \delta(\tau)$. Relate the input points for which input and output are orthogonal. Are the signals uncorrelated at these points? Explain.

68) Consider the signal $Y_n = X_n - \alpha X_{n-1}$, generated by means of white noise X_n, with zero mean and variance σ_X^2. Compute the mean value, the autocorrelation function and the power spectral density of the signal Y_n.

69) Compute the transfer function

$$H(\omega) = \frac{S_{YX}(\omega)}{S_X(\omega)}$$

of the optimum filter for estimating $Y(t)$ from $X(t) = Y(t) + N(t)$, where $Y(t)$ and $N(t)$ are zero mean statistically independent processes.

70) Consider the autoregressive process with moving average $Y_n + \alpha Y_{n-1} = X_n + \beta X_{n-1}$, built from white noise X_n, with zero mean and variance σ_X^2. Calculate the mean value, the autocorrelation and the power spectral density of the signal Y_n.

71) When estimating a time series the following result was obtained by minimizing the mean square error

$$\theta(t+\tau) \approx \frac{R(\tau)}{R(0)}\theta(t) + \frac{R'(\tau)}{R''(0)}\theta'(t),$$

where $R(\tau)$ denotes the autocorrelation of the process $\theta(t)$. By simplifying the above expression, show that the increment $\Delta\theta(t) =$

$\theta(t+\tau) - \theta(t)$ can be written as

$$\Delta\theta(t) = \tau\theta'(t) - \frac{(\tau\omega_M)^2}{2}\theta(t),$$

where ω_M denotes the RMS bandwidth of the process. Justify the intermediate steps used in your proof.

72) Show that the system below, having $h(t) = \frac{1}{T}[u(t) - u(t-T)]$, works as a meter for the autocorrelation of the signal $X(t)$. Is this measure biased?

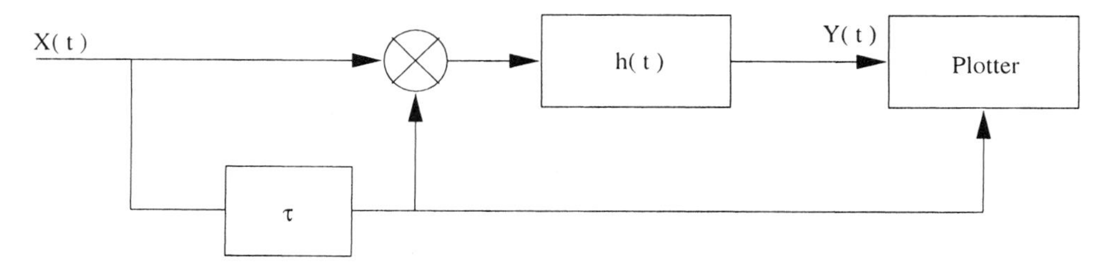

Figure 2.39. Autocorrelation meter.

73) Calculate the transfer function of the optimum filter, for estimating $Y(t)$ from $X(t) = Y(t) + N(t)$, where $Y(t)$ and $N(t)$ are zero mean independent random processes.

74) The stochastic process $Z(t) = X(t)X'(t)$ is built from a Gaussian signal $X(t)$ with zero mean and power spectral density $S_X(\omega)$. Calculate the power spectral density and the autocorrelation of $Z(t)$, knowing that for the Gaussian process

$$S_{X^2}(\omega) = 2\int_{-\infty}^{\infty} S_X(\omega - \phi)S_X(\phi)d\phi.$$

Considering the the signal $X(t)$ has a uniform power spectral density $S_X(\omega) = S_0$, between $-\omega_m$ and ω_M, determine the autocorrelation and the power spectral density of $Z(t)$.

75) It is desired to design a proportional and derivative (PD) control system to act over a signal with autocorrelation

$$R_X(\tau) = \frac{1 - \alpha|\tau|}{1 + \alpha|\tau|}, \ |\tau| \leq 1, \ \alpha > 0.$$

Determine the optimum estimator, in terms of mean square, and estimate the signal value after an interval of $\tau = 1/\alpha$ time units, as a function of the values of $X(t)$ and its derivative $X'(t)$.

Chapter 3

SPEECH CODING*

3.1 Introduction

The fundamental purpose of signal compression techniques is to reduce the number of bits required to represent a signal (speech, image, video, audio, for example), while maintaining an acceptable signal quality. Signal compression is essential for applications which need the minimization of the storage and/or transmission requirements, such as multimedia systems, integrated services digital networks (ISDN), videoconference, voice response systems, voice mail, music broadcast, high resolution facsimile, high definition television (HDTV), mobile telephony, storage of medical images, archiving of finger prints and transmission of remote sensing images obtained from satellites.

Although some systems do not present severe bandwidth limitations, such as optical fiber networks, and although the technological evolution is continuously leading to memories with high storage capacity, signal compression plays an important role, due the many aspects, such as (Jayant and Noll, 1984, Gersho and Gray, 1992):

- The wide use of multimedia systems has led to an increasing demand concerning the storage of speech, audio, image and video in compressed form;

- A larger number of communication channels may be multiplexed in broadband systems, by using compression techniques for reducing the bandwidth requirements of each signal to be multiplexed;

*Chapter 3 contributed by Francisco Madeiro and Waslon Terllizzie Araújo Lopes

- In mobile telephony, the bandwidth is severely limited, which has motivated research and development in speech coding.

3.2 Signal Coding – Preliminaries

Signal coding is the process of representing an information signal with the purpose of reaching a communication target, such as analog to digital conversion, low bit rate transmission or message encryption. In the literature, the terms source coding, signal compression, data compression, digital coding and bandwidth compression are all employed to the techniques for obtaining a compact digital representation of a signal (Jayant, 1992, Jayant et al., 1993).

Figure 3.1 presents a block diagram for a digital communication system. While the source coder aims at minimizing the bit rate needed for adequately representing an input signal, the *mo*dulator-*dem*odulator (*modem*) aims at maximizing the bit rate that may be supported in a given channel or storage device without causing an unacceptable level of bit error probability. In source coding, the bit rate is measured in bits per sample or bits per second (usually denoted by b/s or bps). In modulation, the rate is measured in bits per second per hertz (b/s/Hz). The channel coding blocks add redundancy to the bit stream with the purpose of protection against errors.

Throughout this chapter, the term signal coding is widely used, referring specifically to source coding.

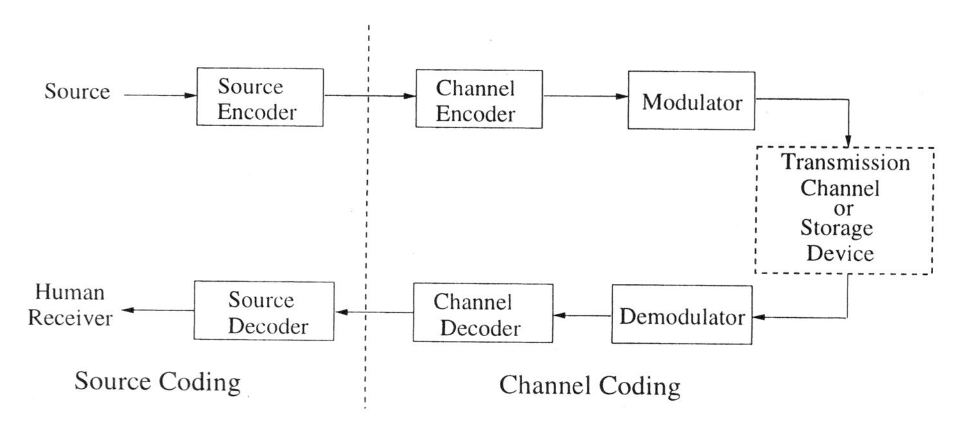

Figure 3.1. Block diagram of a digital communication system.

3.3 The Performance of a Signal Compression System

The purpose of signal compression is to reduce the bit rate in the digital representation of a signal, while maintaining the required levels of signal quality, implementation complexity and communication delay. Each one of those aspects will be discussed in the following.

3.3.1 Quality of the Reconstructed Signals

A great challenge in digital coding of signals is the development of methods for assessing the quality of reconstructed signals (obtained from the application of compression techniques). The measures used for assessing the quality of signals may be classified into two general groups: subjective quality measures and objective quality measures. The subjective measures are based on comparisons (performed by means of listening or visualization tests) between the original signal and the processed signal, performed by a group of human observers which subjectively rank the quality of the processed signal according to a predetermined scale. The objective measures are based on a direct mathematical comparison between the original and processed signals (Deller Jr. et al., 1993).

In order to be useful, the objective quality measures should present two attributes. First, they should have a subjective significance, in the sense that low and high changes in the objective measures correspond, respectively, to low and high changes in the subjective quality of the reconstructed signals. Hence, they should present a strong correlation, positive or negative, with the subjective evaluation results. Second, they should be mathematically tractable and be easily implementable.

Subjective quality measures are used to definitely assess the quality of algorithms and techniques of signal coding. However, subjective tests are difficult to implement, since they require personnel to participate in the evaluation (including lay men, specialists and possible users of the signal coding system), involve a great volume of processed signal and require the availability of laboratories with conditions suitable for the subjective evaluations. Therefore, subjective tests are very time consuming. For those reasons, objective quality measures, which are not time consuming, play an important role in the process of assessing the quality of reconstructed signals and are very useful for the task of adjusting the parameters of algorithms and techniques of compression.

The assessment of speech and image quality has been a research theme for a long time, being addressed by many researchers (Eskicioglu and Fischer, 1995, Dimolitsas, 1991).

In the following, an important subjective measure, the mean opinion score (MOS), is discussed. Three objective quality measures are addressed: the signal to noise ratio (SNR), the segmental signal to noise ratio (SNRseg) and the spectral distortion (SD).

Mean Opinion Score

A subjective quality measure widely used for assessing the performance of speech compression systems is the mean opinion score (MOS). Each listener rates the quality of the reconstructed signal according to a predetermined scale, shown in Table 3.1 (Jayant and Noll, 1984, Deller Jr. et al., 1993). Then the scores are averaged to determine the final value of the evaluation, that is

$$MOS = \frac{1}{L}\sum_{l=1}^{L} s_l, \tag{3.1}$$

where L is the number of listeners in the test and s_l is the score assigned by the l-th listener.

Score (s)	Quality
5	Excellent
4	Good
3	Fair
2	Poor
1	Unsatisfactory

Table 3.1. Mean Opinion Score (MOS) five-point scale.

In speech coding, assessments by means of MOS are well accepted and sometimes complemented by intelligibility measures, such as MRT (modified rhyme test) and DRT (diagnostic rhyme test) (Deller Jr. et al., 1993). Other measures have been used to assess the quality of speech signals, such as IAJ (isometric absolute judgment), QUART (quality acceptance rating test) and DAM (diagnostic acceptability measure) (Deller Jr. et al., 1993).

Signal-to-noise Ratio (SNR)

Let $x(n)$ be the original signal, $y(n)$ the processed signal and $e(n) = x(n) - y(n)$ the error signal at time n.

The energy contained in the original signal is

$$E_x = \sum_n x^2(n). \tag{3.2}$$

The energy contained in the error signal is

$$E_e = \sum_n e^2(n) = \sum_n [x(n) - y(n)]^2. \tag{3.3}$$

The resulting SNR, in dB, is given by (Deller Jr. et al., 1993)

$$\text{SNR} = 10 \log_{10} \frac{E_x}{E_e} = 10 \log_{10} \frac{\sum_n x^2(n)}{\sum_n [x(n) - y(n)]^2}. \tag{3.4}$$

Segmental Signal-to-noise Ratio (SNRseg)

In spite of its mathematical simplicity, the SNR measure has a limitation: it weights all time domain errors equally. This is the reason why a high SNR measure, with undesirable results, can be obtained if the speech utterance presents high concentration of voiced segments (high energy segments), since noise has a greater perceptual effect in low-energy segments, such as unvoiced fricatives (Deller Jr. et al., 1993).

An improved quality measure may be obtained if SNR is measured over short time intervals and the results averaged. The frame-based measure is called the segmental signal to noise ratio (SNRseg) and is expressed as

$$\text{SNRseg} = E[\text{SNR}(j)], \tag{3.5}$$

where SNR(j) denotes the conventional SNR for the j-th frame (time interval) of the signal.

The SNRseg measure is formulated as (Deller Jr. et al., 1993)

$$\text{SNRseg} = \frac{1}{J} \sum_{j=0}^{J-1} 10 \log_{10} \left[\frac{\sum_{n=m_j-N_A-1}^{m_j} x^2(n)}{\sum_{n=m_j-N_A-1}^{m_j} [x(n) - y(n)]^2} \right], \tag{3.6}$$

where $m_0, m_1, ..., m_{J-1}$ are the end-times for the J frames, each of which is length N_A samples, typically 15-25 msec.

Other objective quality measures may be pointed out, such as LAR (log-area ratio), Itakura log-likelihood measure (Deller Jr. et al., 1993) and the spectral distortion.

Spectral Distortion

For applications that require low-rate speech coding, it is essential to quantize suitably the line spectral frequency (LSF) parameters, using a number of bits as low as possible. The development of LSF coding methods has been an intense research area (Paliwal and Atal, 1993, LeBlanc et al., 1993, Eriksson et al., 1999).

The quality of the LSF quantization is measured by the spectral distortion (SD), formulated as

$$\mathrm{SD} = \left[\frac{1}{F_S}\int_0^{F_S} [10\log_{10} S(f) - 10\log_{10}\hat{S}(f)]^2 df\right]^{1/2}, \tag{3.7}$$

where $S(f)$ and $\hat{S}(f)$ denote, respectively, the original and quantized envelope.

3.3.2 Bit Rate

The bit rate of a digital representation may be measured in bits per sample, bits per pixel (bpp) or bits per second, depending on the scenario. The rate in bits per second is simply the product of the sampling rate by the number of bits per sample. The sampling rate is usually slightly higher than twice the signal bandwidth, as stated by the Nyquist Sampling Theorem (Jayant and Noll, 1984, Lathi, 1988).

Table 3.2 shows some formats usually used for audio (Jayant, 1992, Jayant et al., 1993). Typical sampling rates are 8 kHz for telephone speech, 16 kHz for AM radio-grade audio, 32 kHz for FM-audio and 44.1 kHz or 48 kHz for CD (compact disc) audio or DAT (digital audio tape) audio. Notice that the bandwidths are lower than half the corresponding sampling rates, in accordance to Nyquist sampling principle. It is worth mentioning that a PCM signal has a band limited by a low-pass filter. The telephony band usually transmitted is 300 Hz to 3,400 Hz in Europe and Latin America and 200 Hz to 3,400 Hz in United States and Japan.

3.3.3 Complexity

The complexity of a coding algorithm is related to the computational effort required to implement the encoding and decoding processes. Therefore, it concerns to arithmetic capability and memory requirements. The complexity is usually measured in millions of instructions per second (MIPS). Other measures related to the the complexity are the physical size of the encoder, decoder or *codec* (*enco*der plus *dec*oder), the cost and the power consumption (measured, for instance, in milliwatts,

Format	Sampling rate (kHz)	Bandwidth (kHz)	Frequency range (Hz)
Telephony	8	3.2	200-3,400
Teleconferencing	16	7	50-7,000
Compact Disc (CD)	44.1	20	20-20,000
Digital Audio Tape (DAT)	48	20	20-20,000

Table 3.2. Digital audio formats (Jayant, 1992).

mW), which, in turn, is a criterion particularly important for hand-held systems (Jayant, 1992, Jayant et al., 1993).

3.3.4 Communication Delay

The complexity increase of a coding algorithm is generally associated to a processing delay increase in encoder and decoder. The importance of delay in a communication system depends on the application. Depending on the communication environment, the tolerated overall delay may be severely limited, as in network telephony (Jayant, 1992, Jayant et al., 1993). Thus, the delay produced by a *codec* imposes some practical constraints concerning the use in communication systems, since the delay must not be higher than a certain limit. However, for some applications the communication delay in irrelevant.

3.4 Features of Speech Signals

The knowledge of the features of speech signals has been efficiently used in techniques for speech coding, speech synthesis, as well as in speech and speaker recognition systems.

The mechanism of speech production, as occurs in any physical system, presents limited frequency response. The limit, which varies from person to person, is about 10 kHz (Aguiar Neto, 1995). However, in telephony systems the speech signal is limited to the range 200-3,400 Hz, without great damage to speech quality. In other applications, as teleconference, the speech signal is limited to 7 kHz, which is known as broadband transmission.

Speech sounds may be classified in three distinct classes: voiced sounds (such as /a/ and /i/), unvoiced sounds (such as /sh/) and plosive sounds (such as /p/, /t/ and /k/) (Rabiner and Schafer, 1978).

The voiced sounds are quasi-periodic in time domain and harmonically structured in frequency, while the unvoiced sounds have a random nature and broad band. The energy of the voiced sounds is generally higher than that of unvoiced frames. Voiced sounds are produced by quasi-periodic pressure waves exciting the vocal tract, which, acting as a resonator, produces resonance frequencies called formants, which characterize the different voiced sounds. In general there are three to five formants below 5 kHz.

The amplitudes and the location of the first three formants are very important to synthesis and perception of speech (Spanias, 1994). The fundamental frequency of the voiced sounds is typically in the range of 80 Hz (for men) to 350 Hz (for children), with 240 Hz being a typical value for women (Aguiar Neto, 1995). In generation of plosive sounds, the air is completely oriented to the mouth, which is completely closed. With the increase of the pressure, the occlusion is suddenly broken.

One can also cite the mixed excitation sounds, such as unvoiced fricative sounds (such as /j/, /v/ and /z/), which are produced by combining the vocal chords vibration with a turbulent excitation, and the occlusive (or plosive) voiced sounds (such as /d/ and /b/).

Speech signals are non-stationary. However, they may be considered as quasi-stationary in short frames (short time intervals), typically 5-20 ms (Spanias, 1994). In this sense, the speech signals can be considered approximately ergodic.

One of the simplest descriptions of a waveform is given by a graphical representation of amplitude versus time, as shown in Figure 3.2. In the waveform, one can identify high energy frames, low energy frames and inter-syllabic pause frames. These and the frames of pause between words correspond to about 50 to 60 % of the total duration of the speech signal (Aguiar Neto, 1995).

Speech signals present a high variation of signal amplitude, as shown in Figure 3.3. The amplitude variation, which corresponds to about 50 dB, is called the dynamic range of the signal. It is worth mentioning that the acquisition (8.0 bit/sample, 8 kHz sampling rate) of the speech signals shown in Figures 3.2 and 3.3, as well as in Figures 3.4 and 3.5, was performed by using a Sun® workstation, equipped with audio processing tools.

Regarding the probability density function (pdf) for the amplitudes of the speech signal, approximations by the models exponential (two-sided), Laplacian, Gamma and Gaussian have been used (Aguiar Neto, 1995).

Concerning the energy of the speech signal, a concentration in the lowest frequencies of the spectrum is observed, mainly in the range 500 to 800 Hz. However, although presenting low values of energy, the highest frequency components are important since they determine most of

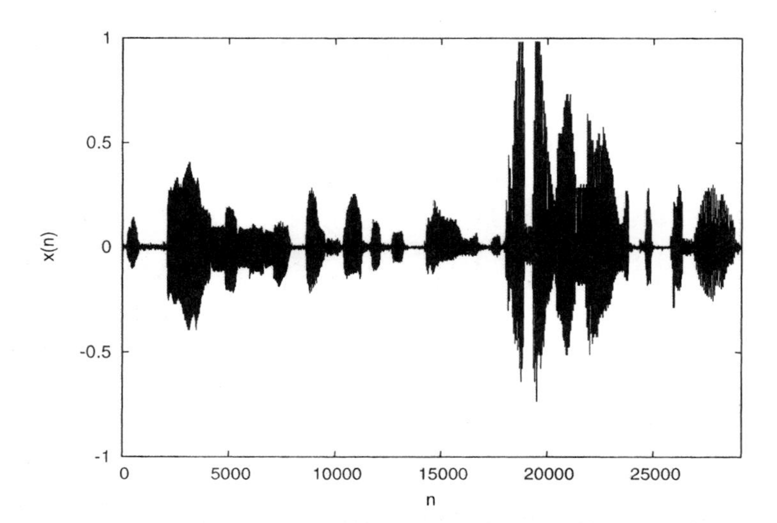

Figure 3.2. Voice waveform corresponding to the Portuguese sentence "*O sol ilumina a fachada de tarde. Trabalhou mais do que podia.*", with 29,120 samples, 3.64 s duration, 8 kHz sampling rate.

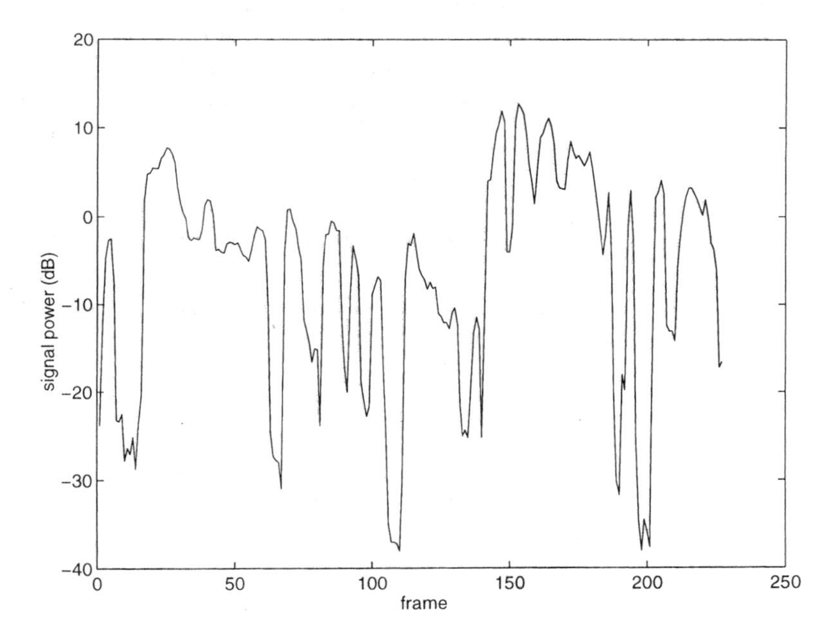

Figure 3.3. Dynamic range of the speech signal "*O sol ilumina a fachada de tarde. Trabalhou mais do que podia.*", corresponding to 29,120 samples, 3.64 s, 8 kHz sampling rate (Madeiro, 1998).

the speech intelligibility. The spectrum decays about 8-10 dB per octave. Frequencies below 500 Hz contribute very little for the speech understanding, but they have a crucial role in the naturalness of the speech (Aguiar Neto, 1995).

Figure 3.4 is a representation, in two-dimensional Euclidean space, of a speech signal (whose histogram is presented in Figure 3.5) corresponding to ten phonetically balanced Portuguese phrases (extracted from (Alcaim et al., 1992) and produced by ten different speakers, five males and five females). Some typical features of the speech signals may be observed in Figures 3.4 and 3.5, such as predominance of small amplitude samples and correlation among consecutive samples (the two-dimensional speech vectors, corresponding to two consecutive samples, are concentrated near the direction corresponding to the principal component of the speech signal, for which $x_2 = x_1$.).

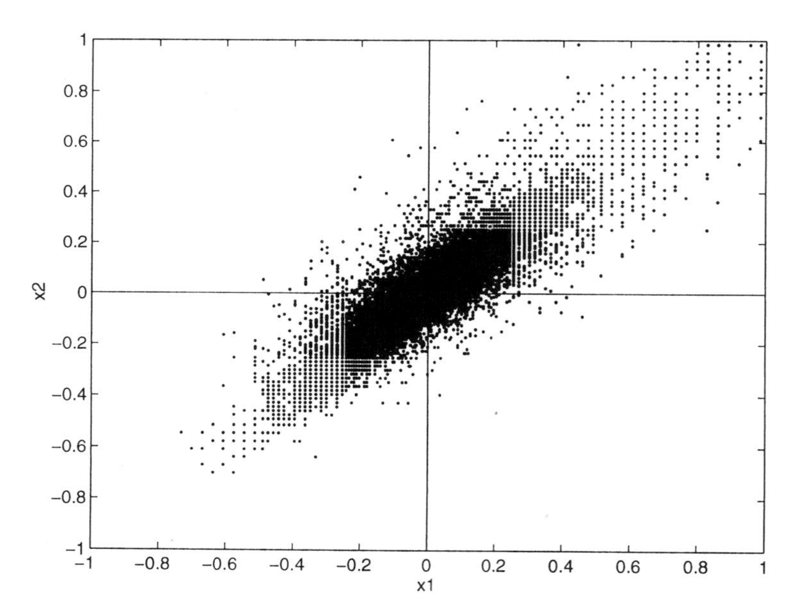

Figure 3.4. Speech signal corresponding to ten phonetically balanced Portuguese sentences (18.76 s, 75,040 vectors). Coordinates x_1 and x_2 represent the first and second components of speech vector $\boldsymbol{x} \in R^2$ respectively.

3.5 Pulse Code Modulation

Pulse code modulation (PCM) provides a method for accomplishing the digital transmission or storage of an original analog signal. In this method of signal coding, the analog message signal is sampled and the amplitude of each sample is mapped (approximated, rounded off) to the

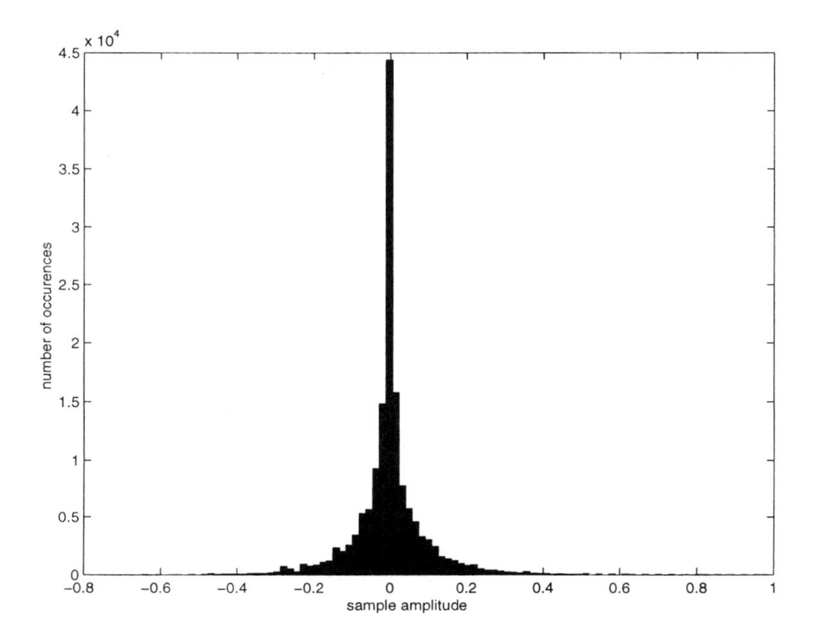

Figure 3.5. Histogram of a speech signal.

nearest one of a finite set of discrete levels (allowed quantization levels), so that both time and amplitude are represented in discrete form. This allows the message to be transmitted by means of a digital (coded) waveform.

The fundamental operations in the transmitter of a PCM system are sampling, quantizing and encoding.

The incoming message waveform is sampled using a train of narrow rectangular pulses so as to closely approximate the instantaneous sampling process. In the sampling process a continuous-time signal is transformed to a discrete-time signal. The sampling, or Nyquist theorem, previously discussed, states that a signal $f(t)$, band-limited to B Hz (that is, $F(w) = 0$ for $|w| > 2\pi B$) may be reconstructed exactly (without error) from its samples taken at a rate $R_S > 2B$ Hz (samples per second). In other words, the minimum sampling frequency for a baseband signal to be recovered without distortion is $f_S = 2B$ Hz (Lathi, 1988).

In practice, a low-pass pre-alias filter is used at the front end of the sampler in order to exclude frequencies greater than B, before sampling.

After sampling, the sample values are still analog because they lie in a continuous range. By means of the quantization process, shown in Figure 3.6, each sample is approximated to the nearest quantization level. Each sample is approximated to the midpoint of the interval in which the sample value falls. The information is thus digitized.

The quantization process introduces a certain amount of error or distortion in the signal samples. This error, known as quantization noise or quantization error, is intrinsic to the process of analog to digital conversion (Sripad and Snyder, 1977). It is minimized by establishing a large number (L) of quantization intervals. To assure the intelligibility of voice signals, for example, $L = 8$ or 16 is sufficient. For commercial use, $L = 32$ is a minimum, and for telephone communication, $L = 128$ or 256 is commonly used (Lathi, 1988).

During each sampling interval, one quantized sample is transmitted, which takes one of the L possible quantized values. To exploit the advantages of sampling and quantizing, it is required the use of an encoding process to translate the discrete set of sample values to a binary signal, because of its simplicity and ease of detection. Figure 3.6 exemplifies the quantization process performed by using $L = 8$ quantization levels. This requires the use of codewords with $log_2 8 = 3$ bits to represent each quantization level.

Several formats (waveforms) can be used to represent the binary sequences produced by the analog-to-digital conversion. Figure 3.7 depicts an example of a format, by which the binary symbol "1" is represented by a pulse of constant amplitude and duration of one bit, and symbol "0" is represented by switching off the pulse for the same duration. This format is called a nonreturn-to-zero unipolar signal, or on-off signal.

The scheme to transmit or store the digitized data using pulses is known as pulse code modulation (PCM) (Shannon, 1948).

The quantization noise can be minimized by increasing the number of quantization levels. This leads to the use codewords with a higher number of bits for representing each quantization level, which increases the transmission rate.

3.5.1 Uniform Quantization

Quantization can be uniform or nonuniform. In uniform quantization, the quantization levels are uniformly spaced. Otherwise, the quantization is nonuniform. The quantizer characteristic can be of midtread or midriser type. Figure 3.8(a) shows the input-output characteristic of a uniform quantizer of the midtread type, which is so called because the origin lies in the middle of a tread of the staircase like figure. Figure 3.9(a) shows the input-output characteristic of a uniform quantizer of the midriser type, in which the origin lies in the middle of a rising part of the staircase like diagram.

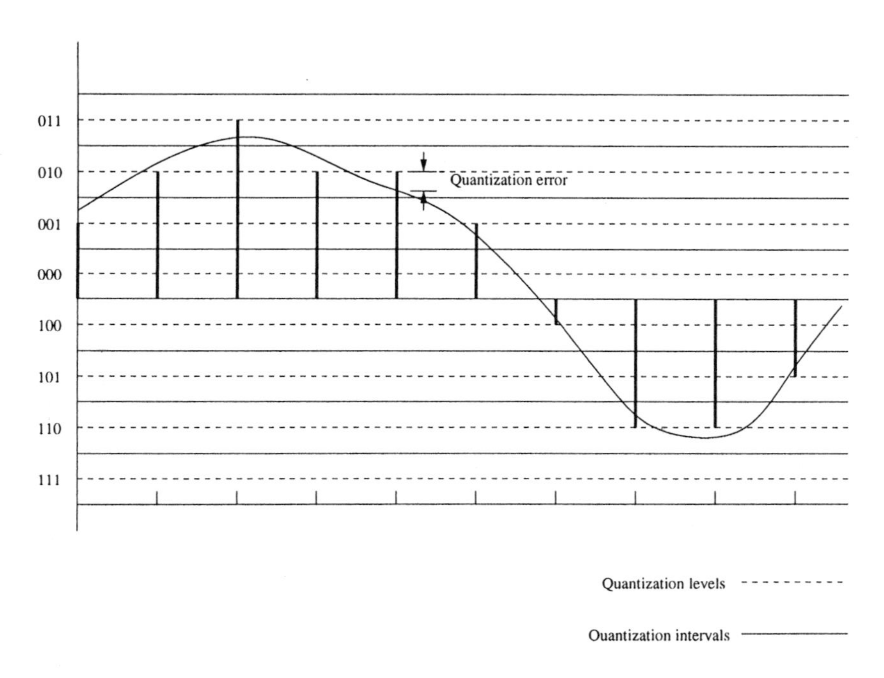

Figure 3.6. Quantization process.

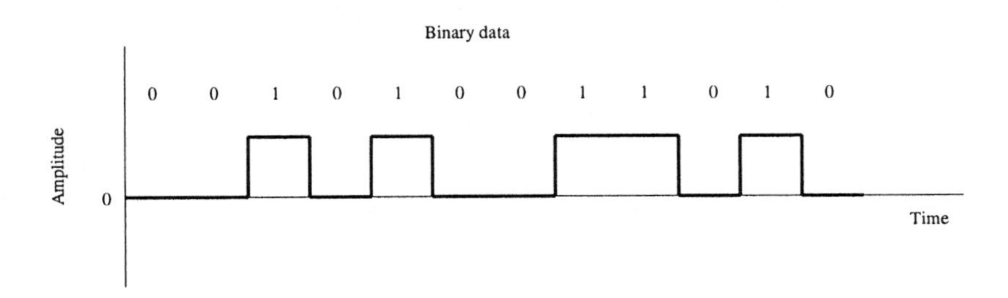

Figure 3.7. Binary waveform: nonreturn-to-zero unipolar (on-off). The waveform corresponds to the first four samples of Figure 3.6.

The staircase-like characteristic of the midtread quantizer in Figure 3.8(a) shows that the decision thresholds of the quantizer are located at $\pm d/2$, $\pm 3d/2$, $\pm 5d/2$, ..., and the representation levels (quantization levels) are located at 0, $\pm d$, $\pm 2d$, ..., where d is the stepsize. The staircase-like characteristic of the midriser quantizer in Figure 3.9(a) shows that the decision thresholds of the quantizer are located at $0, \pm d, \pm 2d, \ldots$, and the representation levels (quantization levels) are located at $\pm d/2$, $\pm 3d/2$, $\pm 5d/2$, ..., where d is the stepsize.

Figures 3.8(b) and 3.9(b) show that the maximum quantization error (which corresponds to the difference between the output and input values of the quantizer) for a sample is half one stepsize. It is also observed that the total range of variation of the quantization noise is from $-d/2$ to $d/2$.

3.5.2 Quantization Noise

The error or quantization noise consists of the difference between the signal at the quantizer input and the signal at the output, $n = x - y$, where $y = q(x)$ and $q(\cdot)$ represents the quantization function. The performance of coding or processing systems is limited by the level of the quantization noise. The channel capacity itself is limited by that noise. As a consequence, the figure of merit which is most used in comparative analysis is the signal to quantization noise ratio (SQNR). Throughout this chapter the notation SNR (signal to noise ratio) is also used to denote the signal to quantization noise ratio (Paez and Glisson, 1972). The mean square error, for a uniform quantizer, is given by $d^2/12$ (Bennett, 1948), supposing a uniform distribution for the quantization noise, where d represents the quantization step. This result is shown in the sequel.

Assuming a uniform probability distribution for the noise in the interval $[-d/2, d/2]$,

$$p_N(n) = \frac{1}{d}, \tag{3.8}$$

the power of the quantization noise is determined by the expression

$$P_N = \int_{-\infty}^{\infty} n^2 p_N(n) dn. \tag{3.9}$$

Taken into account the distribution assumed,

$$P_N = \frac{1}{d} \int_{-d/2}^{d/2} n^2 dn = \frac{1}{d} \frac{d^3}{12} = \frac{d^2}{12}. \tag{3.10}$$

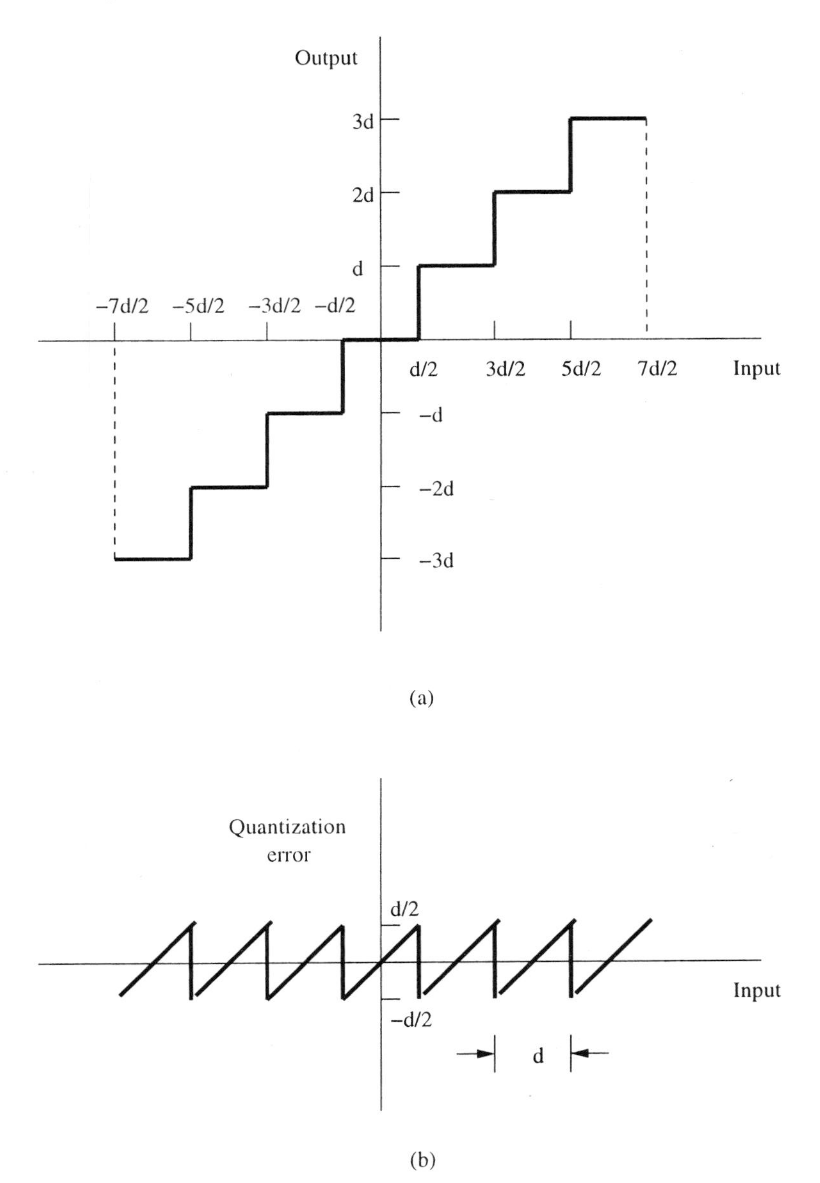

Figure 3.8. (a) Characteristic of the midtread quantizer. (b) Variation of the corresponding quantization error with input.

This result was obtained by the first time by Claude E. Shannon, in 1948 (Shannon, 1948).

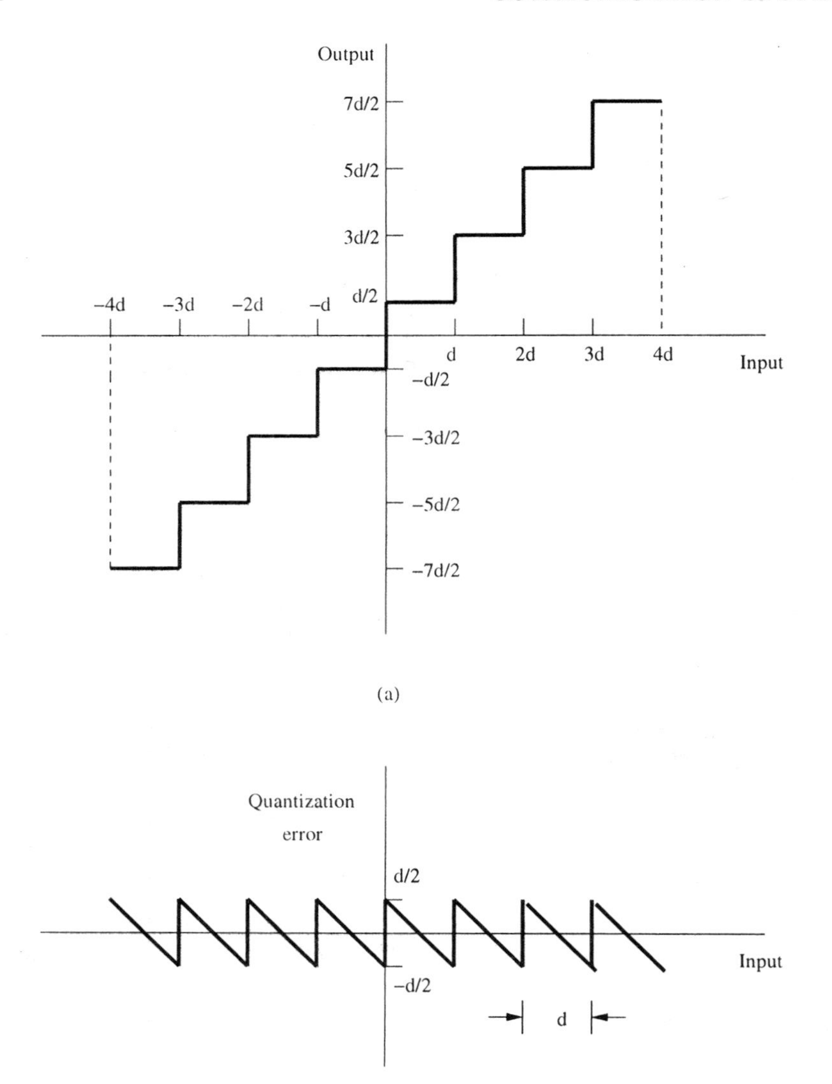

Figure 3.9. (a) Characteristic of the midriser quantizer. (b) Variation of the corresponding quantization error with input.

In what follows it is shown that each additional bit increases the SQNR by about 6 dB. The SQNR is given by

$$\text{SQNR} = 10 \log \frac{P_X}{P_N}, \tag{3.11}$$

where P_X denotes the power of the signal to be quantized and $P_N = d^2/12$ is the power of the quantization noise.

For a signal with dynamic range (that is, the interval corresponding to the signal amplitude variation) equals to $2\sqrt{P_X}$ and quantization with $N = 2^m$ levels, where m denotes the number of coding bits, the quantization step is given by

$$d = \frac{2\sqrt{P_X}}{N} = \frac{2\sqrt{P_X}}{2^m}. \tag{3.12}$$

As a consequence, the power of the quantization noise is given by

$$P_N = \frac{d^2}{12} = \frac{4P_X}{12N^2} = \frac{P_X}{3 \cdot 4^m}. \tag{3.13}$$

Substituting Equation 3.13 in 3.11, it follows that

$$\text{SQNR} = 10 \log 3 \cdot 4^m = 10 \log 3 + 10m \log 4 \approx 5 + 6m \text{ dB}. \tag{3.14}$$

Hence, from Equation 3.14 it is observed that the SNQR increase is about 6 dB for each additional bit, considering uniform quantization.

3.6 Noise Spectrum for the Uniform Quantizer

Quantization noise can be thought as the result of the application of the signal $x(t)$ to a circuit with characteristic $f(x)$, as shown in Figure 3.10 (Alencar, 1998a).

The input signal is assumed stationary in the wide sense, but no restriction is made concerning its probability density function (Alencar, 1993a). A small quantization step, as well as a uniform quantization scheme are considered in the model. The function $f(x)$ is periodic, with period d, therefore it can be written as

$$f(x) = x - m.d$$
$$(m - \frac{1}{2})d < x \leq (m + \frac{1}{2})d, \; m = 0, \pm 1, \pm 2, \cdots \tag{3.15}$$

A Fourier series representation for the preceding function can be obtained in terms of the input signal where d represents the stepsize, $x(t)$ is the input signal and X_M is the quantizer threshold, given by (Lévine, 1973)

$$f(x(t)) = \frac{d}{\pi} \sum_{n=1}^{\infty} \frac{(-1)^n}{n} \sin \left[\frac{2\pi n x(t)}{d} \right], \; |x(t)| \leq X_M. \tag{3.16}$$

The signal is not allowed to exceed the threshold level and so there is no overload noise in the model. Equation 3.16 represents a sum of

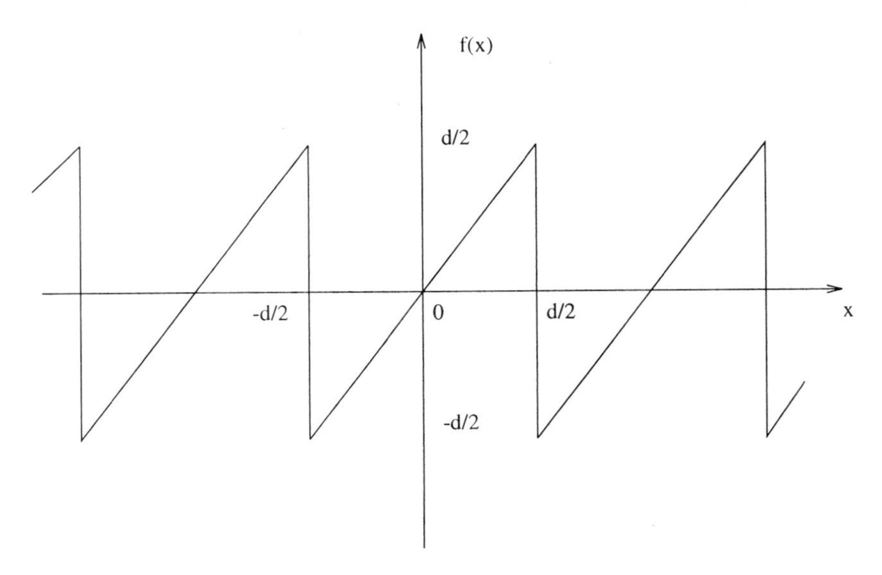

Figure 3.10. Characteristic function of the quantizer.

phase modulated signals, with phase deviation $2\pi n/d$ and amplitude $d/\pi n$. The autocorrelation function of equation 3.16 can be computed, as described in (Alencar, 1993a), giving

$$\begin{aligned} R_N(\tau) &= \frac{d^2}{\pi^2}\sum_{n=1}^{\infty}\sum_{n=1}^{\infty}\frac{(-1)^{n+m}}{n.m} \\ &\cdot\quad E[\sin(\frac{2\pi n x(t)}{d})\sin(\frac{2\pi n x(t+\tau)}{d})]. \end{aligned} \tag{3.17}$$

The phase modulated signals are assumed to form an orthogonal set, which implies that the cross-correlations are zero. Equation 3.17 can be written as

$$R_N(\tau) = \frac{d^2}{2\pi^2}\sum_{n=1}^{\infty}\frac{1}{n^2}E[\cos(\frac{2\pi n}{d}(x(t+\tau)-x(t)))]. \tag{3.18}$$

Equation 3.18 can be re-written, using the exponential formula for the cosine, to emphasize the role of the expectancy operator in that equation

$$R_N(\tau) = \frac{d^2}{4\pi^2}\sum_{n=1}^{\infty}\frac{1}{n^2}E[e^{\pm j\frac{2\pi n}{d}(x(t+\tau)-x(t))}]. \tag{3.19}$$

The linear mean square estimator can be used in the last expression to account for the expression $x(t+\tau)-x(t)$. This estimator can be shown to be unbiased and consistent (Alencar, 1998b). In fact, the mean square error of the estimator is always reduced as $\tau \to 0$.

Finally, the autocorrelation function of Equation 3.16 is evaluated

$$R_N(\tau) = \frac{d^2}{4\pi^2} \sum_{n=1}^{\infty} \frac{1}{n^2} E[e^{-j\frac{2\pi n}{d}\tau x'(t)} + e^{+j\frac{2\pi n}{d}\tau x'(t)}]. \quad (3.20)$$

The previous formula can be easily verified by evaluation at the origin, which gives the well known value for the total noise power,

$$P_N = R_N(0) = \frac{d^2}{2\pi^2} \sum_{n=1}^{\infty} \frac{1}{n^2} = \frac{d^2}{2\pi^2} . \frac{\pi^2}{6} = \frac{d^2}{12}. \quad (3.21)$$

The actual noise power, inside the signal bandwidth, can be much smaller. It worths noting that the last result was obtained without any constraints on the shape of the noise probability density function. This means that the probability distribution for the quantization noise does not have to be uniform in order to verify Equation 3.21.

The power spectrum density of the quantization noise can be obtained by using the Wiener-Khintchin theorem in Equation 3.20 (Alencar and Neto, 1991),

$$S_N(w) = \frac{d^2}{2\pi^2} \sum_{n=1}^{\infty} \frac{1}{n^3} p_{X'}(\frac{wd}{2\pi n}), \quad (3.22)$$

where $p_{X'}(\cdot)$ is the probability density function of the derivative of the input signal.

Equation 3.22 demonstrates that the power spectral density of the quantization noise is related to the probability density function of the derivative of the input signal. The convergence of the noise spectrum to Equation 3.22, as the stepsize decreases, is a result of a previous work (Alencar, 1993b). In fact, the noise spectrum reflects an infinite sum of contributions, each one with the shape of the probability density function, but with decreasing intensity and increasing bandwidth.

In order to give an example of how to use Formula 3.22, suppose the input signal is zero mean Gaussian, with variance $\sigma_X^2 = P_X$

$$p_X(x) = \frac{1}{\sqrt{2\pi P_X}} e^{-\frac{x^2}{2P_X}}, \; P_X = R_X(0). \quad (3.23)$$

For this kind of signal it is easy to compute the probability density function of its derivative, in terms of the second derivative of the autocorrelation function $R_X(\tau)$

$$p_{X'}(x) = \frac{1}{\sqrt{2\pi P_{X'}}} e^{-\frac{x^2}{2P_{X'}}}, \; P_{X'} = -R''_X(0). \quad (3.24)$$

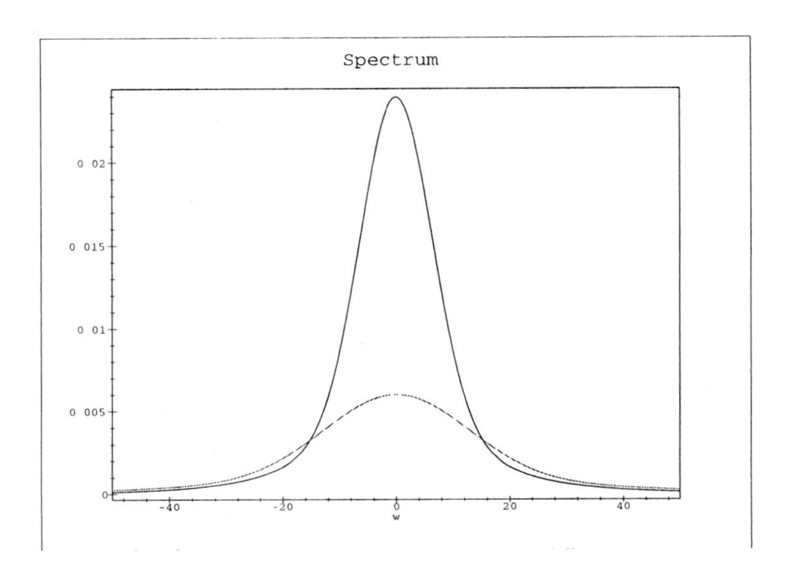

Figure 3.11. Spectrum of the quantization noise for a Gaussian input signal, for two values of the stepsize.

Substituting last equation into Formula 3.22 yields the power spectrum density of the quantization noise, when the input to the quantizer is a stationary Gaussian signal.

$$S_N(w) = \frac{d^2}{2\pi^2} \sum_{n=1}^{\infty} \frac{1}{n^3} \frac{1}{\sqrt{-2\pi R''_X(0)}} e^{\frac{(wd)^2}{8\pi^2 n^2 R''_X(0)}}. \tag{3.25}$$

Figure 3.11 illustrates the application of the formula for a Gaussian input signal. I shows the spectrum of the quantization noise for two values of the stepsize. It is noted that the spectrum broadens for a smaller stepsize.

Another example can be given, to clarify the use of Formula 3.22. Suppose a sine waveform, $x(t) = Asin(\omega_c t + \phi)$, is applied to the input of the quantizer. The pdf for a sine waveform is (Papoulis, 1991)

$$p_X(x) = \frac{1}{\pi\sqrt{2P_X - x^2}}, \quad P_X = R_X(0) = \frac{A^2}{2}. \tag{3.26}$$

Again, for a sinusoidal signal it is easy to compute the pdf of its derivative, in terms of the second derivative of the autocorrelation function $R_X(\tau)$

$$p_{X'}(x) = \frac{1}{\pi\sqrt{(\omega_c A)^2 - x^2}}. \tag{3.27}$$

Substituting this equation into Formula 3.22 yields the power spectrum density of the quantization noise, when the input to the quantizer is a sine waveform

$$S_N(w) = \frac{d^2}{2\pi^2}\sum_{n=1}^{\infty}\frac{1}{n^3}\frac{1}{\pi\sqrt{(\omega_c A)^2 - (\frac{wd}{2\pi n})^2}}. \tag{3.28}$$

3.6.1 Nonuniform Quantization

In uniform quantization, the signal to quantization noise ratio is given by Equation 3.11, and the noise power depends only on the magnitude of the quantization intervals, d, as shown in Equation 3.10. Thus, SQNR depends directly on the power of the input signal. As the dynamic range of speech signals is about 50 dB, the SQNR varies significantly, decreasing with the decrease of the power of the input signal. The signal quality may deteriorate, for instance, in time intervals when the person speaks very softly. Statistically, it is observed a predominance of low amplitude speech samples, meaning that the SQNR is low most of the time.

However, it is ideally desirable to obtain an SQNR that is as constant (same quality) as possible for all the values of power of the message signal (input signal). The use of a nonuniform quantizer attempts to assure the SQNR constance. In this case, the quantization step is not constant, it varies as a function of the signal amplitude. For lower levels of the input signal, the quantization step is small. The width of the quantization step is logarithmically increased as the the level of the message signal increases.

Alternatively, the logarithm of the input signal can be uniformly quantized rather than the input signal itself. Indeed, at the transmitter, the encoder expands the number of levels at low amplitudes and compresses those at high amplitudes, as depicted in Figure 3.12. At the receiver, this procedure is reversed to bring the relative amplitude levels of the signal to their original value. The process of first compressing and then expanding a signal is referred to as *companding*. Figure 3.13 illustrates a block diagram of a communication system based on *companding*.

The method for voice waveform uniform coding, called pulse code modulation (PCM), is defined in CCITT G.711, and AT&T 43801. Basically, the signal is sampled at a rate 8,000 times per second and uniform

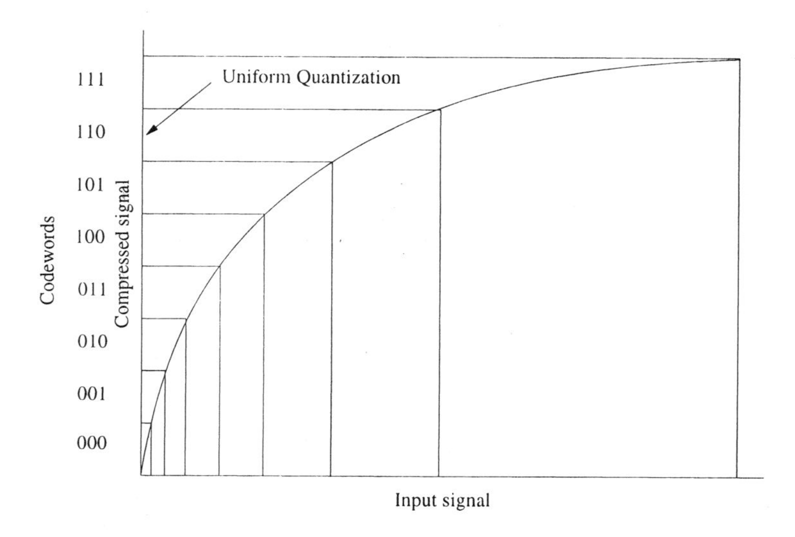

Figure 3.12. Typical compression curve for nonuniform quantization.

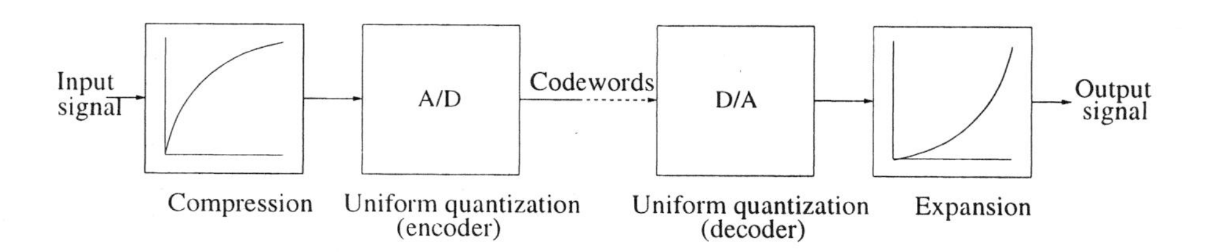

Figure 3.13. Block diagram of a communication system based on *companding*.

coded using 8 bits per sample. For non-uniform coding, two compression laws are recommended by the International Telecommunication Union (ITU-T), formerly know as CCITT: the μ-law used in North America and Japan, and the A-law used in the Europe and most of the rest of the world. The A-law is also used in international routes.

The corresponding expressions for the previous compression laws are given next. Note that the maximum amplitudes of both input and output signals are V.

- μ-law

$$y = \begin{cases} C(x) = \dfrac{V \ln(1 + \mu x/V)}{\ln(1+\mu)}, & x > 0 \\ C(x) = -C(-x), & x \leq 0 \end{cases} \tag{3.29}$$

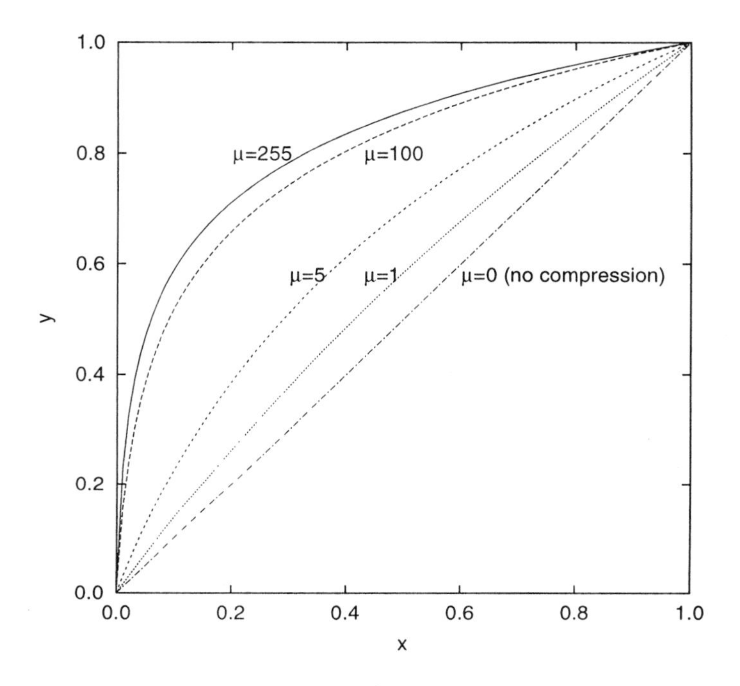

Figure 3.14. Compression curves for μ-law.

- A-law

$$y = \begin{cases} C(x) = \dfrac{Ax}{1+\ln A}, & 0 \leq x \leq V/A \\ C(x) = \dfrac{V(1+\ln(Ax/V))}{1+\ln A}, & V/A \leq x \leq V \\ C(x) = -C(-x), & x \leq 0. \end{cases} \tag{3.30}$$

The corresponding compression curves are shown respectively in Figures 3.14 and 3.15. The curves show the effect of changing the parameters μ and A. An increase on the respective parameter leads to an increase in the curve nonlinearity. For $A = 1$ and $\mu = 0$ the compression curves are linear, i.e., no compression is obtained. The typical values, obtained from subjective tests (*Mean Opinion Score* – MOS), are $\mu = 255$ and $A = 87.6$. The G.711 PCM A-law or μ-law has an MOS of about 4.2.

The curve of signal to quantization noise, as a function of the inverse of the signal amplitude, for nonuniform quantizers is more flat when

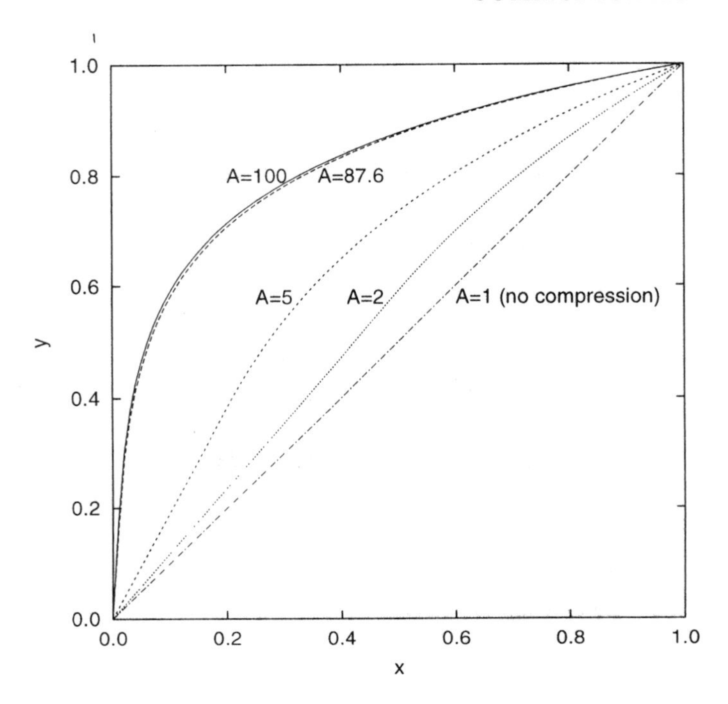

Figure 3.15. Compression curves for A-law.

compared to the uniform quantizer. This way, the input lower levels are preserved.

For the nonuniform case, one can assume that the signal is transformed by a nonlinear function $g(\cdot)$ prior to the quantization process. This gives

$$S_N(w) = \frac{d^2}{2\pi^2} \sum_{n=1}^{\infty} \frac{1}{n^3} p_{g'(x)}(\frac{wd}{2\pi n}), \tag{3.31}$$

where $g'(x)$ is the derivative of the compression function. Its probability density function is usually difficult to compute. For example, using a sine waveform, with amplitude V, as input and the nonlinear $A - Law$ scheme, the resulting pdf is given by

$$p_{g'(x)}(\xi) = \begin{cases} [\pi \sqrt{(\frac{A.V}{1+\ln A})^2 - \xi^2}]^{-1} & |\xi| < \frac{A.V}{1+\ln A} \\ \frac{\xi^2(1+\ln A)}{\pi X_M V} & |\xi| \geq \frac{A.V}{1+\ln A} \end{cases} . \tag{3.32}$$

Careful selection of $g(x)$ can minimize the following expression and maximize the signal to quantization noise ratio. The compression function must be chosen in order to displace the peak of the quantization noise spectrum far outside the signal bandwidth.

$$P'_N = \int_{-\omega_M}^{\omega_M} S_N(\omega) d\omega = \frac{d^2}{2\pi^2} \sum_{n=1}^{\infty} \frac{1}{n^3} \int_{-\omega_M}^{\omega_M} p_{g'(x)}(\frac{\omega d}{2\pi n}) d\omega, \tag{3.33}$$

where P'_N represents the noise power that falls inside the signal bandwidth. This quantity can be made quite small, compared to the total noise P_N in 3.21.

3.7 Vector Quantization

Vector quantization (Gersho and Gray, 1992, Gray, 1984), which may be seen as an extension of scalar quantization to a multidimensional space, is supported by Shannon's Rate Distortion Theory, which states that a better performance is achievable by coding blocks of samples (vectors) instead of individual samples (scalars). Mathematically, vector quantization may be defined as a mapping Q of a vector $\boldsymbol{x}$ belonging to the K-dimensional Euclidean space, $\mathbb{R}^K$, into a vector belonging to a finite subset W of $\mathbb{R}^K$, that is,

$$Q : \mathbb{R}^K \rightarrow W. \tag{3.34}$$

The codebook $W = \{\boldsymbol{w}_i;\ i = 1,\ 2,\ \dots,\ N\}$ is the set of the K-dimensional codevectors (reconstruction vectors). The index of codevector $\boldsymbol{w}_i$ is denoted by $\boldsymbol{i}$. Each index $\boldsymbol{i} \in \{0, 1\}^b$ represents a binary word of b bits. The code rate of a vector quantizer, measuring the number of bits per vector component, is $R = \frac{1}{K} \log_2 N = \frac{b}{K}$. In voice waveform coding, R is expressed in bits/sample. In image coding, R is expressed in bits per pixel (bpp).

In a signal coding systems based on vector quantization, shown in Figure 3.16, the coder and the decoder operate as follows. Given a vector $\boldsymbol{x} \in \mathbb{R}^K$ from the signal to be coded, the coder determines the distortion $d(\boldsymbol{x}, \boldsymbol{w}_i)$ between that vector and each codevector $\boldsymbol{w}_i$, $i = 1,\ 2,\ \dots,\ N$ from the codebook W. The optimum coding rule is the nearest neighbor rule, by which a binary word $\boldsymbol{i}$ is transmitted to the decoder if codevector $\boldsymbol{w}_i$ corresponds to the minimum distortion, that is, if $\boldsymbol{w}_i$ is the vector with greatest similarity to $\boldsymbol{x}$ among all codevectors of the codebook. In other words, the coder employs the coding rule $C(\boldsymbol{x}) = \boldsymbol{i}$ if $d(\boldsymbol{x}, \boldsymbol{w}_i) < d(\boldsymbol{x}, \boldsymbol{w}_j)$, $\forall j \neq i$. The task of the decoder is very simple: upon receiving the index $\boldsymbol{i}$ of b bits, the decoder simply looks for the vector $\boldsymbol{w}_i$ in its copy of the codebook W and outputs $\boldsymbol{w}_i$ as the reproduction (reconstruction) of $\boldsymbol{x}$. Hence, the decoder uses the decoding rule $D(\boldsymbol{i}) = \boldsymbol{w}_i$. The mapping of $\boldsymbol{x}$ in $\boldsymbol{w}_i$ is usually expressed as $\boldsymbol{w}_i = Q(\boldsymbol{x})$.

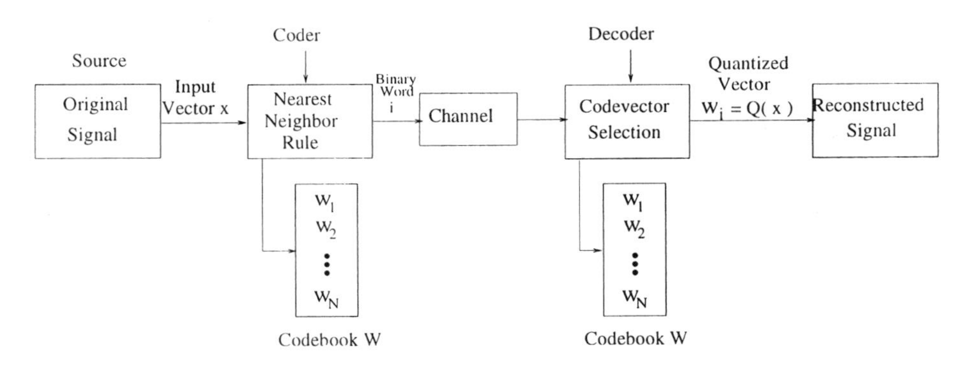

Figure 3.16. Coding system based on vector quantization.

Thus, in the scenario of digital coding of signals, vector quantization is a lossy compression technique, since the reconstructed signal is a distorted version of the original signal. The quantization error, introduced when representing the input signal by its corresponding quantized version, is called quantizer distortion. A key question in vector quantizer design is the tradeoff between rate and distortion. The target is an optimum codebook, which minimizes, for a given code rate, the average distortion introduced by approximating the input vectors by the corresponding codevectors. The LBG (Linde-Buzo-Gray) algorithm (Linde et al., 1980) is the most widely used technique for codebook design. Other methods have been applied for codebook design, such as Kohonen learning algorithm (Kohonen, 1990) and other unsupervised learning algorithms (Krishnamurthy et al., 1990, Chen et al., 1994); stochastic relaxation (Zeger et al., 1992); fuzzy algorithms (Karayiannis and Pai, 1995) and genetic algorithm (Pan et al., 1995).

Codebook design plays an important role for the performance of digital processing systems based on vector quantization (VQ) (Gray, 1984, Gersho and Gray, 1992). In speech and image coding based on VQ (Abut et al., 1982, Ramamurthi and Gersho, 1986), the quality of the reconstructed signals depends on the designed codebook. In speaker identification systems based on VQ (Soong et al., 1987, Fechine, 2000), the recognition rates depend on the codebook of acoustic parameters designed for each speaker registered in the system.

The mapping Q leads to a partitioning of $\mathbb{R}^K$ into N subspaces (cells, called Voronoi regions) S_i, $i = 1, 2, \ldots, N$, for which

$$\bigcup_{i=1}^{N} S_i = \mathbb{R}^K \text{ and } S_i \cap S_j = \emptyset \text{ if } i \neq j. \tag{3.35}$$

Each Voronoi region S_i is defined as

$$S_i = \{\boldsymbol{x} : \ Q(\boldsymbol{x}) = \boldsymbol{w}_i\} = \{\boldsymbol{x} : \ C(\boldsymbol{x}) = \boldsymbol{i}\}. \tag{3.36}$$

Codevector $\boldsymbol{w}_i$ is the representative vector of all input vectors belonging to cell S_i, as shown in Figure 3.17.

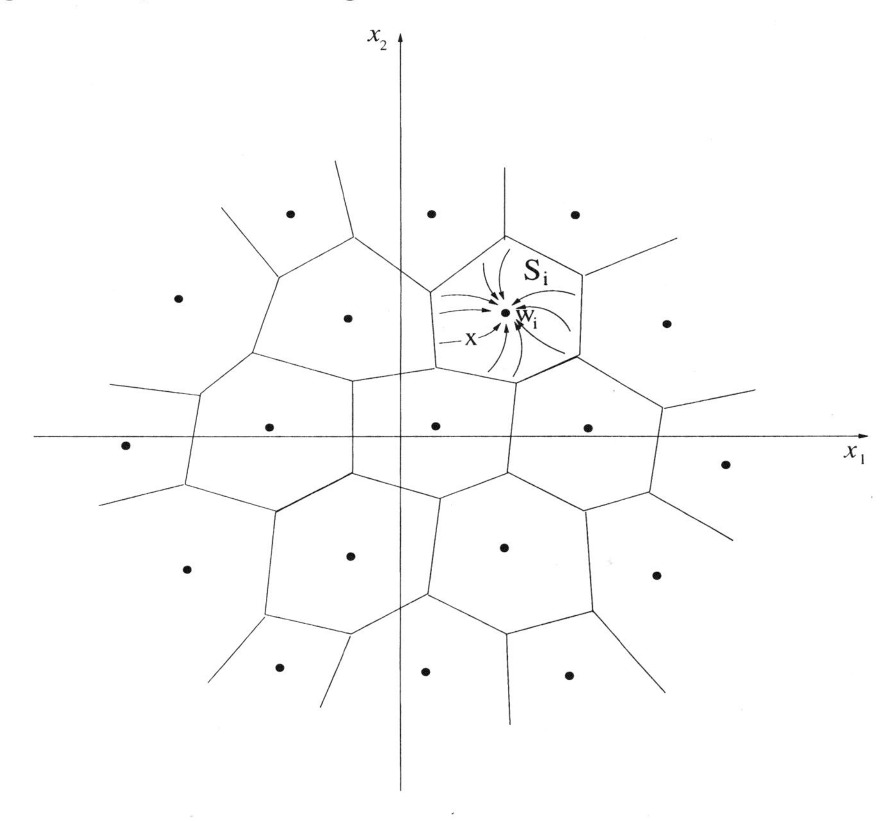

Figure 3.17. Partitioning of the Euclidean space $\mathbb{R}^2$, introduced by mapping input vectors $\boldsymbol{x}$ into codevectors $\boldsymbol{w}_i$. Coordinates x_1 and x_2 represent the first and the second components of vector $\boldsymbol{x} \in \mathbb{R}^2$, respectively.

3.7.1 LBG Algorithm

Let the iteration of the LBG algorithm be denoted by n. Given K, N and a distortion threshold $\epsilon \geq 0$, the LBG algorithm (Linde et al., 1980) consists of the following steps:

- *Step (1)* initialization: given an initial codebook W_0 and a training set $\boldsymbol{X} = \{\boldsymbol{x}_m; \ m = 1, \ 2, \ \ldots, \ M\}$, set $n = 0$ and $D_{-1} = \infty$.

- *Step (2)* partitioning: given W_n (codebook at the n-th iteration), assign each training vector (input vector) in the corresponding class (Voronoi cell) according to the nearest neighbor rule; determine the distortion

$$D_n = \sum_{i=1}^{N} \sum_{\boldsymbol{x}_m \in S_i} d(\boldsymbol{x}_m, \boldsymbol{w}_i). \tag{3.37}$$

- *Step (3)* convergence test (stop criterion): if $(D_{n-1} - D_n)/D_n \leq \epsilon$ then stop, with W_n representing the final codebook (designed codebook); else, continue.

- *Step (4)* codebook updating: calculate the new codevectors as the centroids of the classes; set $W_{n+1} \leftarrow W_n$; set $n \leftarrow n+1$ and go to *Step (2)*.

The distortion decreases monotonically in the LBG algorithm, since the codebook is iteratively updated attempting to satisfy the centroid condition and nearest neighbor condition. In the LBG algorithm the distortion introduced by representing the training vectors by the corresponding codevectors (centroids) is monitored at each iteration. The stopping rule (convergence test) is based on that monitored distortion. The codebook training stops when $(D_{n-1} - D_n)/D_n \leq \epsilon$. The convergence speed of the LBG algorithm depends on the initial codebook.

3.8 LPC Parameters

Most current speech coding algorithms use the source-filter model of speech production. The voice is modeled as the response of a linear time-variant synthesis filter to an input signal called excitation. Examples of speech coders based on that model include the numerous variations of the LPC (linear predictive coding) vocoder and the wide family of analysis by synthesis coders based on linear prediction (LPAS, linear-prediction-based analysis-by-synthesis) – including CELP (code-excited linear prediction). The synthesis filter, which determines the short-term spectral envelope of the synthesized speech, is characterized by the linear prediction coefficients, obtained from a linear prediction (LP) analysis of the input speech signal. Those coefficients are usually called LPC coefficients, and are related generically to each of the many different (but equivalent) sets of parameters that specify the synthesis filter.

Much research has been developed in LPC quantization (Paliwal and Atal, 1993, Kleijn and Paliwal, 1995, Atal et al., 1993). The main purpose of LPC quantization for speech coding is to avoid the introduction of any audible distortion in the coded speech, maintaining the bit rate as low as possible. If this purpose is reached, then we have *transparent quality* or *transparent quantization.*

Due to the cost and difficulty to perform subjective quality tests, the researchers have used objective measures to assess the distortion in spectral envelope caused by the LPC quantization. Specifically, the spectral distortion or spectral distance has been used as a standard performance criterion. Associated to that measure, an objective criterion for transparent LPC quantization has been proposed by Paliwal and Atal (Paliwal and Atal, 1993), based on the role of the *outliers.*

The first study on vector quantization applied to LPC parameters is due to Buzo *et al.* in 1980 (Buzo et al., 1980). Recently, many papers regarding sophisticated schemes of vector quantization have been presented. Paliwal and Atal (Paliwal and Atal, 1993) have efficiently applied split VQ (split vector quantization) for obtaining transparent quality at 24 bits per frame. That work has been a reference usually used as a benchmark for comparing other results. Other researchers have obtained similar or better results.

3.8.1 LPC Quantization

In a speech coder based on the source-filter model, the LPC coefficients $\{a_i\}$ are obtained by using linear predictive analysis (Markel and Gray, 1976) in each speech frame. The coefficients are used to form a synthesis filter given by $H(z) = 1/A(z)$, where $A(z)$ is the inverse filter, given by

$$A(z) = 1 + a_1 z^{-1} + \cdots + a_M z^{-M}, \tag{3.38}$$

where M is typically a number between 10 and 16 called order of the predictor. Due to the quasi stationary nature of the speech, that filter is updated at every frame, whose typical size is 20 msec. This leads to a rate of 50 frames per second. A description of that synthesis filter may be communicated to the receiver at each frame. The quantization of the filter into a finite number of bits per frame is known as LPC spectrum quantization.

The purpose of LPC quantization is to efficiently code the LPC parameters without introducing audible distortion in the coded speech. As stated previously, the difficulties associated to subjective evaluation tests have led to performance assessments of quantization schemes by using the spectral distortion, SD, which is expressed in dB and determined to a speech frame according to

$$\mathrm{SD}^2 = \frac{2}{F_S} \int_0^{F_S} \{20 \log_{10} |H(e^{j\pi f/F_S})| - 20 \log_{10} |\hat{H}(e^{j\pi f/F_S})|\}^2 df, \tag{3.39}$$

where F_S is the sampling frequency and $\hat{H}(z)$ is the quantized synthesis filter transfer function.

Thus, SD^2 is the mean square error related to the log-magnitudes of the synthesis filter frequency responses, quantized and non-quantized, considering that the averaging is taken in the frequency domain. For evaluating the performance of a LPC quantization scheme, the average spectral distortion (average value of SD considering all speech frames) is determined and the percentage of outliers is determined. The outliers are frames whose SD exceeds a determined threshold value. In (Paliwal and Atal, 1993) three conditions based on SD for transparent quantization are presented:

1 The average spectral distortion is lower than 1 dB.

2 There are no outliers with SD greater than 4 dB.

3 The percentage of outliers with SD in the range 2-4 dB is lower than 2 %.

For quantization of the LPC parameters, the prediction coefficients $\{a_i\}$ are mapped into an equivalent representation, with good quantization properties in terms of distribution, stability and spectral sensitivity. Representations such as log-area ratios, arcsines of reflection coefficients and line spectrum frequencies (also called line spectrum pairs, LSPs) have been studied. They give better quantization efficiency and better stability properties when compared to the LPC coefficients. Since the 1980's, the LSF representation has been the dominant form for the purpose of LPC spectrum quantization. According to (Soong and Juang, 1993), the formal result in terms of DRT evaluation of an 800 bps vocoder LSP is only 0.7 worse than the 2,400 bps residual excited vocoder LPC. At a rate 4,800 bps, the DRT score of the vocoder LSP in only 0.7 worse than the 9,600 bps residual excited vocoder LPC .

It is worth mentioning that in speech coding algorithms using linear prediction the transmission of the LPC parameters, usually transformed to the LSF representation, consumes most of the total bit rate of the coder.

In order to define the LSF parameters, the inverse filter polynomial is used to construct two polynomials:

$$P(z) = A(z) + z^{-(M+1)} A(z^{-1}) \tag{3.40}$$

and

$$Q(z) = A(z) - z^{-(M+1)}A(z^{-1}). \tag{3.41}$$

The roots of polynomials $P(z)$ and $Q(z)$ are called LSFs. The polynomials $P(z)$ and $Q(z)$ have the following properties: 1) all zeros of $P(z)$ and $Q(z)$ lie on the unit circle and 2) zeros of $P(z)$ and $Q(z)$ are interlaced with each other, that is, the LSFs w_i are in ascendent order in $(0, \pi)$, in the form (Kim and Oh, 1999)

$$0 < w_1 < w_2 < \ldots < w_p < \pi. \tag{3.42}$$

An important question in LSF coding is that the ordering relation is required for assuring the stability of the synthesis filter (Sugamura and Farvardin, 1988). The properties 1) and 2) previously presented are useful for determining the LSF parameters from $P(z)$ and $Q(z)$.

For a long time, almost all coding schemes have used some form of scalar quantization (Eriksson et al., 1999). Complexity questions (number of operations performed to compare an input vector to each codevector of the codebook) and memory requirements were viewed as limiting factors to the effective use of vector quantization. The first work incorporating VQ was described in (Buzo et al., 1980), but a performance far from the acceptable was obtained with VQ at 10 bits per frame. Hence, due to the prohibitive size of the training set, to the high computational cost of the codebook design and to the computational requirements prohibitive to full search vector quantization, hybrid schemes (scalar and vectorial) (Grass and Kabal, 1991, Laroia et al., 1991) and methods for reducing the computational complexity have been investigated. As an example, one can cite the method proposed by Paliwal and Atal in (Paliwal and Atal, 1993), by which the vector of LSF parameters is divided into two vectors, each one quantized by a different codebook. This procedure is known as split VQ.

In memoryless quantization, each vector of LSF parameters is quantized independently of the past LSF vectors. However, this is not the most efficient method for coding the LSF vectors, which present significant interframe correlation (correlation regarding successive frames). As a consequence, performance gains in coding may be obtained by exploring the interframe correlation, as occurs in predictive vector quantization (Shoham, 1987) and finite-state vector quantization.

Besides the type of quantization (scalar, vectorial, hybrid) and the inclusion or exclusion of memory in the quantization process, many aspects, generally related, affect the performance of an LSF quantizer, such as (Ramachandran et al., 1995): the distortion measure used for the quantizer design, the codebook design, the complexity of the search, the number of bits, the memory requirements for codebook storage and the robustness to channel errors.

Code excited linear prediction (CELP) (Schoroeder and Atal, 1985) is a class of speech coders which present a good strategy for high quality digital speech transmission at low bit rates. An important technique belonging to that class of coders is the vector sum excited linear prediction (VSELP) (Electronic Industries Association (EIA), 1989).

3.9 Overview of Speech Coding

Speech compression has been an intense research area for decades. Almost all works in speech compression regard to lossy compression, by which the numerical representation of the signal samples is never recovered exactly after decoding. There exists a wide range of compromises between the bit rate and the quality of the reconstructed signal, that are of practical interest for telephone speech coding. The cell phone users, for instance, are accustomed to varied levels of signal degradation.

The speech coding algorithms may be divided into two main groups: waveform coders and vocoders (Gersho, 1994).

Historically, the term vocoder originated from the contraction of voice encoder and decoder. In waveform coders, the data transmitted from the coder to the decoder specify a speech representation as a waveform of amplitude *versus* time, such that the reconstructed signal approximates the original waveform and, consequently, provides an approximated recreation of the original sound. On the other hand, vocoders do not produce an approximation of the original waveform. In vocoders, parameters which characterize the segments of sounds are specified and transmitted to the decoder, which reconstructs a new waveform, with a similar sound. Vocoders are also called parametric coders by an obvious reason. Frequently those parameters characterize the short term spectrum of a sound.

Alternatively the parameters specify the mathematical model of human speech production. Either way, the parameters do not provide sufficient information for generating a good approximation of the original waveform, but the information is sufficient for the decoder to synthesize a sound which is perceptually similar to the voice signal. Vocoders operate at rates lower than the ones of waveform coders, but the quality of the reconstructed speech, although intelligible, suffers from loss of *naturalness* and some unique features which identify a speaker may be damaged. Thus, for waveform coders the quantization is performed directly in the signal waveform, while for parametric coders the quantization is performed in the parameters of the model under consideration. Hybrid coders are based on the models of speech production, but they use an excitation for the synthesizer which is more accurate than that one used in parametric coders

Numerous works in speech coding are based on speech in telephony bandwidth, nominally limited to 3.2 kHz (corresponding to the range 200 Hz to 3.4 kHz), sampled at 8 kHz. Speech coding in broadband has gained attention and concerns 7 kHz signals, sampled at 16 kHz. Most work in waveform coding regards modifications and improvements in well established general methods.

A coding technique notable and very popular is the CELP. As coding methods one can cite: ADM (adaptive delta modulation), ADPCM (adaptive differential pulse code modulation), APC (adaptive predictive coding), MP-LPC (multipulse linear predictive coding) and RPE (regular pulse excitation). The coders MP-LPC, RPE and CELP belong to the family of algorithms of analysis by synthesis, which can be seen as hybrid coders, since they combine some features of vocoders and waveform coders. Another analysis by synthesis speech coder is the full rate GSM RPE-LTP 13 kbps, standardized by ETSI in 1988 for the digital cellular mobile system (Hersent et al., 2002).

The LPC (linear predictive coding) vocoder is widely used in secure telephone speech. One can also cite as another vocoder approach the sinusoidal coding and corresponding versions STC (sinusoidal transform coding) and MBE (multiband excitation).

3.10 Waveform Coding

In waveform coders the quantization is performed directly in the voice waveform. The purpose is to reproduce the waveform, sample by sample, as efficiently as possible. In general the coders (as an exception one can cite VQ-based techniques) have low implementation complexity, present low delay and are suitable for use in telephone networks at bit rates $R \geq 16$ kbit/s.

Differential PCM

Differential coding exploits the fact that the speech signal presents significant correlation between successive samples. This means that the speech signal is very redundant. The purpose of differential pulse code modulation (DPCM) is to reduce the redundancy of the speech signal, by quantizing the difference between amplitudes of adjacent samples. As the difference presents lower variance when compared to the original signal, fewer bits may be used for its representation.

Figure 3.18 shows a block diagram of a DPCM encoder. It can be seen that the previous input value is reconstructed by a feedback loop that integrates the encoded sample differences. The advantage of the feedback implementation is that quantization errors do not accumulate indefinitely (Bellamy, 1991). As in PCM systems, the analog-to-digital conversion process can be uniform or companded. By using DPCM, the transmission rate may be decreased to 56 kbps with comparable quality to that of PCM.

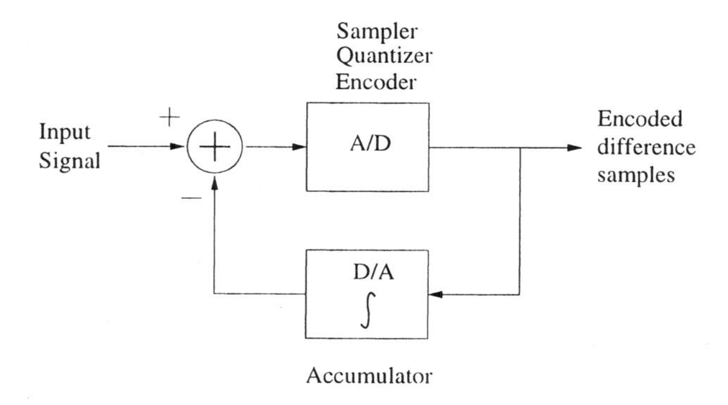

Figure 3.18. Block diagram of differential PCM.

Adaptive Differential PCM

It is interesting to notice that a DPCM coder may increase its performance gain if the quantization and/or prediction is carried out adaptively. In such case the coder is called adaptive differential pulse code modulation (ADPCM).

The key idea behind adaptive techniques for speech coding is to solve the tradeoff between a stepsize large enough to accomodate the maximum peak-to-peak range of the signal and small enough to minimize the quantization noise. So, adaptive quantization consists on adjusting the quantization steps as a function of the signal level. A nonuniform quantization is performed, by which the width of the quantization levels is not predetermined. Adaptive prediction consists on the dynamic adjustment of the predictor coefficients, according to the variations of the speech signal. This way, the nonstationary nature of the speech signal is taken into consideration in the coding process. ADPCM coders present good speech quality for rates between 24 and 48 kbps.

Delta Modulation

Delta modulation (DM) is a special case of DPCM. The variation of amplitude from sample to sample is quantized, by using only two quantization levels. Thus, DM uses only one bit per sample of the difference signal. The output of the two-level quantizer is related to the input by the expression $y = 2du(x) - d$, where $u(\cdot)$ is the unit step function, d denotes the quantization step and x is the difference signal between sucessives samples of the original signal (Alencar, 2002). The single bit is used to increment or decrement, by a constant stepsize, the coded signal when the difference signal is positive or negative, respectively. Thus, the coded signal resembles a staircase, as illustrated in Figure 3.19.

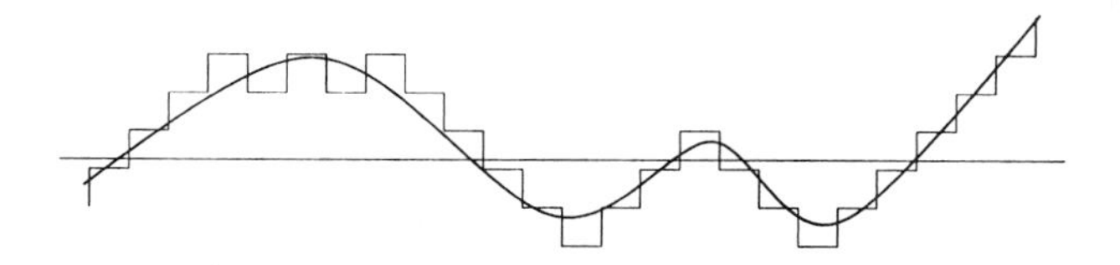

Figure 3.19. Example of the delta modulation.

The practical advantages of using a small number of quantization levels, including quantization with only one bit, have been confirmed by the popularity of *sigma-delta* ($\Sigma - \Delta$) coding schemes. Those schemes have wide acceptance due to their robustness to circuitry imperfections and suitability to VLSI (very large scale integration) implementation (Gray, 1987, Zamir and Feder, 1995).

The quantization noise, caused by the quantization stage in the $\Sigma - \Delta$ modulator, has been analyzed by many techniques (Gray, 1989, Galton, 1993). The use of pseudo random noise (*dither*) and the sampling at high rates have been studied, demonstrating the viability of recovering a signal with two levels (Chou and Gray, 1991, Shamai, 1994).

3.11 Parametric and Hybrid Coding

Parametric coders, or vocoders, are based on the model of speech production. This model is represented by a set of parameters periodically updated. For determining those parameters, the signal is segmented at periodic intervals, called frames. The parameters are usually updated at each frame. The rate required by the vocoders is low (lower than 4.8 kbps), but the delay and the complexity are high and the signal sounds synthetic.

Different from waveform coders, in parametric coders the quantization is carried out on the parameters of the speech production model, which are used in speech synthesis. The parameters of the speech model are determined over short frames, in which the speech signal may be considered stationary, and transmitted to the synthesizer in the receiver.

Parametric coders do not provide speech quality required by the telephone network. They are more used in military applications.

Hybrid coders combine the quality of waveform coders with the efficiency of parametric coders. Hybrid coders, more sophisticated than vocoders, are based on models of speech production and use a better excitation to the synthesizer. The excitation improvement is responsible for the improvement in the quality of the synthesized speech, which is more intelligible than the speech from conventional vocoders. The improvement is due to quantization and coding of the parameters that define the excitation as well as to parameters of the synthesis filter. A process known as analysis by synthesis is responsible for obtaining the parameters used in short frames. Hybrid coders are usually complex and lead, for rates from 4 to 16 kbps, to a better quality when compared to the one obtained from waveform coding with higher rates.

Channel Vocoder

Most of the coding processes of the channel coders involve the determination of the sampled spectrum of the speech signal as a time function. In channel vocoder, a bank of bandpass filters is used to separate the energy of the speech signal in sub-bands, which are completely rectified and filtered for determining the relative power levels. The individual power levels are coded and transmitted to the destination.

In addition to measuring the signal spectrum, modern channel vocoders also determine the nature of the speech excitation.

Most of the difficulties encountered in the great majority of vocoder implementations regard the determination of the speech harmonics. Without an accurate information of excitation, the quality in the coder output is poor and many times it depends on the person who talks. Some channel vocoders have produced high intelligibility, in spite of producing a synthetic sound at a rate 2.4 kbps (Bayless et al., 1973).

Formant Vocoder

A formant vocoder determines the localization and the amplitude of the formants (in which the energy of the speech signal is concentrated) and transmits that information instead of transmitting the whole spectral envelope. Hence, the formant vocoder produces a low bit rate by coding only the most important pulses of the speech spectrum (Alencar, 2002).

The most important point in formant vocoder is the accurate tracking of the formants. Once that requirement is satisfied, the formant vocoder may provide speech intelligibility at a rate lower than 1 kbps (Flanagan et al., 1979).

Linear Predictive Coder

Linear predictive coder (LPC) is a widely used vocoder. It extracts the perceptually important features of the speech directly from the waveform in the time domain. The effect is better than that one obtained from the frequency spectrum, such as in channel vocoder and formant vocoder. Fundamentally, an LPC analyses the voice waveform for producing a model of the time varying vocal tract and for producing the transfer function of the vocal model. A synthesizer at the receiver terminal recreates the speech signal by using a mathematical model of the vocal tract. By the periodic updating of the parameters of the model and the specification of the excitation, the synthesizer is adapted to the changes performed. Figure 3.20 presents the basic model of speech generation of the linear predictive coding. The figure is also a model of an LPC coder/synthesizer.

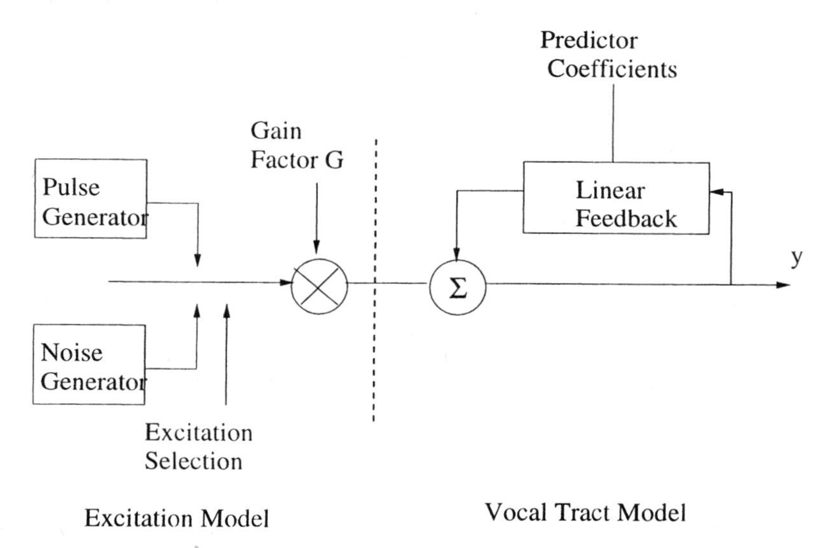

Figure 3.20. Model of speech generation in linear predictive coding.

The equation of the model of the vocal tract is defined by

$$y(n) = \sum_{k=1}^{p} a_k y(n-k) + Gx(n), \tag{3.43}$$

where $y(n)$ denotes de n-th output sample, a_k denotes the k-th predictor coefficient, G is the gain factor, $x(n)$ denotes the input sampled at time n and p is the order of the model.

The speech output in Equation 3.43 is represented as the actual input of the system, summed to a linear combination of the predicted outputs of the vocal tract. The model is adaptive and the coder determines, periodically, a new set of parameters corresponding to the successive frames of speech. A basic LPC does not perform the measurement and coding of the differences of waveforms. Instead, the error signals are minimized in a mean square sense, when the predictor coefficients are determined (Alencar, 2002).

The information that the LPC coder/analyzer determines and transmits to the decoder/synthesizer consists of:

1 Nature of excitation, if voiced or not.

2 Counting of the period for the excitation of the speech.

3 Gain factor.

4 Predictor coefficients, that is, parameters of the vocal tract model.

An LPC coder is capable of performing gradual changes in the spectral envelope. As a final result, a more natural speech representation is obtained when compared to that of vocoders based purely in the frequency domain (Bayless et al., 1973). Most LPCs have concentrated speech coding in the range 1.4 to 2.4 kbps.

Code-excited Linear Prediction – CELP

The basic LPC algorithm synthesizes the speech in the decoder by using a very simple excitation model, which requires only 10% of the data rate (Alencar, 2002). The simplicity of the model leads to speech that sounds synthetic. To overcome this problem, many techniques have been developed to improve the excitation. Three methods may be cited: Multipulse excitation LPC (MLPC), Residual excitation LPC (RELP) and code-excited linear prediction (CELP).

The CELP coders look like the multipulse coders. The synthesis filter and the predictors are the same. The difference resides in the fact that in CELP coders the sequence of excitation pulses is selected from a set of random vectors, previously stored, forming a kind of codebook. The vectors stored in the codebook have Gaussian distribution and zero mean, as an attempt to match the short term characteristics of the speech signal (Aguiar Neto, 1995).

The choice of the ideal sequence to be used as the excitation of the synthesis filter at the decoder is performed by a search in a codebook, using the analysis by synthesis technique. As in MLPC, a short term correlation analysis and a long term correlation analysis are carried out, leading to the spectral envelope (formants) and the periodicity (pitch) of the speech signal. By using those parameters, the speech signal is synthesized.

The CELP coders lead to a good speech quality at rates ranging from 4.8 kbps to 16 kbps and present a better performance in terms of quality versus bit rate when compared to other coders, as can be seen in Figure 3.21. However, the CELP coding demands a high computational effort, which was seen as an obstacle, for a long time, to real-time implementation.

Simplifications in the basic structure of CELP coder (introduction of efficient search methods in the codebook or use of algebraic codebooks) and the increase of MIPS in digital signal processors (DSPs) made the coding method feasible. It is worth mentioning that the decoder is much simpler than the coder (since there is no analysis by synthesis search), and a posterior filtering may be optionally performed (Hersent et al., 2002).

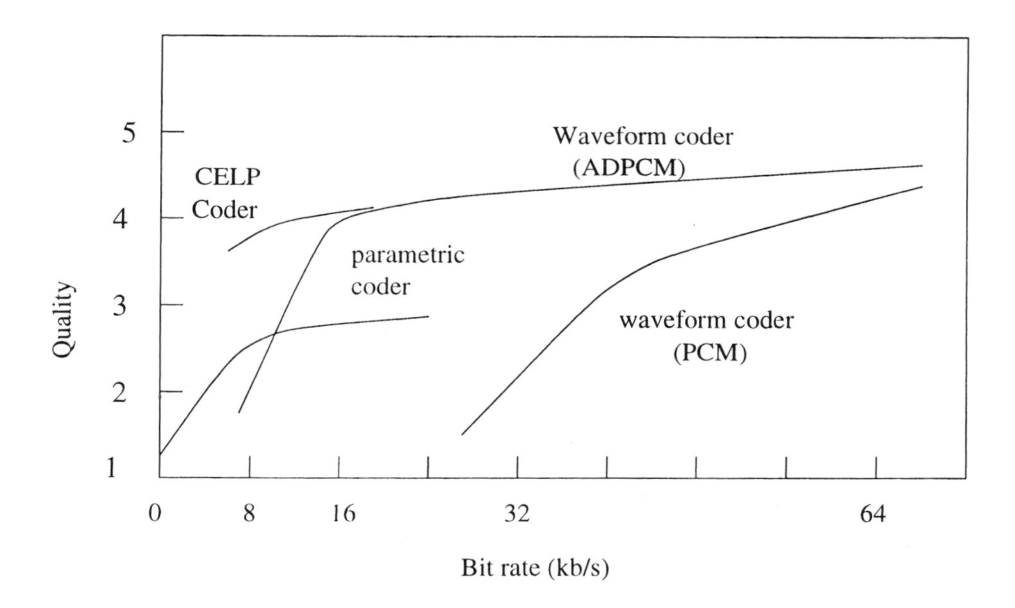

Figure 3.21. Comparing coders in terms of bit rate and speech quality.

A modification introduced in the CELP coder, such that the delay was lowered from about 20-40 ms to about 2 ms, has originated the LD-CELP (low delay CELP).

Many international standards in the range 4.8 kbps to 16 kbps are CELP coders or CELP based coders, such as: ITU-T G.729, multimedia standard ITU-T G.723.1, full-rate GSM ETSI and half-rate GSM.

3.12 Speech Coder Attributes

Speech coders have attributes classified into four groups: bit rate, quality, complexity and delay. For a given application, some attributes are pre-specified, while tradeoffs can be made among the others. As an example, the communication channel may establish a limit in the bit rate; cost constraints may set limits in the complexity.

Quality can be generally improved by increasing the bit rate or the complexity, and sometimes by increasing the delay. Requirements or goals may be set for all of these speech coders attributes. For the purposes of standards, two additional attributes are the method of specification and conformance validation and the schedule in which the work plan is to be accomplished (Cox, 1995).

Bit Rate

As mentioned earlier, telephone bandwidth speech signals have a bandwidth of about 300-3,400 Hz and are generally sampled at 8,000 Hz. The speech coders that have been standardized in recent years have bit rates ranging from 800 bps to 16 kbps (Cox, 1995).

Some of these coders, notably the ones that have been standardized for cellular telephony, also have a channel coder associated with them. In such case, the bit rate increases (for instance, to 22.8 kpbs). The lower bit rates (800 to 4,800 bps) are associated mainly to secure telephony, while cellular telephony speech coding rates range from 3.3 kbps to 14 kbps, for regular voice services (Cox, 1995).

Wideband speech signals have a bandwidth of 50-7,000 Hz and are sampled at 16 kHz. Due to the wider bandwidth, intelligibility and naturalness are improved and listener fatigue during long conversations are lessened. Among the wideband telephony standards (7 kHz), one can cite G.722, with rates 64, 56 and 48 kbps.

Delay

Low rate speech coders can be considered block coders. They encode a block (frame) of speech at a time. Depending on the application, the total speech coding delay is some multiple of the frame size. The minimum delay is generally about three to four times the frame size. For purposes of standardization, the delay is a factor in real-time conversational systems (Cox, 1995).

Complexity

Regarding speech coder implementation, speed and random access memory (RAM) usage are the two most important aspects to complexity. The faster the chip (DSP) or the greater the chip size, the greater the cost. The same attributes also influence the power consumption, which is a critical attribute to handheld applications.

Hence, complexity is determinant for both cost and power consumption. For the perspective of standardization, complexity is determined by the application (Cox, 1995). Speech coders that require less than 15 MIPS are considered low complexity coders. The ones that require 30 MIPS or more are considered high complexity coders (Rao et al., 2002).

Quality

The Speech Quality Experts Group (SQEG) of the International Telecommunications Union (ITU) has a very strict view of what constitutes network toll quality.

The first digital speech coding standard was CCITT Recommendation G.711 for 64 kbps PCM speech. The distortion introduced by a G.711 codec is considered one QDU (quantization distortion unit). SQEG uses QDU for network planning purposes. The second digital speech coding standard was G.721 32 kbps ADPCM. This coder was standardized to be used in combination with a G.711 codec as its input and output. The resulting distortion is considered to be 3.5 QDU (Cox, 1995).

SQEG network planning guidelines call for a maximum of 14 QDU for an end-to-end international connection and less than 4 QDU for a domestic connection (Cox, 1995). G.728 16 kbps LD-CELP codec is considered to have the same QDU of G.721. Both are ITU coders which present toll quality.

Regarding quality, some aspects must be considered, such as intelligibility, naturalness and coder performance when the input contains other signals (such as background noise or music) besides speech. Speech quality can be measured by subjective tests.

Other dimensions of speech quality can be considered. The channel error sensitivity can be regarded as one aspect of quality. In the case of digital cellular standards, additional bits for channel coding are provisioned to protect the information bearing bits.

3.13 Problems

1 Explain signal compression (definition, types, importance, purpose, applications).

2 Discuss the function of source coding, channel coding and modulation, in a typical digital communication system.

3 Enunciate the Sampling Theorem, presenting a brief mathematical analysis of the sampling process. Comment on the following: if the sampling frequency is lower than the Nyquist frequency, then the signal can not be completely recovered.

4 Discuss the aspects used for assessing the performance of a signal compression system.

5 Present the main features of speech signals. Some features are efficiently explored in speech coding systems? Explain.

6 Pulse code modulation (PCM) transforms an analog signal into a series of binary pulses. What is the minimum sampling frequency for a signal with bandwidth of 3.4 kHz?

7 The minimum square error of a uniform scalar quantizer is given by $d^2/12$, where d is the quantization step. What are the reasons by which this result is obtained?

8 The scalar quantization noise leads to a loss in signal quality. What is the improvement, in terms of signal to quantization noise ratio (SQNR), obtained by using two additional bits in the coding process? Explain.

9 Define mathematically vector quantization. Discuss its importance in the scenario of signal compression.

10 Regarding LPC quantization, what is transparent quantization?

11 Describe what is meant by waveform coders, parametric coders and hybrid coders. Present a comparison concerning reconstructed signal quality *versus* bit rate.

12 Discuss the following attributes of speech coders: bit rate, delay, complexity and quality.

Chapter 4

AMPLITUDE MODULATION

4.1 Introduction

The problem of information transmission between two points is of fundamental relevance for communications. This problem occurs in radio broadcast, television transmission, telephony, satellite communication and in many other instances in various systems.

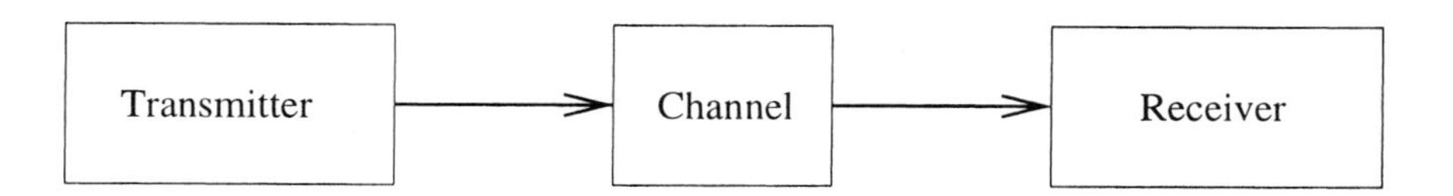

Figure 4.1. Transmission system model.

Information is carried from one point to another by means of signals, using a propagation media, as illustrated in block diagram in Figure 4.1 by the channel block. The propagation media, as for example the telephone lines, is usually used for confining and guiding the signals, but that is not always the case since in radio broadcast the propagation media is the atmosphere or free space.

In order for the signal energy to flow conveniently from the transmitter to the propagation media, and from the propagation media to the receiver, it is necessary to feed the signal through transition elements working either as energy transmitters or as energy receivers. In many cases the system characteristics are determined by the realizability of such elements (Schwartz, 1970). For example, when using acoustic signals the radiating element is the loud-speaker the physical dimensions of which depend on the frequency band and on the amount of energy to be radiated. When using radio transmitters it would be impractical to transmit signals in the human voice frequency band directly since, by the theory of electromagnetic waves, efficient signal radiation can only be achieved with a radiating antenna size of the same order as the wavelength of the signals to be radiated. It follows in this case that the radiating antennas would require sizes of hundreds of kilometers in order to radiate efficiently.

Therefore, in order to allow efficient radiation of radio signals carrier waves are employed the parameters of which are compatible with the dimensions of the radiating element and with the media characteristics. Such waves provide a support for carrying information. The information is also represented by an electrical signal, called modulating signal, used to modify one or more parameters of the carrier. Figure 4.2 presents the electromagnetic spectrum available for signal transmission, where R.F. stands for radiofrequency, M.W. is microwave, M.M. is millimeter wave, I.R. is infrared and U.V. is ultraviolet.

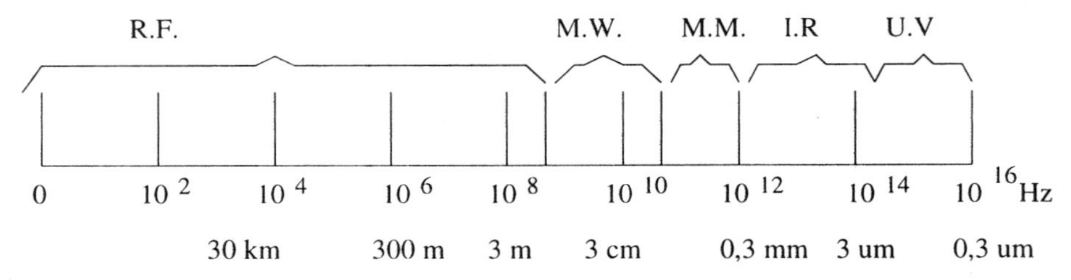

Figure 4.2. Electromagnetic spectrum.

As briefly mentioned earlier, modulation consists of the variation of one or more characteristics of the carrier waveform as a function of the modulating signal. The sinusoidal waveform is traditionally used as the carrier, and the modulation is performed in general in three ways.

- Amplitude Modulation (AM), if amplitude is the carrier parameter that is varied;
- Angle Modulation, if either the phase (PM) or the frequency (FM) is the carrier parameter that is varied;
- Quadrature Modulation (QAM), if both the amplitude and the phase of the carrier are varied simultaneously.

4.2 Amplitude Modulation

Amplitude modulation is a modulation system where the carrier wave has its instantaneous amplitude varied in accordance with the modulating signal (Alencar, 1999). Amplitude modulation (AM), also known as double-sideband amplitude modulation (AM-DSB), is by far the most widely used modulation system and was adopted for commercial broadcast. This is a consequence of some practical advantages as, for example, economy, simplicity of receiver design and easy maintenance. The AM carrier is a sinusoid represented as

$$c(t) = A\cos(\omega_c t + \phi), \tag{4.1}$$

where A denotes the carrier amplitude, ω_c denotes the angular frequency (measured in radians per second, i.e., rad/s) and ϕ denotes the carrier phase. Figure 4.3 illustrates an AM modulator.

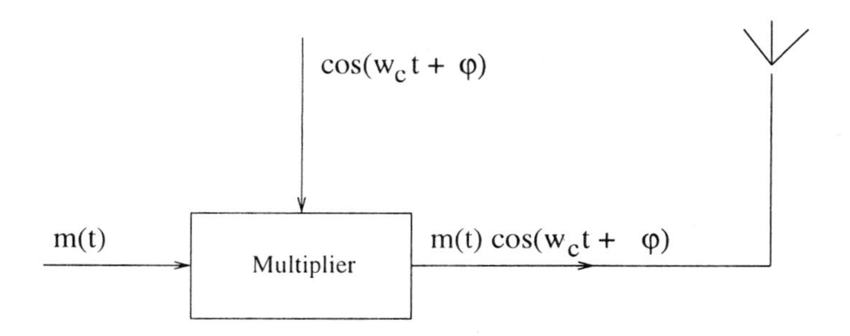

Figure 4.3. AM modulator.

The message signal, or modulating signal, denoted usually as $m(t)$, must be such that its largest frequency is still much less than ω_c. The available modulation bandwidth is restricted by the linear region of operation of the transmitter amplifiers, and varies from 0,1% to 1% of the carrier frequency (Gagliardi, 1978). The modulation process is illustrated in Figure 4.4.

In order to have the carrier amplitude varying proportionally to $m(t)$, the *instantaneous carrier amplitude* needs to have the form

$$a(t) = A + Bm(t) = A[1 + \Delta_{AM} m(t)], \tag{4.2}$$

where $\Delta_{AM} = B/A$ is called the *AM modulation index*. The instantaneous amplitude is responsible for producing the modulated waveform as follows

$$s(t) = a(t)\cos(\omega_c t + \phi), \tag{4.3}$$

or

$$s(t) = A[1 + \Delta_{AM} m(t)]\cos(\omega_c t + \phi). \tag{4.4}$$

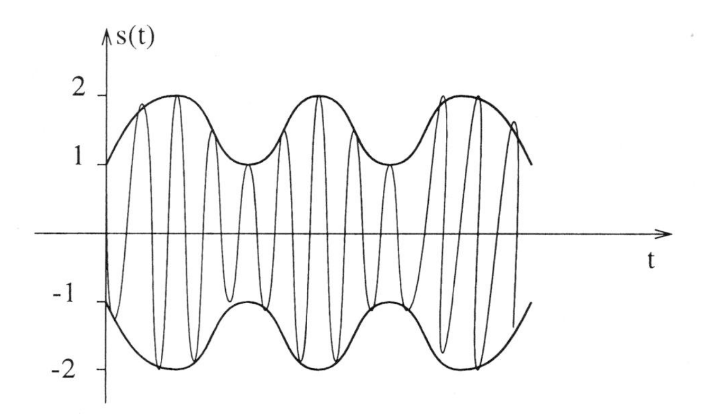

Figure 4.4. Amplitude modulation.

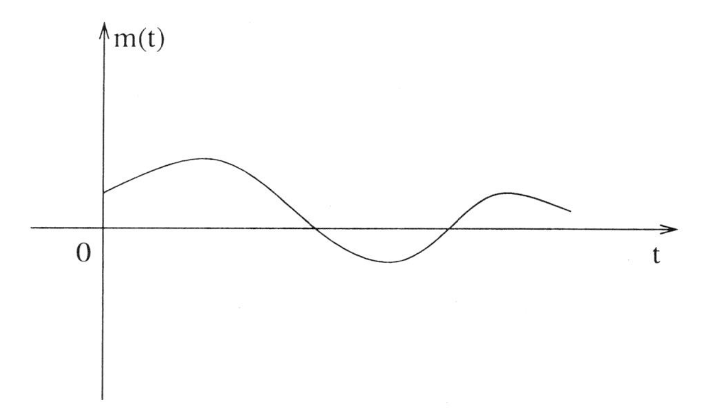

Figure 4.5. Modulating signal $m(t)$.

Figure 4.5 shows the modulating waveform, which alters the carrier amplitude in order to produce the modulated waveforms seen in Figures 4.6, 4.7 and 4.8, for various values of the modulation index.

It is noticed in Figure 4.6 that the envelope of the modulated waveform has the same format as the modulating signal. This follows because $A + Bm(t) > 0$. In Figure 4.7 the modulation index is equal to 1, and is usually referred as a 100% modulated carrier. In Figure 4.8 the modulation index is greater than 1, and causes phase inversion or phase rotation, in the modulated carrier, which is said to be over modulated.

The modulation index indicates how strong the modulating signal is with respect to the carrier. The modulation index should not exceed 100% in order to avoid distortion in a demodulated signal whenever an envelope detector is employed.

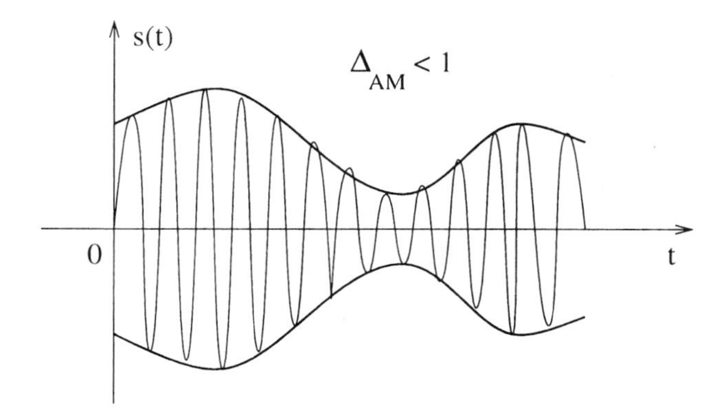

Figure 4.6. Effect of variation of the modulation index, for $\Delta_{AM} < 1$.

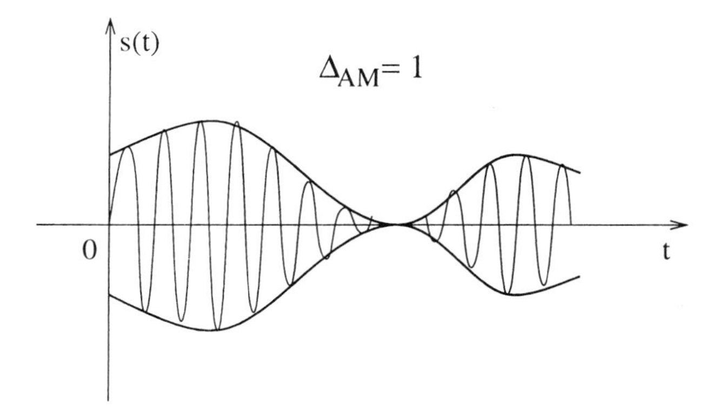

Figure 4.7. Effect of variation of the modulation index, for $\Delta_{AM} = 1$.

Depending on the manner by which the instantaneous amplitude varies as a function of Δ_{AM}, assuming $|m(t)| = 1$, the following terminology is applicable.

- $\Delta_{AM} = 1$, means 100% carrier modulation and allows envelope detection of the modulating signal;

- $\Delta_{AM} > 1$, means carrier overmodulation, causes phase rotation of the carrier and requires synchronous demodulation for recovering the modulating signal;

- $\Delta_{AM} < 1$, means carrier undermodulation, allows envelope detection of the modulating signal but does not make efficient use of the carrier power.

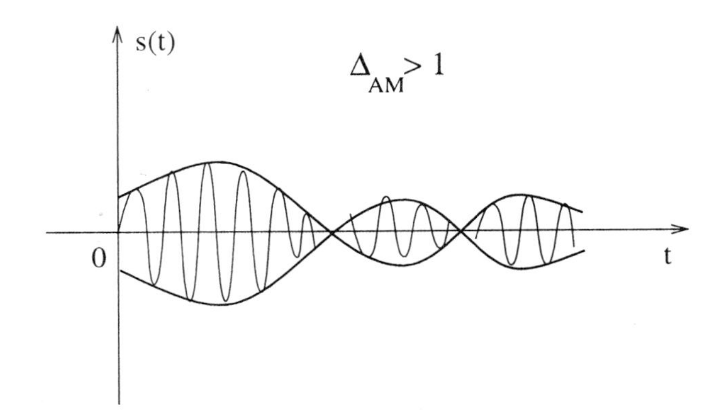

Figure 4.8. Effect of variation of the modulation index, for $\Delta_{AM} > 1$.

4.3 Amplitude Modulation by Random Signals

The mathematical treatment used in this section involves concepts of a random process, or random signal, which was developed in Chapter 3. The results derived here by using an analysis based on random processes by far compensate for the extra effort required to understand the theory of stochastic processes. The theoretical development that follows, besides being particularly more elegant than the standard treatment based on deterministic signals, is closer to real life signals of a true random nature.

Consider the modulated carrier given by $s(t) = a(t)\cos(\omega_c t + \phi)$, where $a(t) = A[1 + \Delta_{AM} m(t)]$. Let the carrier phase ϕ be a random variable with a uniform distribution in the interval $[0, 2\pi]$. The signal $m(t)$ is assumed to be a stationary random process with zero mean and statistically independent of ϕ.

The carrier is a random process because it possesses a random phase and because it was modulated by a stochastic process. The stationarity of the carrier is thus guaranteed. The autocorrelation of a stationary random process, which was discussed in Chapter 3, is given by

$$R_S(\tau) = E[s(t)s(t + \tau)]. \tag{4.5}$$

By plugging (4.3) into (4.5) it follows that

$$R_S(\tau) = E[(a(t)\cos(\omega_c t + \phi))(a(t + \tau)\cos(\omega_c(t + \tau) + \phi))], \tag{4.6}$$

or

$$R_S(\tau) = E[a(t)a(t+\tau)\cos(\omega_c t + \phi)\cos(\omega_c(t+\tau) + \phi)], \quad (4.7)$$

and by replacing a sum of cosines for a product of cosines it follows that

$$R_S(\tau) = \frac{1}{2}E[a(t)a(t+\tau)\left(\cos(\omega_c\tau) + \cos(2\omega_c t + \omega_c\tau + 2\phi)\right)]. \quad (4.8)$$

By making use of properties of the mean, considering that $a(t)$ and ϕ are independent random variables and that the mean value of the carrier is zero, it follows that

$$R_S(\tau) = \frac{1}{2}R_A(\tau)\cos(\omega_c\tau), \quad (4.9)$$

where the autocorrelation of the modulating signal $R_A(\tau)$ is defined as

$$R_A(\tau) = E[a(t)a(t+\tau)]. \quad (4.10)$$

Replacing the expression (4.2) for $a(t)$ in (4.10) it follows that

$$\begin{aligned} R_A(\tau) &= E[A(1+\Delta_{AM}m(t))A(1+\Delta_{AM}m(t+\tau))] \quad (4.11)\\ &= A^2E[1+\Delta_{AM}m(t)+\Delta_{AM}m(t+\tau)+{\Delta_{AM}}^2m(t)m(t+\tau)]. \end{aligned}$$

Again using properties of the mean and recalling that $m(t)$ is stationary and zero mean, i.e., that $E[m(t)] = E[m(t+\tau)] = 0$, it follows that

$$R_A(\tau) = A^2[1 + \Delta_{AM}^2 R_M(\tau)], \quad (4.12)$$

where $R_M = E[m(t)m(t+\tau)]$ represents the autocorrelation of the message signal. Finally, the autocorrelation of the amplitude modulated carrier is given by

$$R_S(\tau) = \frac{A^2}{2}[1 + \Delta_{AM}^2 R_M(\tau)]\cos\omega_c\tau. \quad (4.13)$$

4.3.1 Total Power of an AM Carrier

The total power of an AM carrier is given by the value of its autocorrelation for $\tau = 0$, thus

$$P_S = R_S(0) = \frac{A^2}{2}(1 + \Delta_{AM}^2 P_M), \tag{4.14}$$

where $P_M = R_M(0)$ represents the power in the message signal $m(t)$. The power in the unmodulated carrier is given by $\frac{A^2}{2}$, as can be easily checked, and represents a significant portion of the total transmitted power.

4.3.2 Power Spectral Density

The power spectral density of the AM modulated carrier is obtained as the Fourier transform of the autocorrelation function $R_S(\tau)$. This result is known as the Wiener-Khintchin theorem, seen in Chapter 3, i.e.,

$$S_S(\omega) = \mathcal{F}[R_S(\tau)],$$

where $R_S(\tau) = \frac{1}{2}R_A(\tau)\cos(\omega_c\tau)$.

The function $R_S(\tau)$ can be seen as the product of two functions: $\frac{1}{2}R_A(\tau)$ and $\cos(\omega_c\tau)$. Using this line of reasoning, the Fourier transform of $R_s(\tau)$ is calculated with the application of the convolution theorem, i.e.,

$$\begin{aligned} S_S(\omega) &= \mathcal{F}[R_S(\tau)] = \frac{1}{2\pi}\left(\mathcal{F}[\frac{1}{2}R_A(\tau)] * \mathcal{F}[\cos(\omega_c\tau)]\right) \\ &= \frac{1}{2\pi}\left[\frac{1}{2}S_A(\omega) * (\pi\delta(\omega+\omega_c) + \pi\delta(\omega-\omega_c))\right], \end{aligned}$$

and applying the impulse filtering property it follows that

$$S_S(\omega) = \frac{1}{4}\left[S_A(\omega+\omega_c) + S_A(\omega-\omega_c)\right], \tag{4.15}$$

where $S_A(\omega) = \mathcal{F}[R_A(\tau)]$.

The power spectral density of the modulating signal can be derived by writing the expression for $R_A(\tau)$ and then calculating its Fourier transform. Thus

$$\begin{aligned} S_A(\omega) &= \mathcal{F}[A^2(1 + \Delta_{AM}{}^2 R_M(\tau)], \\ &= A^2[2\pi\delta(\omega) + \Delta_{AM}{}^2 S_M(\omega)]. \end{aligned}$$

where $S_M(\omega)$ is the power spectral density of the message signal.

Finally, the power spectral density of the modulated AM carrier is given by

$$S_S(\omega) = \frac{A^2}{4}\left[2\pi(\delta(\omega+\omega_c)+\delta(\omega-\omega_c)) + {\Delta_{AM}}^2(S_M(\omega+\omega_c)+S_M(\omega-\omega_c))\right]. \quad (4.16)$$

The power spectral densities of a message signal, of the modulating signal (carrier amplitude) and of the modulated carrier are illustrated in Figure 4.9a, b and c, respectively.

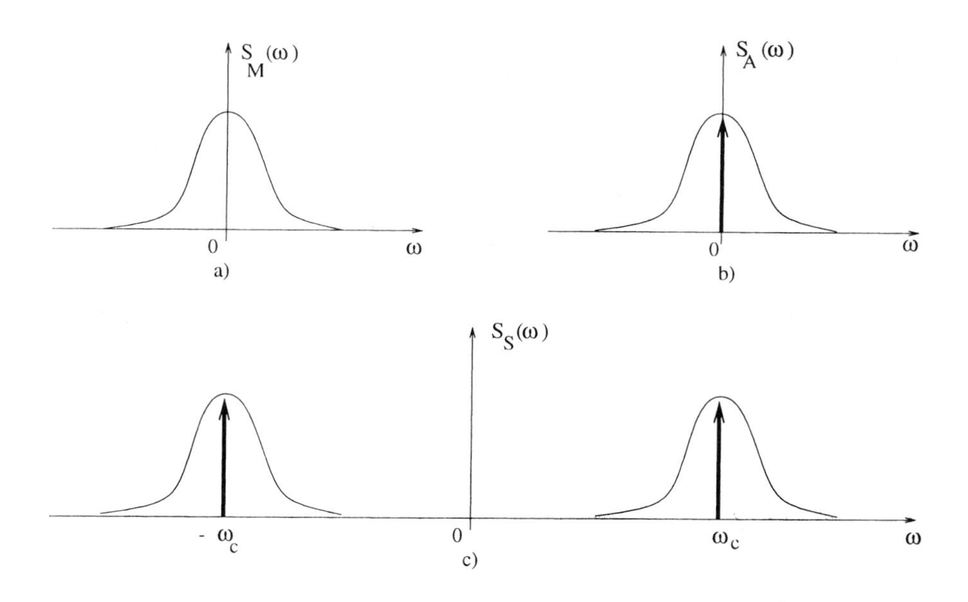

Figure 4.9. Power spectral density of an AM signal.

The bandwidth required for the transmission of an AM signal is precisely twice the bandwidth of the message signal. In AM radio broadcast the maximum frequency of the message signal is limited to 5 kHz, and consequently the AM bandwidth for commercial radio broadcast is 10 kHz.

4.4 Amplitude Modulators

Modulators are devices that allow, from a modulating signal and the carrier, to produce the modulated carrier waveform. In general, the AM modulators can be classified in two basic categories as follows.

1 Quadratic modulators,

2 Synchronous modulators.

4.4.1 Quadratic Modulator

Every nonlinear device having in its characteristic curve a term of degree two can be used as a modulator. In case the characteristic curve also contains terms of order higher than two, the undesirable effect of such terms may be filtered out at a subsequent stage.

The quadratic modulator must possess an element of nonlinear characteristic containing predominantly a quadratic term. Let $x(t)$ denote an input signal and let $y(t)$ denote the corresponding output signal associated with a nonlinear device, i.e., let

$$y(t) = ax(t) + bx^2(t) + \cdots . \tag{4.17}$$

By considering only the first two terms in (4.17) and an input signal given by

$$x(t) = A\cos(\omega_c t + \phi) + m(t), \tag{4.18}$$

where the function $\cos(\omega_c t+\phi)$ represents the carrier and $m(t)$ represents the modulating signal, the output signal is given by

$$y(t) = a[A\cos(w_c t + \phi) + m(t)] + b[A\cos(w_c t + \phi) + m(t)]^2,$$

i.e.,

$$\begin{aligned} y(t) &= aA\cos(\omega_c t + \phi) + am(t) + bA^2\cos^2(\omega_c t + \phi) + \\ &\quad 2bAm(t)\cos(\omega_c t + \phi) + bm^2(t). \end{aligned}$$

By filtering the signal $y(t)$ with a bandpass filter centered in ω_c it follows that

$$s(t) = aA\cos(\omega_c t + \phi) + 2bAm(t)\cos(\omega_c t + \phi),$$

which is the expression of the amplitude modulated signal.

4.4.2 Synchronous Modulator

This type of modulator is based on the principle of sampling the signal resulting from the sum of the modulator signal and a DC level. The sampling operation is performed by means of a switch synchronous with

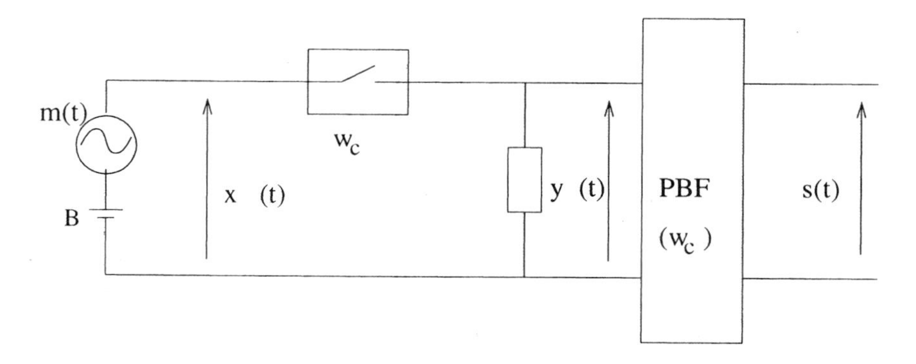

Figure 4.10. Synchronous modulator.

the carrier, as illustrated in Figure 4.10, where PBF stands for passband filter.

The voltage $y(t)$ can be characterized as $y(t) = c(t)x(t)$, where $c(t)$ is a function of the switching operation and can be written as

$$c(t) = c_0 + \sum_{n=1}^{\infty} c_n \cos(n\omega_c t) \tag{4.19}$$

and $x(t) = B + m(t)$. Therefore,

$$y(t) = (B + m(t))(c_0 + \sum_{n=2}^{\infty} c_n \cos(n\omega_c t)). \tag{4.20}$$

Filtering $y(t)$ with a bandpass filter centered in ω_c, it follows that $s(t) = (B + m(t))(c_1 \cos(\omega_c t))$, which is the expression for the amplitude modulated signal.

4.4.3 Digital AM Signal

The digital AM signal, also called *Amplitude Shift Keying* (ASK), can be generated by the process of multiplying the digital modulating signal by the carrier. The ASK signal is shown in Figure 4.11.

The ASK signal can be represented in a different manner, using a phase diagram, which consists of representing the modulated signal in axes which are in phase (I axis) and in quadrature (with a phase lag of $\pi/2$, or Q axis) with respect to the carrier phase.

This diagram is also known as a constellation because it represents signal points as stars on a plane. The digital signal which amplitude modulates the carrier can be written as

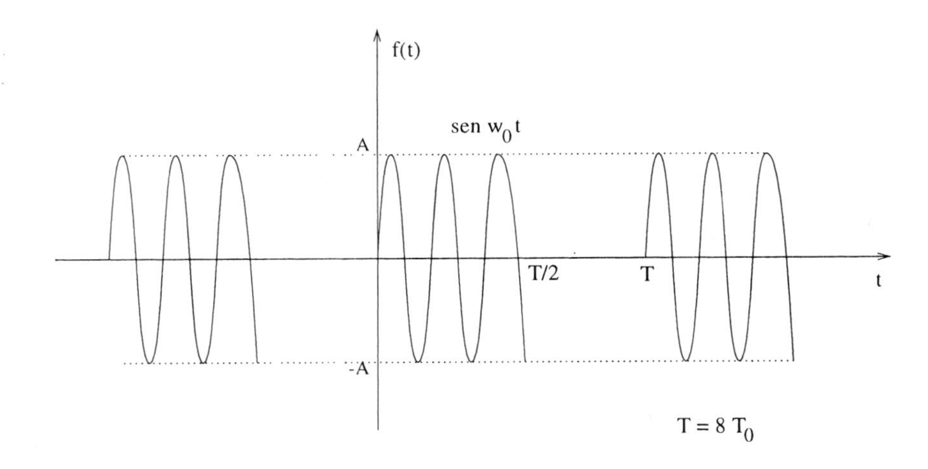

Figure 4.11. Binary ASK signal.

$$m(t) = \sum_{k=-\infty}^{k=\infty} m_k p(t - kT_b), \tag{4.21}$$

where m_k represents the k-th randomly generated symbol, from a discrete alphabet, $p(t)$ is the pulse shape of the transmitted digital signal and T_b is the bit interval.

The modulated signal without carrier is then given by

$$s(t) = \sum_{k=-\infty}^{\infty} m_k p(t - kT_b) \cos(\omega_c t + \phi). \tag{4.22}$$

As an example, Figure 4.12 shows the constellation diagram of a 4ASK signal, the symbols of which are $m_k \in \{-3A, -A, A, 3A\}$. All the signal points are on the cosine axis (in phase with the carrier), since there is no quadrature component in this case.

When the modulating signal is a digital signal, the transmitted power is calculated by considering the average power per symbol. In the ASK case, by considering equiprobable symbols symbols it follows that the transmitted power is given by

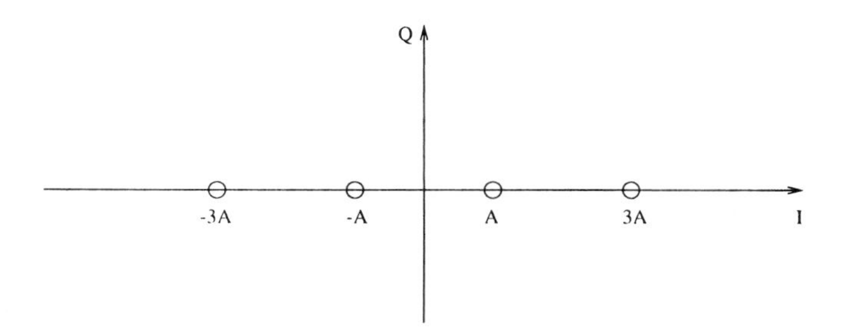

Figure 4.12. Constellation for a 4ASK signal.

$$P_S = \frac{1}{2}\sum_{k=1}^{4} m_k^2 p(m_k) = \frac{1}{2}\left[\frac{(-3A)^2 + (-A)^2 + (A)^2 + (3A)^2}{4}\right] = \frac{5}{2}A^2. \tag{4.23}$$

The probability of error for the coherent binary 2ASK is (Haykin, 1988)

$$P_e = \frac{1}{2}\text{erfc}\left(\sqrt{\frac{E_b}{N_0}}\right). \tag{4.24}$$

where, E_b is the binary pulse energy, N_0 represents the noise power spectral density and erfc$(\cdot)$ is the complementary error function,

$$\text{erfc}(x) = \frac{2}{\sqrt{\pi}}\int_x^{\infty} e^{-t^2}dt. \tag{4.25}$$

For a rectangular pulse, $E_b = A^2T_b$, where A is the pulse amplitude and T_b is the pulse duration.

4.4.4 AM Transmitter

Figure 4.13 shows a block diagram of an AM transmitter, normally used for radio broadcast. The transmitter consists of the following stages.

1 Oscillator, to generate the carrier;

2 RF amplifier, to amplify the carrier power to a level adequate to feed the modulator stage;

3 Audio amplifier, to amplify the modulating signal;

4 Class C modulator, produces the AM signal when excited by the audio signal and by the carrier.

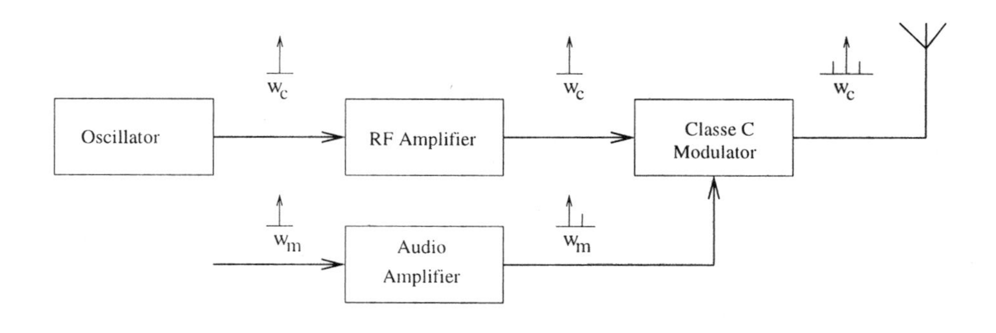

Figure 4.13. AM transmitter.

4.5 Suppressed Carrier Amplitude Modulation

It has been seen that in standard AM systems most of the power is spent in transmitting the carrier. In principle, this is a waste of power if one reasons that the carrier by itself conveys no information.

In a suppressed carrier AM system (AM-SC) the carrier is not sent as part of the AM signal and all of the transmitter power is available for transmitting information over the two sidebands. Obviously, in terms of efficient use of transmitter power the AM-SC system is more efficient than an AM system. However, a receiver for an AM-SC signal is significantly more complex than an AM receiver.

The AM-SC technique essentially translates the frequency spectrum of the modulating signal by multiplying it by a sinusoid the frequency of which has a value equal to the desired frequency translation. In other words the original message, or modulating, signal $m(t)$ becomes $m(t)\cos(\omega_c t + \phi)$, after multiplication by the carrier.

4.6 Spectrum of the AM-SC Signal

Let $s(t) = m(t)\cos(\omega_c t + \phi)$, where $m(t)$ is a stationary random process with zero mean, ϕ is a random variable uniformly distributed in the interval $[0, 2\pi]$ and statistically independent of $m(t)$.

$$R_S(\tau) = E[s(t)s(t+\tau)] \tag{4.26}$$

$$R_S(\tau) = E\left[m(t)m(t+\tau)\cos(\omega_c t+\phi)\cos(\omega_c(t+\tau)+\phi)\right] \tag{4.27}$$

$$R_S(\tau) = E\left[m(t)m(t+\tau)]\cdot E[\frac{1}{2}(\cos\omega_c\tau) + \cos(\omega_c\tau + 2\omega_c t + 2\phi)\right] \tag{4.28}$$

$$R_S(\tau) = E\left[m(t)m(t+\tau)\right]\,(\frac{1}{2}\cos\omega_c\tau) \tag{4.29}$$

$$R_S(\tau) = \frac{1}{2}R_M(\tau)\cos\omega_c(\tau) \tag{4.30}$$

The power of the AM-SC signal can be derived from the autocorrelation function computed for $\tau = 0$.

$$P_S = R_S(0) = \frac{1}{2}P_M. \tag{4.31}$$

4.6.1 Power Spectral Density

The power spectral density is obtained by means of the Fourier transform of the autocorrelation function, as follows

$$S_S(\omega) = \int_{-\infty}^{\infty} R_S(\tau)e^{-j\omega\tau}d\tau, \tag{4.32}$$

$$S_S(\omega) = \frac{1}{4}S_M(\omega) * \delta[(\omega+\omega_c) + \delta(\omega-\omega_c)], \tag{4.33}$$

$$S_S(\omega) = \frac{1}{4}[S_M(\omega+\omega_c) + S_M(\omega-\omega_c)]. \tag{4.34}$$

4.6.2 The AM-SC Modulator

A common type of AM-SC modulator is called a balanced modulator, illustrated in Figure 4.14, which can also be implemented with a diode bridge as shown in Figure 4.15. In Figure 4.14, the output current is composed by the two input currents, produced by the excitation of the nonlinear elements with a voltage consisting of the sum of the modulating signal and the carrier. In Figure 4.15, the diodes act as switches to produce the effect of multiplication of the signal $m(t)$ by a square wave.

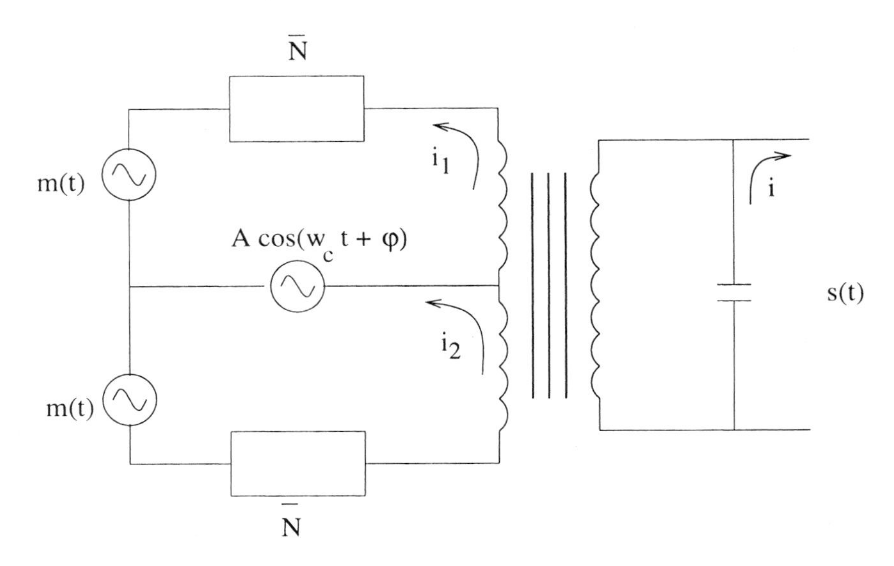

Figure 4.14. Balanced modulator.

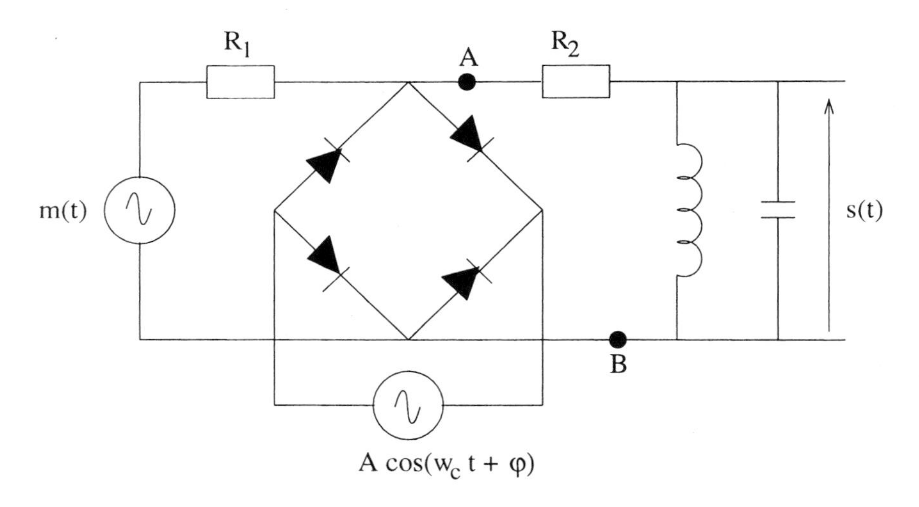

Figure 4.15. Diode bridge modulator.

4.7 AM-VSB Modulation

A special case in the study of amplitude modulation is the *amplitude modulation vestigial sideband* (AM-VSB). The process of generating a single sideband signal (SSB), in order to make efficient use of the available bandwidth, has some practical problems when performed by means of a sideband filter because in some applications a very sharp filter cutoff characteristic is required. The SSB modulation scheme will be studied later on. In order to overcome this filter design difficulty a compromise solution was established between AM-DSB and SSB, called AM-VSB (Taub and Schilling, 1971). Figures 4.16 and 4.17 show a two types of AM modulation.

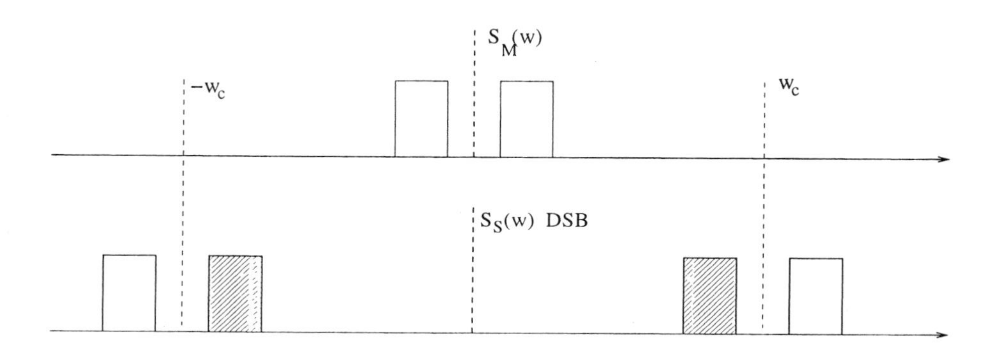

Figure 4.16. Spectra of modulating signal $S_M(\omega)$ and AM-SC modulated signal $S_S(\omega)$ em AM-SC.

Figure 4.16 represents the spectrum of the modulating signal as well as the spectrum of the modulated AM-SC signal. Figure 4.17 presents the spectrum of a modulated SSB signal, to be discussed in the next chapter, and the spectrum of a modulated AM-VSB signal. As shown in Figure 4.17, the AM-VSB signal can be obtained by the partial suppression of the upper (lower) sideband of an AM-SC. The AM-VSB scheme is used for transmitting TV video signals. The main reasons for the choice of an AM-VSB system are the following.

- Bandwidth reduction;
- Practical implementation advantages;
- Lower cost.

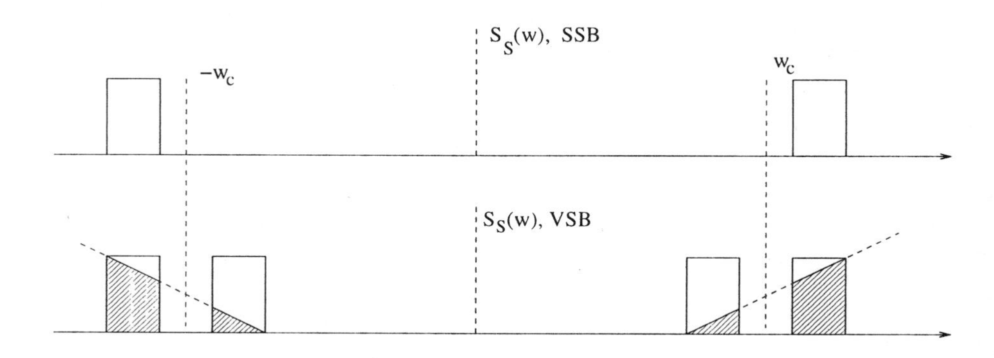

Figure 4.17. Spectra of SSB and VSB modulated signals.

4.8 Amplitude Demodulation

The modulated carrier can be synchronously demodulated to recover the message signal, as shown in Figure 4.18. The incoming signal

$$s(t) = A\left[1 + \Delta_{AM} m(t)\right] \cos(\omega_c t + \phi) \tag{4.35}$$

is multiplied (mixed) by a locally generated sinusoidal signal

$$c(t) = \cos(\omega_c t + \phi) \tag{4.36}$$

and low-pass filtered, which results in

$$s(t) = \frac{A}{2}\left[1 + \Delta_{AM} m(t)\right]. \tag{4.37}$$

The DC level then is blocked, to give the original signal $m(t)$ multiplied by a constant.

In case the phase of the local oscillator is not the same as the received signal phase, the signal is attenuated (fading occurs) by a term proportional to the cosine of the phase difference. If the frequency of the local oscillator is different from the frequency of the received carrier, the demodulated signal can be frequency translated.

The signal can also be recovered using an envelope detector (noncoherent detector), as shown in Figure 4.19. The received carrier is first rectified by the diode and filtered by the low-pass filter formed by R_1 and C_1. The DC component is then blocked by capacitor C_2 to yield the message signal $m(t)$.

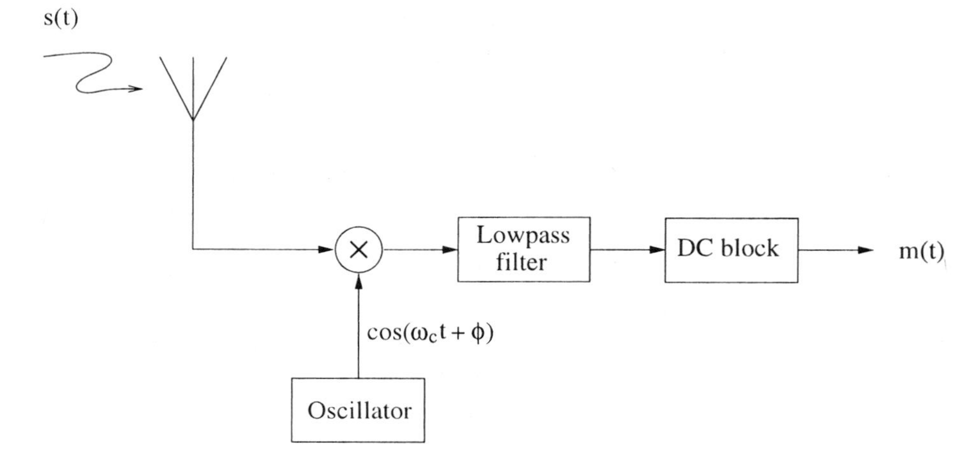

Figure 4.18. Block diagram of an AM demodulator.

4.9 Performance of Amplitude Modulation

When analyzing the performance of modulation systems it is common practice to employ the so called quadrature representation for the noise. The quadrature representation for the noise $n(t)$ is presented next, as a function of its in phase $n_I(t)$ and quadrature $n_Q(t)$ components.

$$n(t) = n_I(t)\cos(\omega_c t + \phi) + n_Q(t)\sin(\omega_c t + \phi). \quad (4.38)$$

It can be shown that the in phase and quadrature components have the same mean and variance as the noise $n(t)$, i.e., $\sigma_I^2 = \sigma_Q^2 = \sigma^2$, where σ^2 represents the variance of the noise $n(t)$. Assuming that an AM-SC was transmitted and was corrupted by additive noise, the received signal is given by $r(t)$ as follows.

$$r(t) = Bm(t)\cos(\omega_c t + \phi) + n(t). \quad (4.39)$$

The signal to noise ratio (SNR_I) at the demodulator input is given by

$$SNR_I = \frac{B^2/2}{\sigma^2}, \quad (4.40)$$

where $\sigma^2 = \frac{2\omega_M N_0}{2\pi}$, for a modulating signal with highest frequency equal to ω_M. As a consequence of synchronous demodulation, the demodulated signal after filtering is given by

$$\hat{m}(t) = Bm(t) + n_I(t), \quad (4.41)$$

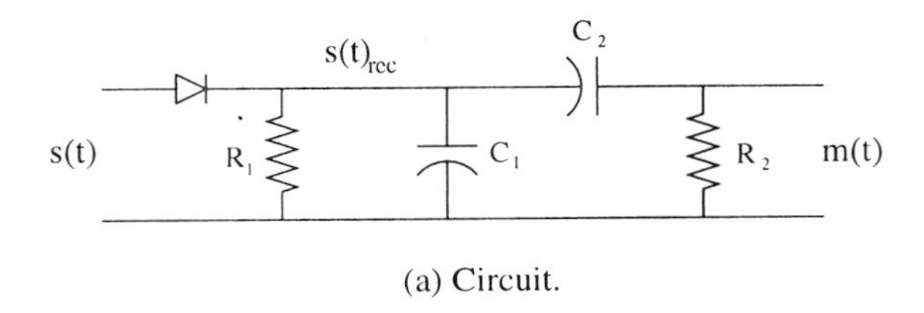

(a) Circuit.

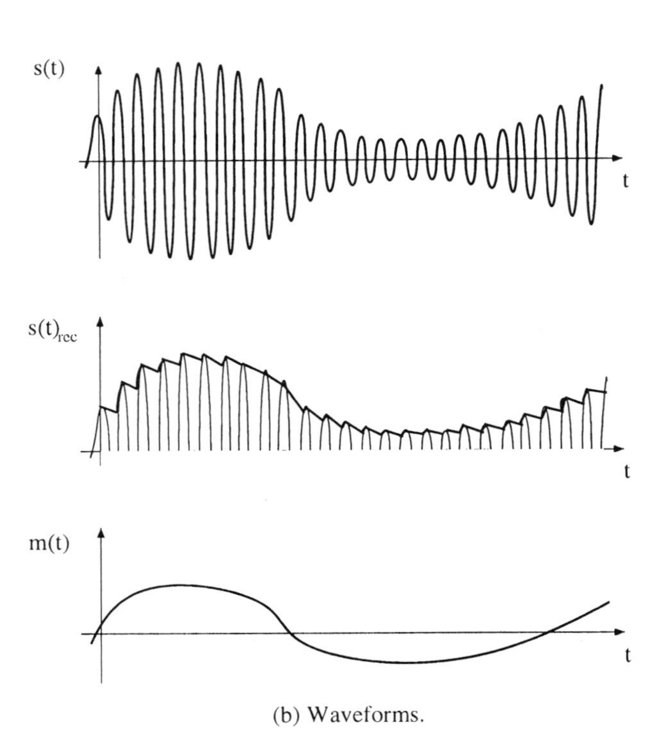

(b) Waveforms.

Figure 4.19. AM envelope demodulation.

thus producing the following output signal to noise ratio

$$SNR_O = \frac{B^2}{\sigma^2}. \tag{4.42}$$

In this manner, the demodulation gain η, for an AM-SC system is equal to

$$\eta = \frac{SNR_O}{SNR_I} = 2. \tag{4.43}$$

The AM-VSB modulation has a demodulating gain approximately equal to that of the AM-SC modulation.

Following a similar procedure it can be shown that the demodulation gain of ordinary AM, employing coherent demodulation, is given by

$$\eta = \frac{2\Delta_{AM}^2 P_M}{1 + \Delta_{AM}^2 P_M} \tag{4.44}$$

It is noticed that the demodulation gain of ordinary AM is at least 3 dB inferior the corresponding AM-SC demodulation gain, eventually reaching up to 6 dB in practice (Lathi, 1989).

4.10 Problems

1 An AM signal is given next, where $m(t) = \cos(\omega_M t + \varphi)$ and where φ and ϕ are independent random variables, with a uniform distribution in the interval $[0, 2\pi]$. Calculate and sketch the modulated signal, in the time domain and in the frequency domain, for the following values of the modulation index $\Delta_{AM} = 0,\ 1$ and 2.

$$s(t)_{AM} = [1 + \Delta_{AM} m(t)] \cos(\omega_c t + \phi).$$

2 The ASK signal ASK

$$s(t)_{ASK} = \sum_{n=-\infty}^{n=\infty} s_n p(t - nT_b) \cdot \cos(\omega_c t + \phi),$$

with $s_n \in \{-3,\ -1,\ 1,\ 3\}$, passes through a nonlinear channel with response $r_n = -\sqrt{-s_n} u(-s_n) + \sqrt{s_n} u(s_n)$. Sketch a diagram for the constellation for $s(t)$ and for $r(t)$. Comment the result in terms of noise immunity of the symbols employed.

3 A modulating signal has autocorrelation function given by $R_A(\tau) = A^2[1 + e^{-|\tau|}]$. Calculate the autocorrelation and the power spectral density of the amplitude modulated carrier for the given modulating signal. Plot the autocorrelation of the modulated signal. Calculate the total power, the DC power and the AC power of the modulated signal.

4 A signal $X(t)$, with autocorrelation function

$$R_M(\tau) = \frac{\omega_M S_0}{\pi} \frac{\sin(\omega_M \tau)}{\omega_M \tau}$$

amplitude modulates a carrier, with modulation index $\Delta_{AM} = 1$. Calculate the power in the modulated carrier. Determine the power spectral density of the modulated carrier and draw its corresponding graph.

5 The signal $s(t) = m(t)\cos(\omega_c t + \theta)$ is synchronously demodulated with a locally generated carrier. After being mixed with the local carrier the signal is filtered to eliminate frequencies above ω_M, where $\omega_M \ll \omega_c$. Determine the demodulated signal $\hat{m}(t)$. Analyze and sketch the effect, on the demodulated signal, of the following alterations in the local carrier.

   (a) The local carrier has a frequency deviation $\omega_c + \Delta\omega$. What happens with the recovered signal if $\Delta\omega = \omega_M$?

   (b) The local carrier has a phase shift $\theta + \Delta\theta$. What effect has this phase shift on the recovered signal? What if $\Delta\theta = \pi/2$?

6 Show that it is possible to amplitude modulate a signal by using a sampling circuit, with sampling function $\delta_T(t) = \sum_{n=-\infty}^{\infty} \delta(t - nT)$. Given that the signal has a maximum frequency ω_M, and that the carrier frequency is $10\omega_0$ (ω_0 is the Nyquist frequency), determine the optimum carrier frequency and the sampling rate as a function of ω_M.

7 A carrier is amplitude modulated by a signal with the following autocorrelation

$$R_M(\tau) = \frac{1}{1+\tau^2}.$$

Calculate the total power and the AC and DC powers of the message signal. Calculate the power of the modulated signal for the following values of the modulation index: $\Delta_{AM} = 0$, $\Delta_{AM} = 0.5$ and $\Delta_{AM} = 1$. Sketch the autocorrelation of the modulated signal in each case.

8 For the AM signal

$$\begin{aligned} s(t) &= \cos(\omega_c t + \phi) + \frac{1}{4}\sin\left[(\omega_c + \omega_M)t + \phi + \theta\right] \\ &- \frac{1}{4}\sin\left[(\omega_c - \omega_M)t + \phi - \theta\right], \end{aligned} \tag{4.45}$$

determine the modulation index, draw the phase diagram and calculate the power spectral density. Draw the time and frequency diagrams for the modulated signal.

9 A random signal $m(t)$, with autocorrelation $R_M(\tau)$, is applied to the input of a conventional AM modulator, with modulation index

$\Delta_{AM} = 1$ and carrier amplitude $A = 1$. Calculate the autocorrelation of the modulated signal $s(t)$. Calculate the power spectral density of $s(t)$ from its autocorrelation. Draw graphs for the autocorrelation and power spectral density for $R_M(\tau) = \frac{1}{1+\tau^2}$.

10 The digital signal message $m(t)$ has the following autocorrelation

$$R_M(\tau) = A^2[1 - \frac{|\tau|}{T_b}][u(\tau + T_b) - u(\tau - T_b)],$$

where T_b is the bit interval. Determine and sketch the autocorrelation function for the amplitude modulated signal with suppressed carrier (AM-SC). Calculate the transmitted power. Calculate and sketch the power spectral density of the modulated signal. Sketch, in the time domain, the modulated signal and discuss the possibility of demodulating this signal with an envelope detector.

11 The message signal $m(t)$, with an arbitrary autocorrelation, modulates a carrier $c(t) = A\cos(\omega_c t + \phi)$ as an AM-SC signal. This modulated signal is transmitted through a communication channel which introduces a random phase variation $\Delta\phi = \{0, \pi/2\}$ in the carrier, where the two phase values are equally likely. Discuss whether or not complete message signal recovery is possible by means of a synchronous detector.

12 The signal $m(t)$ having a maximum frequency ω_M is amplitude modulated and then transmitted. The modulated carrier is received and demodulated by a system using a local carrier with a frequency shift ω_M with respect to that of the transmitted carrier. At the reception, the demodulated signal passes through a filter $H(\omega) = [u(\omega + \omega_M) - u(\omega - \omega_M)]$. Determine and sketch the power spectral density at the filter output. Explain what has happened to the signal. Show that it is possible to recover the original modulating signal.

13 The message signal $m(t)$ with power spectral density given by

$$S_M(\omega) = S_0[u(\omega + \omega_M) - u(\omega - \omega_M)]$$

is amplitude modulated, with a 100% modulation index, and is transmitted through a channel with impulse response $h(t) = \delta(t) + \delta(t - \sigma)$. Determine and sketch the spectrum of the message signal, the spectrum of the modulated signal, the spectrum of the received signal and the spectrum of the demodulated signal.

14 A given signal with power spectral density $S_M(\omega) = M_o$ is applied to the input of a filter with transfer function

$$H(\omega) = \frac{\sqrt{2}}{1 + j\omega},$$

and then amplitude modulates a carrier, with a modulation index of $\Delta_{AM} = 0.5$. Calculate the power spectral density, the autocorrelation and the power of the modulated signal. Draw graphs for the autocorrelation and power spectral density.

15 Calculate the mean value, the autocorrelation and power of the AM signal

$$s(t) = [1 + \Delta_B(l(t) + r(t))]\sin(\omega_c t + \theta),$$

where $l(t)$ and $r(t)$ are stationary processes, with zero mean, uncorrelated and θ is a random variable uniformly distributed in the interval $[0, 2\pi]$.

16 Consider the modulated signal $s(t) = [1 + \Delta_{AM} m(t)]\cos(\omega_c t + \pi + \phi)$, where $m(t) = \cos(\omega_M t + \theta)$. Sketch this signal in the time domain, sketch its phase diagram, its autocorrelation and power spectral density, for the following values of the modulation index: $\Delta_{AM} = 1/2, 1$ and 2.

17 A modulating signal has the following autocorrelation function

$$R_A(\tau) = A\frac{\sin^2 \omega_M \tau}{(\omega_M \tau)^2} + B.$$

This signal amplitude modulates a carrier. Determine the power of the modulated signal. Calculate the power spectral density of the modulated carrier. Plot the graphs for the autocorrelation and the power spectral density of the modulated carrier.

18 The AM-SC signal $s(t) = m(t)\cos(\omega_c t + \phi)$, is transmitted through a communication channel producing the signal $r(t) = \alpha s(t) - \beta s(t - \sigma)$ at the receiver input. The spectrum of the modulating signal is given by $S_M(\omega) = S_M$, for $-\omega_M \leq \omega \leq \omega_M$ and is null outside this interval. Determine the power spectral density of the received signal, as a function of that of the transmitted signal. Calculate the frequencies for which the signal reception reaches a peak (find a relationship between ω_c and σ), assuming that the signal bandwidth is narrower than the peak bandwidth of the channel signature (frequency response). Design a demodulator for the transmitted signal, employing a filter that equalizes the channel effects.

19 Consider the modulated signal, $s(t) = A[1 - \Delta_{AM} m(t)] \cos(\omega_c t + \phi)$, where $m(t) = \cos(\omega_M t + \theta)$. Sketch this signal in the time domain, sketch its phase diagram and its autocorrelation and power spectral density, for a modulation index $\Delta_{AM} = 1$.

20 An AM signal, $s(t) = a(t) \cos(\omega_c t + \phi)$, for which $S_A(\omega) = S_A[1 - \frac{|\omega|}{\omega_M}]$, for $-\omega_M \leq \omega \leq \omega_M$, is transmitted through a channel with additive white Gaussian noise, with $S_N(\omega) = S_0$. The demodulator filter has the same bandwidth as the transmitted signal. Calculate the signal to noise ratio (SNR) for the received signal. Plot the corresponding diagrams.

21 The current through a diode is given by the formula $i(t) = i_o[e^{\alpha v(t)} - 1]$, for an applied voltage $v(t)$, where $\alpha > 0$ is a diode parameter and i_o is the reverse current. Show that this diode can be employed to demodulate, as an envelope detector, the signal

$$s(t) = A[1 + \Delta_{AM} m(t)] \cos(\omega_c t + \phi),$$

with the help of a low-pass filter. Plot the associated diagrams.

22 Two ASK signals are transmitted without carrier. The first has a modulating signal with amplitude levels $m_j \in \{A, -A\}$, and the second has amplitude levels $m_k \in \{-2B, -B, B, 2B\}$. Determine the power for each one of the two modulated signals. Draw the constellation diagrams for the two modulated signals. What is the value of B for which the two modulated signals have the same power? For which values of m_k the modulated signals have similar probability of error? Which of the two signals carries more information per unit bandwidth, if both employ the same symbol interval T_b?

23 The AM signal $s(t) = a(t) \cos(\omega_c t + \phi)$, where $a(t)$ has a maximum frequency ω_M, is synchronously demodulated by a receiver which simultaneously presents a frequency error $\Delta\omega << \omega_c$ and a phase error $\Delta\phi$ in the local oscillator. Write the expression for the demodulated signal after filtering the higher frequency terms. Explain the result obtained using graphs. Analyze what happens if: *a)* $\Delta\omega = 0$; *b)* $\Delta\phi = 0$; *c)* $\Delta\omega = \omega_M$; *d)* $\Delta\phi = \pi$.

24 A quadratic modulator is defined by the following formula

$$y(t) = ax(t) + bx^2(t),$$

where $x(t)$ denotes the input signal and $y(t)$ denotes the output signal. Calculate the modulation index as a function of the modulator parameters, for an input signal $x(t) = A\cos(\omega_c t + \phi) + m(t)$.

25 Describe the main characteristics of white noise, of bandlimited noise, of narrowband noise and of impulse noise. Comment on the autocorrelation and power spectral density for each one of the noises above mentioned.

26 The quadrature representation of the noise $n(t)$ is given by

$$n(t) = n_I(t)\cos(\omega_c t + \phi) + n_Q(t)\sin(\omega_c t + \phi).$$

Determine the resulting amplitude $r(t)$ and phase $\theta(t)$ for the noise as functions of the in phase $n_I(t)$ and quadrature $n_Q(t)$ components.

27 Show that the in phase and quadrature components have the same power spectral density

$$S_I(\omega) = S_Q(\omega) = S_N(\omega + \omega_c) + S_N(\omega - \omega_c), \text{ for } |\omega| \leq \omega_N.$$

28 Prove that the in phase and quadrature components have mean and variance identical to those of the noise $n(t)$.

29 Calculate the correlation between $n_I(t)$ and $n_Q(t)$, and show that the in phase and quadrature components are uncorrelated.

30 Prove that if $n(t)$ is Gaussian, then the in phase and quadrature components are Gaussian distributed.

31 Show that if the noise $n(t)$ is Gaussian, then for an arbitrary t the resulting amplitude, or noise envelope $v(t)$, is a random variable with a Rayleigh distribution and the phase $\theta(t)$ is a random variable uniformly distributed in the interval $[0, 2\pi]$.

32 Explain why the performance in the presence of noise of an AM system with carrier (meant for envelope detection) is at least 3 dB inferior to that of an AM-SC system.

33 An AM-SC signal with additive noise is demodulated by a synchronous detector presenting a phase error θ. Assume that the local oscillator generates a carrier $2\cos(\omega_c t + \theta)$ and calculate the signal to noise ratio and the demodulation efficiency of the detector.

34 Consider a baseband communication channel with additive noise, the power spectral density of which is equal to S_N. The channel has a frequency response given by

$$H(\omega) = \frac{B}{B + j\omega}.$$

The distortion is equalized by a filter with transfer function

$$G(\omega) = \frac{1}{H(\omega)}, \quad 0 \leq |\omega| \leq B.$$

Obtain an expression for the output signal to noise ratio.

35 Show that AM envelope detection provides a detection gain equivalent to that of synchronous detection, for high signal to noise ratios.

36 The threshold level of an AM system which employs envelope detection is defined for that input SNR value for which $V \ll A$ with probability 0.99. Here, V and A are narrowband noise and carrier amplitudes respectively. Show that if $\Delta_{AM} = 1$ and $P_M = 1$ W, then $P\{V \ll A\} = 0.99$ requires $\mathrm{SNR}_I \approx 10\,\mathrm{dB}$. For that purpose, consider the Rayleigh distribution for the noise amplitude

$$p_V(v) = \frac{v}{\sigma_N^2} e^{-\frac{v^2}{2\sigma_N^2}}.$$

37 A digital signal $x(t)$ with amplitude levels A and $-A$ modulates a carrier in amplitude (ASK) and is transmitted by a communication system affected by additive white Gaussian noise $n(t)$, with variance σ_N^2. Assuming that the two transmitted amplitude levels are equiprobable, i.e., that $P\{A\} = P\{-A\} = 0.5$, determine the error probability in the received signal.

38 The signal $m(t)$, with autocorrelation

$$R_M(\tau) = 2e^{-4000|\tau|} \cos 1000\pi\tau,$$

modulates the AM carrier

$$c(t) = 2\cos(10^6 \pi t + \phi)$$

with a 100% modulation index and is subsequently applied to an amplifier with a 30 dB gain. What is the required 3 dB bandwidth for the amplifier? What is the output power?

39 A conventional AM signal (AM with carrier) is received in the presence of additive noise and is demodulated by a synchronous detector with a phase error $\Delta\theta$. Assume that the local oscillator generates a carrier $2\cos(\omega_c t + \phi + \Delta\theta)$ and calculate the signal to noise ratio and the demodulation efficiency of the detector employed.

40 Show that envelope detection of conventional AM, at high signal to noise ratio, produces a demodulation gain equivalent to that of a

coherent demodulation. What happens when the signal to noise ratio is low?

41 Calculate the input signal to noise ratio, the output signal to noise ratio and the modulation gain for the synchronous detection of a signal $s(t) = m(t)\cos(\omega_c t + \phi)$, using a local carrier $c(t) = \sin(\omega_c t + \phi)$. Explain the result obtained.

42 An ASK signal, modulated by a signal with equally likely amplitude levels $\{3A, A, -A, -3A\}$, is transmitted through a communications channel with additive noise. Consider that the modulating signal has the form

$$m(t) = \sum_{k=-\infty}^{\infty} m_k p(t - kT_b),$$

and that the noise has power spectral density $\eta/2$. What is the signal to noise ratio at the input to the detector? Assume that 10% of the signal power is lost in a bandpass filter that allows only the main spectral lobe through.

43 A digital signal $a(t)$ with amplitude probability distribution $p_A(a) = p\delta(a) + (1-p)\delta(a - V)$, amplitude modulates a carrier (ASK). Calculate the AC power and the DC power of the modulating signal. Calculate the modulation index Δ_{AM} and the power at the modulator output. Sketch a diagram for the constellation and for the modulated signal in the time domain.

44 The signal in the previous problem is received contaminated by additive Gaussian noise with probability distribution

$$p_N(x) = \frac{1}{\sigma_N\sqrt{2\pi}} e^{-\frac{x^2}{2\sigma_N^2}},$$

and is synchronously demodulated. Calculate the signal to noise ratio (SNR) both at the input and at the output of the demodulator as well as the demodulation efficiency η.

45 A modulation system employing binary on-off keying (OOK) modulation, used in optical communications, uses a modulating signal with power $P_A = \frac{V^2}{2}$. Plot the autocorrelation of the modulating signal, illustrating the AC and DC power levels. Sketch the autocorrelation of the modulated carrier. Draw a diagram for the power spectral density of the modulated signal.

46 A digital signal has amplitude probability distribution $p_X(x) = p\delta(x+A) + q\delta(x-A)$, where $p+q=1$. Draw a diagram for the autocorrelation of this signal, assuming that the symbols are statistically independent, with symbol interval T_b, for $p = 1/2$ and $p = 1/3$. The signal is used to amplitude modulate a carrier. Determine analytical expressions for the autocorrelation and for the power spectral density of the modulated signal. Sketch the power spectral density for the modulated signal assuming the same values for the symbol probabilities.

47 Two signals with identical autocorrelation, given by

$$R_M(\tau) = A^2 e^{-\alpha|\tau|},$$

are used to modulate carriers with distinct frequencies in the AM-SC mode. What is the minimum frequency separation between carriers for the interference between the two signals not to exceed 3 dB?

48 Calculate the demodulation gain for an AM signal with carrier, non-synchronously demodulated, for both low signal to noise ratio and high signal to noise ratio. Compare the result with that for synchronous demodulation.

49 An operator (carrier company) wants to replace the ASK modulation system of Figure 4.20 with an OOK system with the same error rate. What is the power to be transmitted by the new system? Plot the constellation diagram for the OOK and compare the two systems with respect to power, bandwidth and complexity.

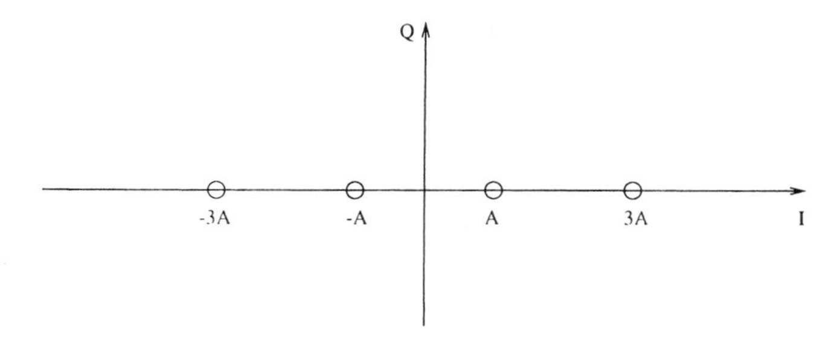

Figure 4.20. ASK constellation.

50 The digital signal $a(t)$ with autocorrelation

$$R_A(\tau) = A^2[1 - \frac{|\tau|}{T_b}][u(\tau + T_b) - u(\tau - T_b)],$$

where T_b denotes the symbol interval, is used to amplitude modulate a carrier. Determine and sketch the autocorrelation function of the modulated signal. Calculate and sketch the power spectral density of this signal.

51 Consider the ASK signal $s(t)$ with levels $a_n \in \{-3A, \ -A, \ A, \ 3A\}$. Draw a constellation diagram for $s(t)$ and comment on the result from the point of view of symbol immunity to noise. What would be the error rate if the constellation were rotated? What would be the effect on the error rate caused by the addition of a DC level?

$$s(t)_{ASK} = \sum_{n=-\infty}^{n=\infty} a_n p(t - nT_b) \cdot \sin(\omega_c t + \phi).$$

52 Two ASK signals are transmitted without carrier. The first has a modulating signal with levels $m_j \in \{A, -A\}$ and the second has levels $m_k \in \{-B, 0, B\}$. Determine the power in each modulated ASK signal. Plot the constellation diagrams for the two modulated signals. What is the value of B for which the two modulated signals have the same power? For which values of m_k the two modulated signals have similar error probabilities? Which of the two signals transmits more information per unit bandwidth if both have the symbol time interval T_b? Sketch a time diagram for the two modulated signals.

Chapter 5

QUADRATURE AMPLITUDE MODULATION

5.1 Quadrature Modulation with Random Signals

The quadrature modulation scheme (QUAM) uses sine and cosine orthogonality properties to allow the transmission of two different signals in the same carrier, which occupies a bandwidth that is equivalent to the AM signal. The QUAM modulator can be assembled using two DSB-SC modulators. The information is transmitted by both the carrier amplitude and phase.

The quadrature modulated signal $s(t)$ can be written as

$$s(t) = b(t)\cos(\omega_c t + \phi) + d(t)\sin(\omega_c t + \phi), \tag{5.1}$$

where the random modulating signals $b(t)$ e $d(t)$ can be correlated or uncorrelated.

It is also possible to write the modulated signal as

$$s(t) = \sqrt{b^2(t) + d^2(t)}\cos\left(\omega_c t - \tan^{-1}\frac{d(t)}{b(t)} + \phi\right), \tag{5.2}$$

in which the modulating signal, or amplitude resultant, can be expressed as $a(t) = \sqrt{b^2(t) + d^2(t)}$ and the phase resultant is $\theta(t) = -\tan^{-1}\left[\frac{d(t)}{b(t)}\right]$.

The autocorrelation function for the quadrature modulated carrier can be computed from the definition

$$R_S(\tau) = E[s(t) \cdot s(t+\tau)]. \tag{5.3}$$

Substituting Equation 5.1 into 5.3, gives

$$\begin{aligned} R_S(\tau) &= E\left[[b(t)\cos(\omega_c t+\phi)+d(t)\sin(\omega_c t+\phi)]\right. \\ &\cdot\ [b(t+\tau)\cos(\omega_c(t+\tau)+\phi) \\ &+\ \left. d(t+\tau)\sin(\omega_c(t+\tau)+\phi)]\right]. \end{aligned} \tag{5.4}$$

Expanding the product, gives

$$\begin{aligned} R_S(\tau) &= E[b(t)b(t+\tau)\cos(\omega_c t+\phi)\cos(\omega_c(t+\tau)+\phi) \\ &+ d(t)d(t+\tau)\sin(\omega_c t+\phi)\sin mega_c(t+\tau)+\phi) \\ &+ b(t)d(t+\tau)\cos(\omega_c t+\phi)\sin(\omega_c(t+\tau)+\phi) \\ &+ b(t+\tau)d(t)\cos(\omega_c(t+\tau)+\phi)\sin(\omega_c t+\phi)]. \end{aligned} \tag{5.5}$$

Using trigonometric properties and collecting terms which represent known autocorrelation functions, it follows that

$$\begin{aligned} R_S(\tau) &= \frac{R_B(\tau)}{2}\cos\omega_c\tau+\frac{R_D(\tau)}{2}\cos\omega_c\tau \\ &+ \frac{R_{DB}(\tau)}{2}\sin\omega_c\tau-\frac{R_{BD}(\tau)}{2}\sin\omega_c\tau. \end{aligned} \tag{5.6}$$

Which can be simplified to

$$R_S(\tau)=\left[\frac{R_B(\tau)+R_D(\tau)}{2}\right]\cos\omega_c\tau+\left[\frac{R_{DB}(\tau)-R_{BD}(\tau)}{2}\right]\sin\omega_c\tau. \tag{5.7}$$

It is observed that the QUAM signal suffers both amplitude and phase modulation, and its autocorrelation function can be written as

$$R_S(\tau)=\frac{R(\tau)}{2}\cos\left[\omega_c\tau+\theta(\tau)\right], \tag{5.8}$$

where

$$R(\tau)=\sqrt{[R_B(\tau)+R_D(\tau)]^2+[R_{DB}(\tau)-R_{BD}(\tau)]^2} \tag{5.9}$$

$$\theta(\tau)=-\tan^{-1}\left[\frac{R_{DB}(\tau)-R_{BD}(\tau)}{R_B(\tau)+R_D(\tau)}\right] \tag{5.10}$$

Considering zero mean uncorrelated modulating signals, $R_{BD}(\tau)=E[b(t)d(t+\tau)]=0$ and $R_{DB}(\tau)=E[b(t+\tau)d(t)]=0$. The resulting autocorrelation is then given by

$$R_S(\tau) = \frac{R_B(\tau)}{2} \cos \omega_c \tau + \frac{R_D(\tau)}{2} \cos \omega_c \tau. \tag{5.11}$$

The carrier power is given by the following formula

$$P_S = R_S(0) = \frac{P_B + P_D}{2}. \tag{5.12}$$

The power spectrum density is obtained by applying the Fourier transform to the autocorrelation function (Wiener-Khintchin theorem), which gives

$$\begin{aligned} S_S(\omega) &= \frac{1}{4}\left[S_B(\omega+\omega_c) + S_B(\omega-\omega_c) + S_D(\omega+\omega_c) + S_D(\omega-\omega_c)\right] \\ &= \frac{j}{4}\left[S_{BD}(\omega-\omega_c) + S_{BD}(\omega+\omega_c)\right. \\ &+ \left. S_{DB}(\omega+\omega_c) - S_{DB}(\omega-\omega_c)\right], \end{aligned} \tag{5.13}$$

where $S_B(\omega)$ and $S_D(\omega)$ represent the respective power spectrum densities for $b(t)$ and $d(t)$; $S_{BD}(\omega)$ is the cross-spectrum density between $b(t)$ and $d(t)$; $S_{DB}(\omega)$ is the cross-spectrum density between $d(t)$ and $b(t)$.

For uncorrelated signals, the previous formula can be simplified to

$$S_S(\omega) = \frac{1}{4}\left[S_B(\omega+\omega_c) + S_B(\omega-\omega_c) + S_D(\omega+\omega_c) + S_D(\omega-\omega_c)\right]. \tag{5.14}$$

The quadrature amplitude modulation scheme is shown in Figure 5.1.

5.2 Single Sideband Amplitude Modulation

Single sideband amplitude modulation, or AM-SSB, is, in reality, a type of quadrature amplitude modulation which uses either the lower or upper AM side band for transmission. The AM-SSB signal can be obtained by filtering out the undesired side band of an AM-SC signal.

The SSB signal saves bandwidth and power, as compared to other systems, but needs frequency and phase synchronization to be recovered. (Carlson, 1975). In order to understand the process of generating the SSB signal, the Hilbert transform is introduced in the following section.

5.2.1 Hilbert Transform

The Hilbert transform of a signal $f(t)$ is written $\hat{f}(t)$, and is defined as (Gagliardi, 1978)

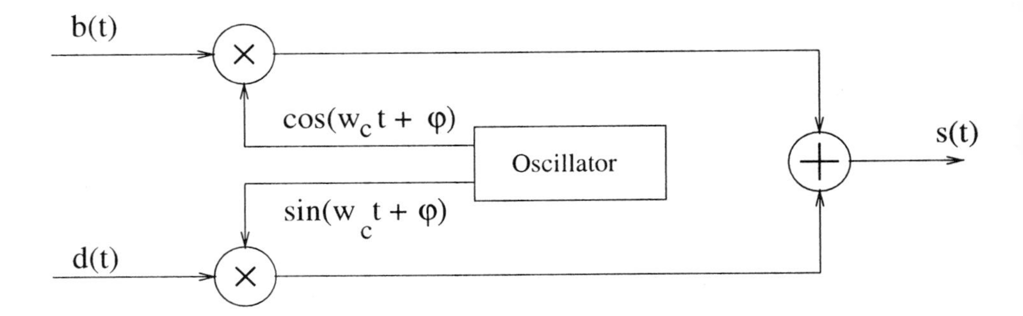

Figure 5.1. Block diagram for the quadrature modulator.

$$\hat{f}(t) = \frac{1}{\pi} \int_{-\infty}^{\infty} \frac{f(\tau)}{t-\tau} d\tau. \tag{5.15}$$

The Hilbert transform is a linear operation and its inverse transform is given by

$$f(t) = -\frac{1}{\pi} \int_{-\infty}^{\infty} \frac{\hat{f}(t)}{t-\tau} d\tau. \tag{5.16}$$

Functions $f(t)$ and $\hat{f}(t)$ form a pair of Hilbert transforms. This transform shifts all frequency components of an input signal by $\pm 90^0$. The positive frequency components are shifted by -90^0, and the negative frequency components are shifted by $+90^0$. The spectral amplitudes are not affected.

From the definition, $\hat{f}(t)$ can be interpreted as the convolution of $f(t)$ and $\frac{1}{\pi t}$. As already known, the Fourier transform of the convolution of two signals is given by the product of the transforms of each signal. Therefore,

$$G(\omega) = \mathcal{F}[f(t) * \frac{1}{\pi t}] = \mathcal{F}[f(t)] \cdot \mathcal{F}[\frac{1}{\pi t}]. \tag{5.17}$$

5.2.2 Fourier transform of $1/\pi t$

To obtain the Fourier transform of $\frac{1}{\pi t}$, one can resort to the following rationale:

- Consider the signal function in Figure 5.2.

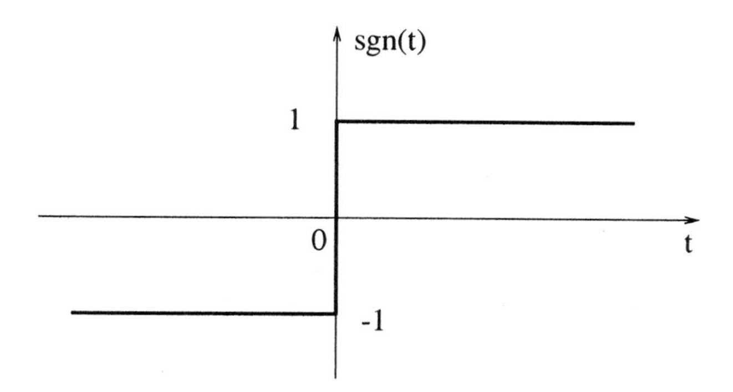

Figure 5.2. Signal function.

- The derivative of this function is an impulse centered at the origin, whose Fourier transform is the constant 2.
- Using the Fourier integral property,

$$u(t) - u(-t) \leftrightarrow \frac{2}{j\omega}.$$

- Finally, using the symmetry property, it follows that

$$\frac{1}{\pi t} \longleftrightarrow j(u(-\omega) - u(\omega)).$$

Therefore,

$$G(\omega) = j[u(-\omega) - u(\omega)] \cdot F(\omega). \tag{5.18}$$

5.2.3 Properties of the Hilbert Transform

The Hilbert transform has the following properties

- Signal $f(t)$ and its transform $\hat{f}(t)$ have the same power spectrum density;
- Signal $f(t)$ and its transform $\hat{f}(t)$ havethe same power autocorrelation;
- Signal $f(t)$ and its transform $\hat{f}(t)$ are orthogonal;
- If $\hat{f}(t)$ the Hilbert transform of $f(t)$, then the Hilbert transform of $\hat{f}(t)$ is $-f(t)$.

5.2.4 Producing the SSB Signal

The usual process to obtain the SSB signal is by filtering one of the AM-SC sidebands. Another method is to use the properties of the Hilbert transform, discussed previously. Consider a sinusoidal signal $m(t) = \cos(\omega_M)t$, which has a Fourier spectrum $S_M(\omega)$ represented by two impulses at $\pm\omega_M$.

The modulated signal is given by $\cos(\omega_M t)\cos(\omega_c t)$, whose Fourier spectrum is that of $S_M(\omega)$ shifted to $\pm\omega_c$. The carrier spectrum is formed by two impulses at $\pm\omega_c$. Therefore, producing an SSB signal, for the special case, $m(t) = \cos(\omega_M t)$ is equivalent to generating the signal $\cos(\omega_c - \omega_M)t$, or the signal $\cos(\omega_c + \omega_M)t$.

By trigonometry, it follows that

$$\cos(\omega_c - \omega_M)t = \cos\omega_M t\cos\omega_c t + \sin\omega_M t\sin\omega_c t. \tag{5.19}$$

Thus, the desired SSB signal can be produced adding $\cos\omega_M t\cos\omega_c t$ and $\sin\omega_M t\sin, \omega_c t$. Signal $\cos\omega_M t\cos\omega_c t$ can be produced by a balanced modulator. Signal $\sin\omega_M t\sin\omega_c t$ can be written as $\cos(\omega_M t - \frac{\pi}{2})\cos(\omega_c t - \frac{\pi}{2})$.

This signal can also be generated by a balanced modulator, as long as $\cos\omega_M t$ and carrier $\cos\omega_c t$ be phase shifted by $\frac{\pi}{2}$. Although this result has been derived for the special case $m(t) = \cos\omega_M t$, it is valid to any waveform, because of the Fourier series properties, that allow the representation of any signal by a sum of sine and cosine functions.

The SSB associated with $m(t)$ is thus

$$s_{SSB}(t) = m(t)\cos(\omega_c t + \phi) + \hat{m}(t)\sin(\omega_c t + \phi), \tag{5.20}$$

where $\hat{m}(t)$ is obtained by phase shifting all frequency components of $m(t)$ by $\frac{\pi}{2}$. The block diagram of the modulator is shown in Figure 5.3.

5.2.5 Lower Sideband SSB – Random Signal

If the modulating signal $m(t)$ is a zero mean, stationary, stochastic process, the usual path to follow, in order to obtain the power spectrum density, is to compute its autocorrelation function. The modulated SSB signal is given by

$$s_{SSB}(t) = m(t)\cos(\omega_c t + \phi) + \hat{m}(t)\sin(\omega_c t + \phi) \tag{5.21}$$

The autocorrelation function is calculated, as usual, by the formula

$$R_S(\tau) = E[s(t)s(t + \tau)] \tag{5.22}$$

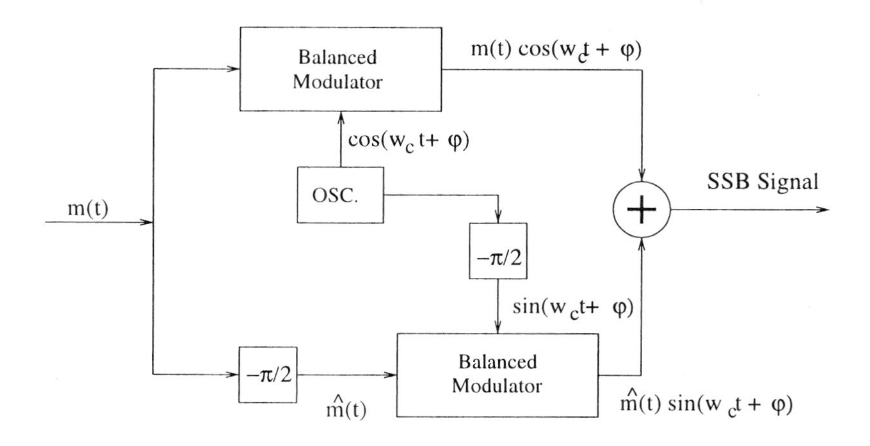

Figure 5.3. Block diagram for the SSB modulator.

Substituting $s(t)$, given by Equation 5.21

$$\begin{aligned} R_S(\tau) &= E\left[(m(t)\cos(\omega_c t + \phi) + \hat{m}(t)\sin(\omega_c t + \phi))\right. \\ &\cdot (m(t+\tau)\cos(\omega_c(t+\tau) + \phi) \\ &+ \left.\hat{m}(t+\tau)\sin(\omega_c(t+\tau) + \phi))\right]. \end{aligned} \tag{5.23}$$

The previous equation can be split into

$$\begin{aligned} R_S(\tau) &= E\left[m(t)m(t+\tau)\cos(\omega_c t + \phi)\cos(\omega_c(t+\tau) + \phi)\right] \\ &+ E\left[m(t)\hat{m}(t+\tau)\cos(\omega_c t + \phi)\sin(\omega_c(t+\tau) + \phi)\right] \\ &+ E\left[\hat{m}(t)m(t+\tau)\sin(\omega_c t + \phi)\cos(\omega_c(t+\tau) + \phi)\right] \\ &+ E\left[\hat{m}(t)\hat{m}(t+\tau)\sin(\omega_c t + \phi)\sin(\omega_c(t+\tau) + \phi)\right]. \end{aligned} \tag{5.24}$$

After the corresponding simplifications, one obtains

$$\begin{aligned} R_S(\tau) &= \frac{1}{2}R_{MM}(\tau)\cos\omega_c\tau + \frac{1}{2}R_{M\hat{M}}(\tau)\sin\omega_c\tau \\ &- \frac{1}{2}R_{\hat{M}M}(\tau)\sin\omega_c\tau + \frac{1}{2}R_{\hat{M}\hat{M}}(\tau)\cos\omega_c\tau. \end{aligned} \tag{5.25}$$

It is known that $R_{MM}(\tau) = R_{\hat{M}\hat{M}}(\tau)$ and $R_{M\hat{M}}(\tau) = -R_{\hat{M}M}(\tau)$. Therefore, the power spectrum density can be computed as $S_S(\omega) = \mathcal{F}[R_S(\tau)]$, by the use of previous relations and using the equation for the Hilbert filter

$$H(\omega) = \begin{cases} -j & \text{se } \omega \geq 0 \\ +j & \text{se } \omega < 0 \end{cases} \tag{5.26}$$

which leads to

$$S_{\hat{M}\hat{M}}(\omega) = S_{MM}(\omega) \tag{5.27}$$
$$S_{M\hat{M}}(\omega) = j[u(-\omega) - u(\omega)] \cdot S_{MM}(\omega) \tag{5.28}$$
$$S_{\hat{M}M}(\omega) = j[u(\omega) - u(-\omega)] \cdot S_{MM}(\omega), \tag{5.29}$$

Figure 5.4 illustrates the procedure.

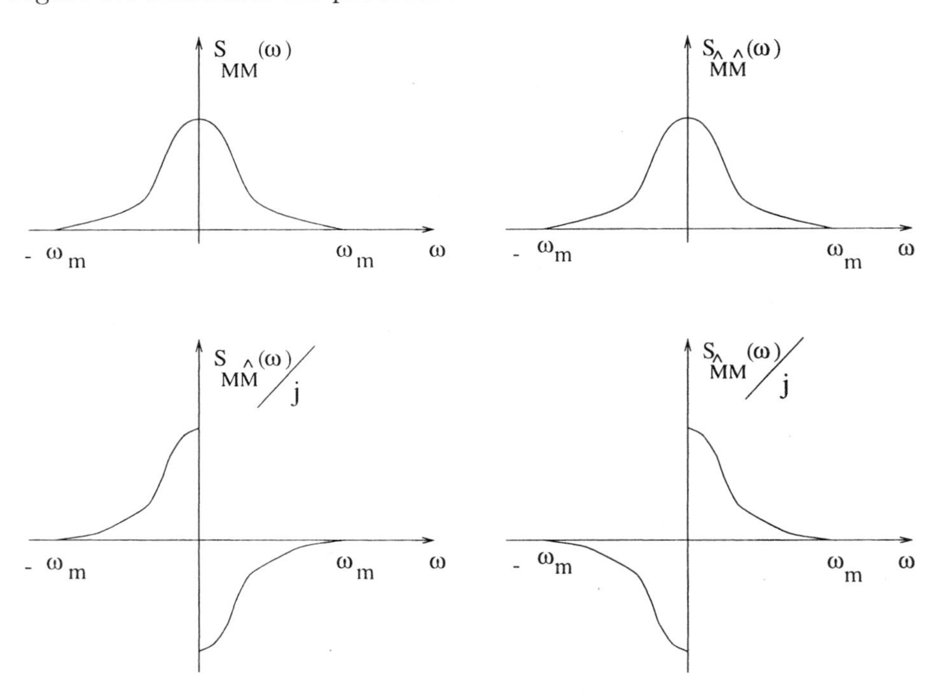

Figure 5.4. SSB Modulation.

The following power spectral density results for the SSB signal

$$S(\omega)_{SSB} = S_M(\omega - \omega_c)u(-\omega + \omega_c) + S_M(\omega + \omega_c)u(\omega + \omega_c), \tag{5.30}$$

which represents the lower sideband SSB signal, obtained from the original spectrum $S_M(\omega)$.

Example: For the following SSB modulation scheme, determine the power of the modulated signal, considering that the modulating signal autocorrelation is given by $R_M(\tau) = e^{-\tau^2}$.

$$s(t) = [1 + m(t)]\cos(\omega_c t + \phi) - \hat{m}(t)\sin(\omega_c t + \phi).$$

Solution: The power can be computed using the formula

$$P_S = \frac{P_A}{2},$$

where $P_A = E[a^2(t)]$ and

$$a(t) = \sqrt{(1 + m(t))^2 + \hat{m}^2(t)}$$

Considering that $m(t)$ has zero mean and that $P_M = P_{\hat{M}}$, it follows that

$$P_A = 1 + 2P_M$$

Because $P_M = R_M(0) = e^{-0^2} = 1$, then

$$P_A = 1 + 2 \cdot 1 = 3\,W$$

and

$$P_S = \frac{P_A}{2} = \frac{3}{2}\,W. \tag{5.31}$$

5.3 ISB Modulation

The process of independent sideband (ISB) modulation consists in sending two distinct signals in separate sidebands. The ISB signal can be obtained by the addition of two SSB signals, as shown in Figure 5.5.

First, consider the expression for the upper sideband signal

$$s_L(t) = l(t)\cos(\omega_c t + \phi) - \hat{l}(t)\sin(\omega_c t + \phi) \tag{5.32}$$

Second, the expression for the lower sideband signal

$$s_R(t) = r(t)\cos(\omega_c t + \phi) + \hat{r}(t)\sin(\omega_c t + \phi) \tag{5.33}$$

The expression for the ISB signal is, then, the sum of the two signals

$$s(t) = s_L(t) + s_R(t) \tag{5.34}$$

or

$$s(t) = [l(t) + r(t)] \cos(\omega_c t + \phi) + [\hat{r}(t) - \hat{l}(t)] \sin(\omega_c t + \phi). \qquad (5.35)$$

From the previous expression, it is noticed that the ISB signal can be produced using just one QUAM modulator. To implement such a modulator, it is necessary to use a device to generate the sum and difference of the input signals. The sum signal is applied to the in-phase modulator and the difference signal is applied to the quadrature modulator, after being phase shifted by the Hilbert filter.

The power spectrum density for the ISB modulated signal can be observed in Figure 5.5.

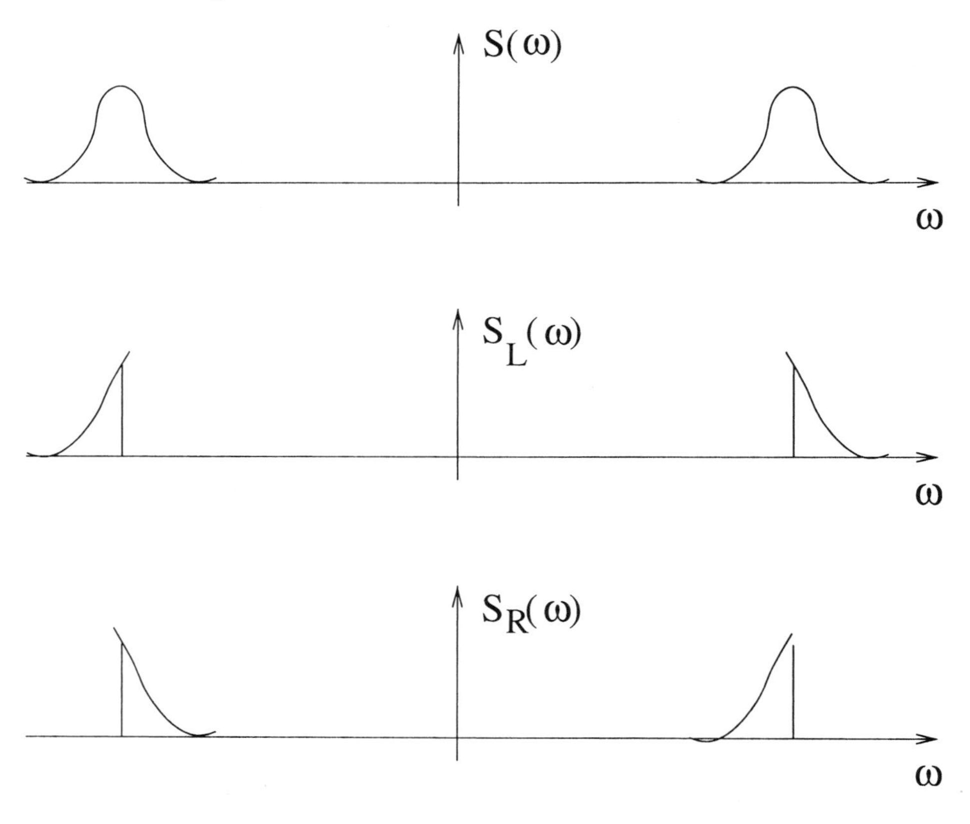

Figure 5.5. Generating the ISB signal.

Example: Compute the power spectrum density and the ISB modulated signal average power, assuming the following power spectrum densities for the modulating signals

$$S_L(\omega) = S_0[u(\omega + \omega_M) - u(\omega - \omega_M)]$$

and

$$S_R(\omega) = S_0(1 - \frac{|\omega|}{\omega_M})[u(\omega + \omega_M) - u(\omega - \omega_M)].$$

Solution: The modulated signal power can be computed by

$$P_S = P_L + P_R$$

where

$$P_L = \frac{1}{2\pi} \int_{-\infty}^{\infty} S_L(\omega) d\omega$$

and

$$P_R = \frac{1}{2\pi} \int_{-\infty}^{\infty} S_R(\omega) d\omega.$$

It is also possible to calculate the average power by integrating the power spectrum density function. Therefore, it follows that

$$P_L = \frac{\omega_M S_0}{\pi}$$

and

$$P_R = \frac{\omega_M S_0}{2\pi}.$$

Then,

$$P_S = \frac{3}{2} \frac{\omega_M S_0}{\pi} \text{ W}.$$

5.4 AM-Stereo

There are five main systems which use the AM-stereo technique. Actually, the modulation technique is QUAM, but commercially it is called AM-stereo. The systems were introduced by the following companies: Belar, Magnavox, Kahn, Harris and Motorola.

The Motorola AM-stereo system, dubbed C-QUAM, is described as follows. It has been developed with the objective of being compatible with the legacy AM mono receivers, which should be able to detect the sum of the stereo channels.

The stereo signal has a left channel l and a right channel r. The compatible AM-stereo envelope is proportional to the sum $l + r$ and its phase is a function of the difference signal $l - r$. Of course, the sum signal must be limited to 100% modulation index, at least for the negative peaks, otherwise the phase modulation is lost.

Figure 5.6 shows the block diagram of the driver circuit to the system. The crystal oscillator, which produces the carrier, is commonplace in conventional AM transmitters. The driver circuit has a phase-shifter to generate a second carrier, which lags the first one by ninety degrees.

The audio program channels are input to a matrix to produce the sum $l + r$ and difference signals $l - r$ which modulate the in-phase and quadrature carriers, respectively, using balanced modulators.

A low level 25 Hz pilot signal is added to the $l - r$ signal, before modulation. The sidebands coming out of the modulators are added to the carrier, producing a QUAM signal, which is both amplitude and angle modulated. A clipper circuit removes the amplitude modulation, leaving only the angle modulation. This signal is the fed, as a carrier, to a conventional amplitude modulator. The expression for the C-QUAM signal is

$$s(t)_{AM-EST} = A[1 + \Delta_B b(t)] \cos(\omega_c t + \theta(t)), \tag{5.36}$$

in which

$$\theta(t) = \arctan \left[\frac{\Delta_D d(t) + \sin \omega_P t}{1 + \Delta_B b(t)} \right], \tag{5.37}$$

where $b(t) = l(t) + r(t)$ and $d(t) = l(t) - r(t)$.

5.5 Quadrature Amplitude Demodulation

The modulated carrier is synchronously demodulated in order to recover the message signal, as shown in Figure 5.7. The incoming signal

$$s(t) = b(t) \cos(\omega_c t + \phi) + d(t) \sin(\omega_c t + \phi) \tag{5.38}$$

is multiplied by two locally generated sinusoidal signals $\cos(\omega_c t + \phi)$ and $\sin(\omega_c t + \phi)$ and low-pass filtered. This results in the recovering of the original $b(t)$ and $d(t)$ signals, multiplied by a constant.

Fading can occur if the local oscillator phase is different from the received carrier phase. The signal is attenuated by a term proportional to the cosine of the phase difference. As in the synchronous AM demodulation, a frequency drift of the local oscillator can shift the demodulated signal frequency and disturb the reception.

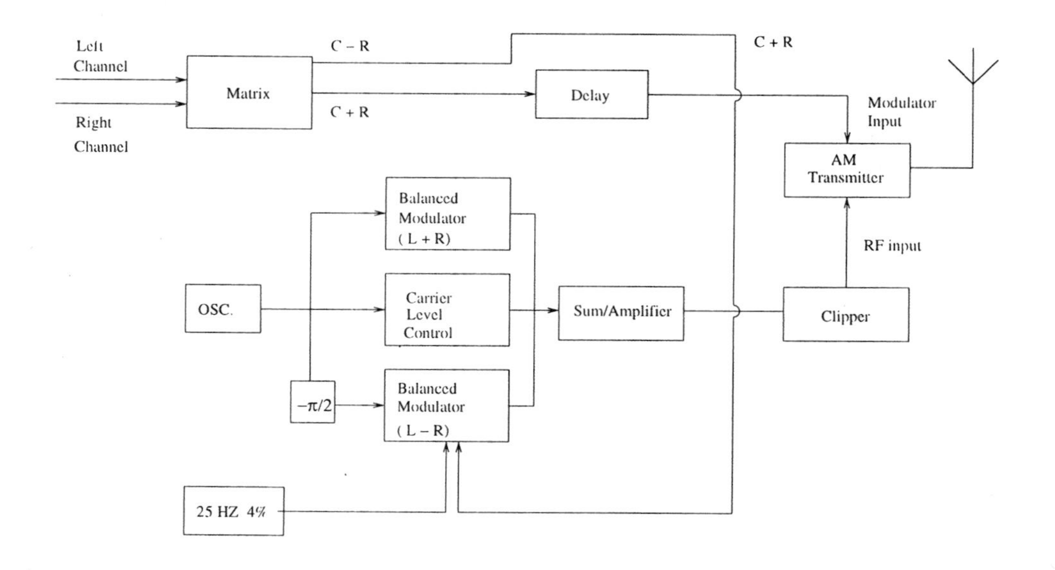

Figure 5.6. Block diagram of C-QUAM modulator.

5.6 Performance Evaluation of SSB

The single sideband modulated signal (SSB), presented previously, can be written as

$$s(t) = m(t)\cos(\omega_c t + \phi) + \hat{m}(t)\sin(\omega_c t + \phi), \tag{5.39}$$

where $m(t)$ represents the message signal and $\hat{m}(t)$ its Hilbert transform.

The SSB signal spectral analysis shows that its bandwidth is the same as that of the baseband signal (ω_M). The received signal is expressed as

$$r(t) = [m(t) + n_I(t)]\cos(\omega_c t + \phi) + [\hat{m}(t) + n_Q(t)]\sin(\omega_c t + \phi). \tag{5.40}$$

Because the noise occupies a bandwidth ω_M and demodulation is synchronous, the demodulation gain η for the SSB signal is the same as the one obtained for the AM-SC, i.e.

$$\eta = \frac{SNR_O}{SNR_I} = 2. \tag{5.41}$$

5.7 Quadrature Modulation with Digital Signal

Quadrature amplitude modulation using a digital signal is usually called QAM. The computation of the autocorrelation function and the power spectrum density follows the same rules previously established.

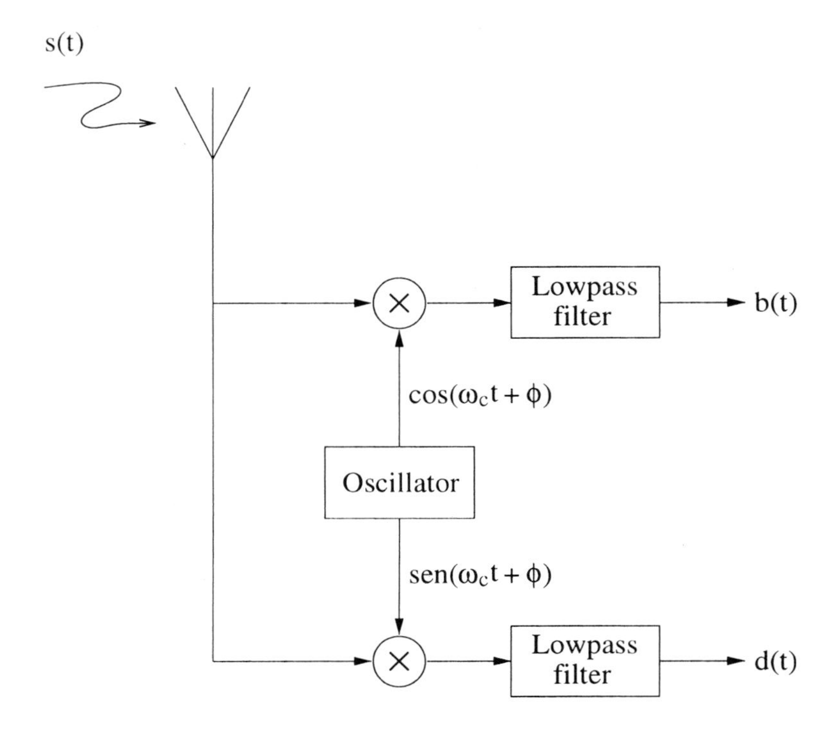

Figure 5.7. Block diagram of a QAM demodulator.

The modulating signals for the QAM scheme are

$$b(t) = \sum_{n=-\infty}^{\infty} b_n p(t - nT_b) \tag{5.42}$$

and

$$d(t) = \sum_{n=-\infty}^{\infty} d_n p(t - nT_b). \tag{5.43}$$

The modulated carrier is similar to the QUAM's

$$s(t)_{QAM} = b(t)\cos(\omega_c t + \phi) + d(t)\text{sin}\,(\omega_c t + \phi). \tag{5.44}$$

The constellation diagram for a 4QAM signal is shown in Figura 5.8, where the symbols $b_n = \{A, -A\}$ and $d_n = \{A, -A\}$. It can be shown, using the same previous methods, that the modulated signal power for this special case is given by $P_S = A^2/2$. This signal is also known as $\pi/4$-QPSK and is largely used in schemes for mobile cellular communication systems. If the constellation is rotated by $\pi/4$ it produces another

QAM modulation scheme, which presents more ripple than the previous one, because one of the carriers if shut off whenever the other carrier is transmitted.

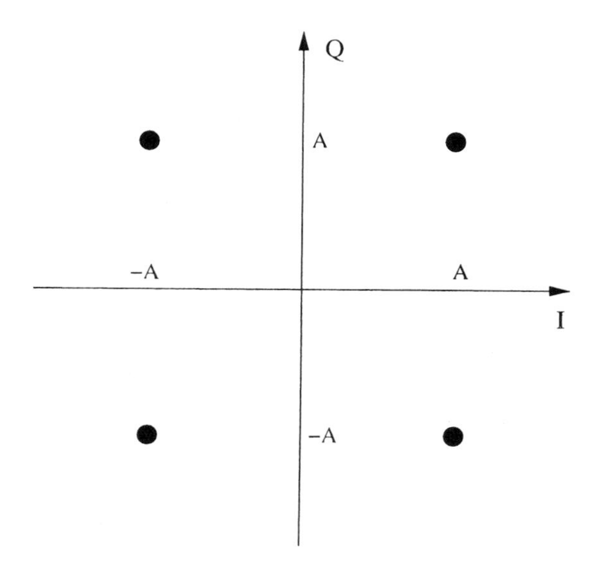

Figure 5.8. Constellation diagram for the 4QAM signal.

Figure 5.9 shows the constellation diagram for a 16QAM signal, whose points have coordinates $b_n = \{-3A, -A, A, 3A\}$, for the abscissa, and $d_n = \{-3A, -A, A, 3A\}$, for the ordinate.

The efficient occupation of the signal space renders the QAM technique more efficient than ASK and PSK, in terms of bit error probability versus transmission rate. On the other hand, the scheme is vulnerable to non-linear distortion, which occurs in power amplifiers on board satellite systems as shown, for example, in Figure 5.10.

It is evident that other problems can occur and degrade the transmitted signal. It is important that the communication engineer could identify them, in order to establish the best combat strategy.

Figure 5.11 shows a constellation diagram where the symbols are contaminated by additive Gaussian noise. The circles illustrate, ideally, the uncertainty region surrounding the signal symbols.

An interesting analogy can be made between the noise effect and the heating process. Heating is the result of an increase in the kinetic energy of the system. If the points on the constellation diagram are heated, they acquire kinetic energy and move from their original position. Because the movement is rapid and random is it more difficult to perfectly identify their positions.

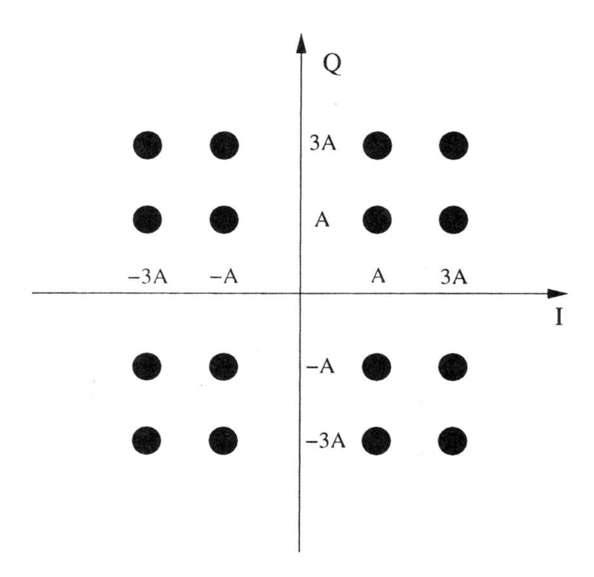

Figure 5.9. Constellation diagram for the 16QAM signal.

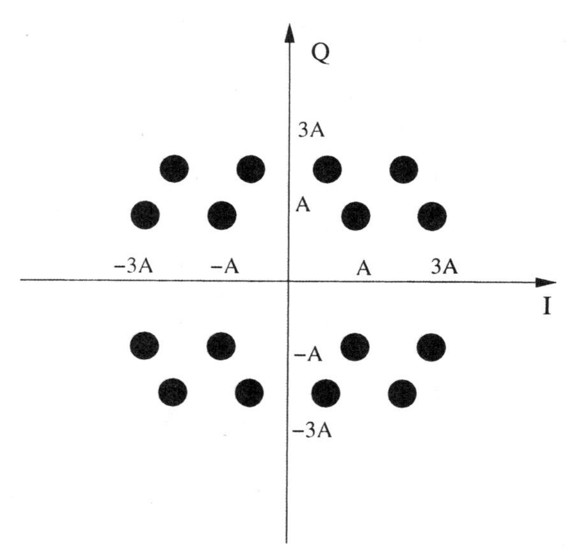

Figure 5.10. Constellation diagram for the 16QAM signal, subject to distortion caused by a non-linear channel.

Figure 5.12 shows the QAM signal affected by amplitude fading. The main fading effect is to decrease the distance between the symbols, which implies an increase in the bit error rate (BER) of the system.

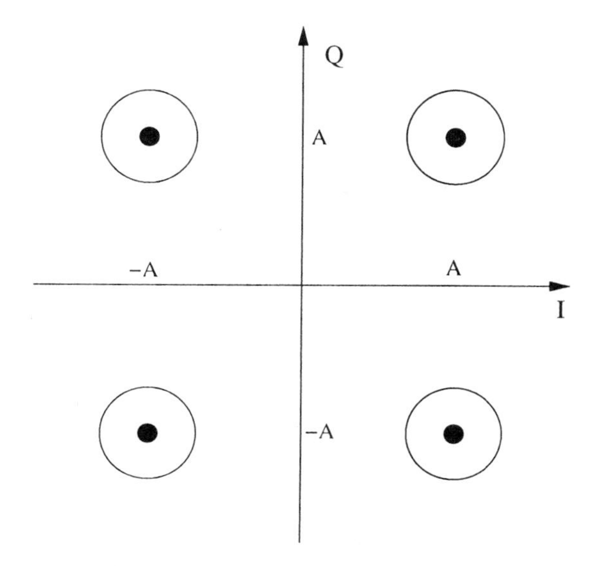

Figure 5.11. Constellation diagram for a 4QAM signal with additive noise.

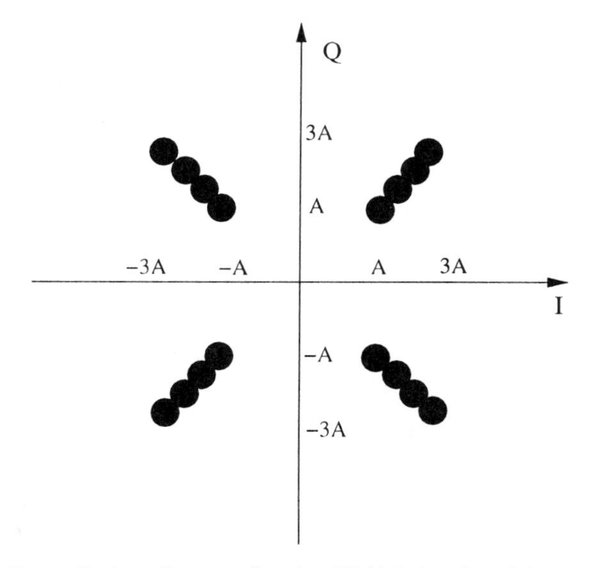

Figure 5.12. Constellation diagram for the 4QAM signal, subject to fading.

The symbol error probability is, usually, controlled by the smaller distance between the constellation symbols. The formulas for the error probability produce curves that decrease exponentially, or quasi-exponentially, with the signal to noise ratio.

The random phase variation effect, or jitter, is shown in Figure 5.13. This occurs when the local synchronization system is not able to perfectly follow the received signal phase variations. This introduces randomness in the detection process, which increases the bit symbol error probability.

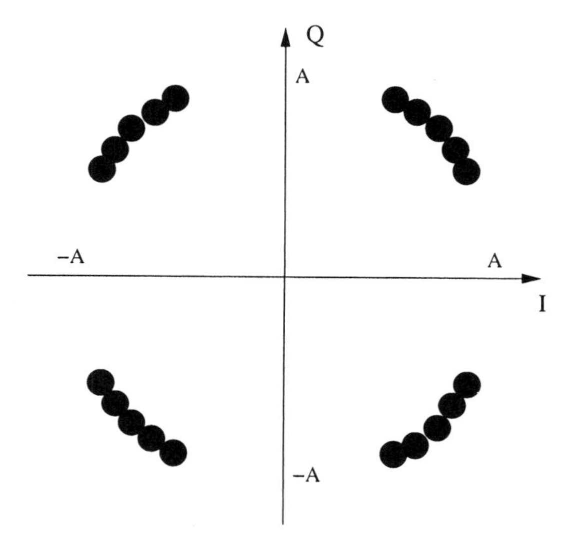

Figure 5.13. Constellation diagram for a 4QAM signal, subject to jitter.

On the other hand Figure 5.14 illustrates a 4QAM signal affected by a total loss of synchronization. The constellation diagram seems to be a circle because the symbols rotate rapidly. In this case detection is impossible.

The probability of error for the 4QAM signal is (Haykin, 1988)

$$P_e \approx \text{erfc}\left(\sqrt{\frac{E_b}{N_0}}\right). \tag{5.45}$$

where, E_b is the pulse energy and N_0 represents the noise power spectral density.

5.8 Problems

1 Compute the modulated carrier power, considering that $l(t)$ and $r(t)$ are zero mean uncorrelated stationary random processes, for the sig-

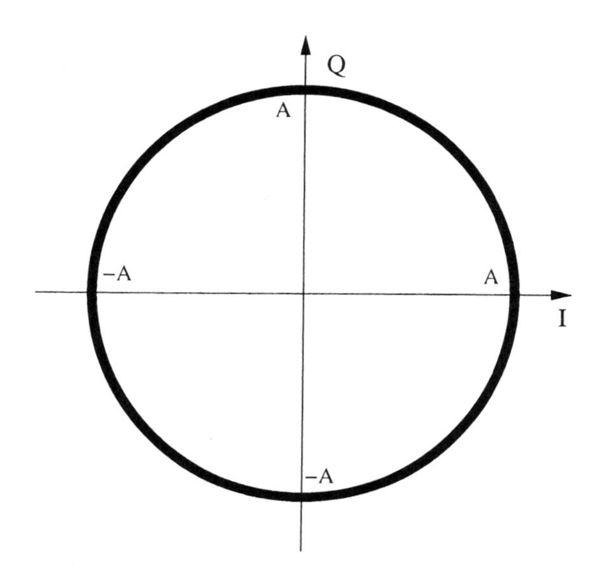

Figure 5.14. Constellation diagram for the 4QAM, subject to synchronization loss.

nal

$$\begin{aligned} s(t)_{QUAM-ST} &= \sin(\omega_c t + \phi) \\ &+ [l(t) + r(t)] \cos(\omega_c t + \phi) + [l(t) - r(t)] \sin(\omega_c t + \phi). \end{aligned}$$

2 Fr the following signal determine and sketch the modulated carrier autocorrelation, considering that $R_B(\tau) = \cos \omega_M \tau$ and $R_D(\tau) = \cos \omega_N \tau$. What is the power of the modulated carrier?

$$s(t)_{QUAM} = b(t) \cos(\omega_c t + \phi) + d(t) \sin(\omega_c t + \phi).$$

3 Draw the constellation diagram for the following QUAM signal, where $b(t) = \sum_{n=-\infty}^{n=\infty} b_n p(t - nT_b)$ and $d(t) = \sum_{n=-\infty}^{n=\infty} d_n p(t - nT_b)$, with $b_n = \pm\sqrt{2}$ and $d_n = \pm\sqrt{2}$. Compute the amplitude and phase resultants.

$$s(t)_{QAM} = b(t) \cos(\omega_c t + \phi) + d(t) \sin(\omega_c t + \phi).$$

4 For the following SSB modulation scheme, determine the modulated carrier power, considering that the modulating signal autocorrelation is given by $R_M(\tau) = e^{-\tau^2}$.

$$s(t) = [1 + m(t)] \cos(\omega_c t + \phi) - \hat{m}(t) \sin(\omega_c t + \phi)$$

5 Sketch the spectrum and compute the power of the ISB signal, assuming that the modulating signals power spectrum densities are given as follows.

$$S_L(\omega) = S_o[u(\omega + \omega_M) - u(\omega - \omega_M)],$$

$$S_R(\omega) = S_o\left(1 - \frac{|\omega|}{\omega_M}\right)[u(\omega + \omega_M) - u(\omega - \omega_M)]$$

6 For the following SSB modulation scheme, determine the power of the modulated carrier. Consider that the modulating signal $m(t)$ autocorrelation is given by $R_M(\tau) = \cos(\omega_M \tau)$. Sketch the spectrum of the modulated carrier.

$$s(t) = m(t)\cos(\omega_c t + \phi) + \hat{m}(t)\sin(\omega_c t + \phi)$$

7 The simplified formula for the C-QUAM modulation scheme is given as follows. Compute the modulated carrier power. Consider that signals $l(t)$ and $r(t)$ are uncorrelated with power P_L and P_R, respectively.

$$s(t) = [1 + l(t) + r(t)]\cos\left[\omega_c t + \tan^{-1}\left(\frac{l(t) - r(t)}{l(t) + r(t)}\right) + \phi\right].$$

8 Given the following SSB lower sideband signal, write the expressions for the amplitude and phase resultants. Compute the modulated carrier power and draw the phasor diagram.

$$s(t) = m(t)\cos(w_c t + \phi) + \hat{m}(t)\sin(w_c t + \phi).$$

9 Given that $l(t)$ and $r(t)$ are zero mean uncorrelated signals, with power P_L and P_R, respectively, compute the power for the following modulated signal.

$$s(t) = [1 + l(t) + r(t)]\cos(w_c t + \theta) + [l(t) - r(t)]\sin(w_c t + \theta),$$

10 Show that the SSB modulation is a linear operation.

$$s(t) = m(t)\cos(\omega_c t + \phi) + \hat{m}(t)\sin(\omega_c t + \phi),$$

where

$$\hat{m}(t) = \frac{1}{\pi}\int_{-\infty}^{\infty} \frac{m(\tau)}{t - \tau} d\tau.$$

11 Signals $b(t)$ and $d(t)$ have the following autocorrelation functions:

$$R_B(\tau) = \frac{A^2}{2} e^{-\alpha|\tau|}$$

and

$$R_D(\tau) = A^2[1 - \frac{|\tau|}{T_b}][u(\tau + T_b) - u(\tau - T_b)].$$

Determine the power of signal $s(t)$, which is QUAM modulated with these signals. Compute the power spectrum density for the modulated signal. Sketch the corresponding diagrams.

12 Compute the modulated carrier power, the amplitude and phase resultants $a(t)$ and phase $\theta(t)$, and draw the phasor diagram for the following stereo signal.

$$s(t) = [l(t) + r(t)] \cos(\omega_c t + \phi) + [1 + l(t) - r(t)] \sin(\omega_c t + \phi).$$

13 The following signal, obtained at the output of a serial to parallel converter, is used to excite a quadrature amplitude modulator. Draw its block diagram, explain its operation and sketch the constellation diagram for the resultant 16QAM signal.

$$b(t) = \sum_{k=-\infty}^{\infty} b_k p(t - kT_b)$$

and

$$d(t) = \sum_{k=-\infty}^{\infty} d_k p(t - kT_b),$$

where $b_k = d_k = \{-3, -\sqrt{3}, \sqrt{3}, 3\}$ and T_b represents the bit interval. Compute the amplitude and phase resultants.

14 Show that the scheme that generates the following modulated carrier operates as an SSB modulator. Compute the modulator output power.

$$s(t) = \hat{m}(t) \cos(\omega_c t + \phi) + m(t) \sin(\omega_c t + \phi)$$

15 The message signals $b(t)$ and $d(t)$, with given autocorrelations, modulate a carrier in quadrature (QUAM). Suppose the signals are uncorrelated and compute the modulated carrier autocorrelation, along

with its power spectrum density. Sketch the corresponding diagrams, considering that $\omega_B = 2\omega_D$.

$$R_B(\tau) = \frac{\omega_B S_0}{\pi} \frac{\sin(\omega_B \tau)}{\omega_B \tau}$$

and

$$R_D(\tau) = \frac{\omega_D S_0}{\pi} \frac{\sin(\omega_D \tau)}{\omega_D \tau}.$$

16 Draw the constellation diagram for the following QAM signal, where $b(t) = \sum_{n=-\infty}^{n=\infty} b_n p(t - nT_b)$ and $d(t) = \sum_{n=-\infty}^{n=\infty} d_n p(t - nT_b)$. Assume, for the first case, $b_n = -1, 0, 1$ and $d_n = -1, 0, 1$. For the second case, consider $b_n = -1, 0, 1$ and $d_n = \pm 1$. Compute the amplitude and phase resultants for both cases. Which system is more efficient? Why?

$$s(t)_{QAM} = b(t)\cos(\omega_c t + \phi) + d(t)\sin(\omega_c t + \phi).$$

17 A random signal, $m(t)$, with autocorrelation $R_M(\tau) = \frac{1}{1+\tau^2}$, is applied to the input of an SSB modulator. Compute the output signal autocorrelation and power spectrum density. Draw the respective diagrams.

18 A certain company designed the following stereo modulation system

$$s(t)_{EST} = b(t)\cos(\omega_c t + \phi) + d(t)\sin(\omega_c t + \phi),$$

where $b(t) = l(t) \cdot r(t)$ and $d(t) = l(t)/r(t)$. Consider $l(t)$ and $r(t)$ independent. Does the system work properly? Compute the output power. Design a demodulator to recover $l(t)$ and $r(t)$. Is the proposed system linear?

19 Determine the power spectrum density, the average power and the amplitude and phase resultants for the following modulation scheme.

$$s(t) = m(t)\cos(\omega_c t + \phi) + \hat{m}(t)\sin(\omega_c t + \phi),$$

where

$$m(t) = \cos(\omega_M t) + \sin(\omega_M t).$$

20 Compute the mean value and average power for the signal

$$\begin{aligned} s(t) &= [1 + \Delta_B(l(t) + r(t))]\sin(\omega_c t + \theta) \\ &- [1 + \Delta_D(l(t) - r(t))]\cos(\omega_c t + \theta), \end{aligned} \quad (5.46)$$

where $l(t)$ and $r(t)$ are zero mean stationary uncorrelated random processes and θ is a random variable, uniformly distributed in the interval $[0, 2\pi)$.

21 The following SSB signal is demodulated using a local carrier $c(t) = \cos(\omega_c t + \theta)$. The high frequency terms are filtered out. Write the expression for the demodulated signal as a function of $\alpha = \phi - \theta$. Explain the phase shift effect on the reception of the signal. Compute the mean value and average power for this signal. Show that the power is independent of α. Consider $R_{M\hat{M}}(0) = R_{\hat{M}M}(0) = 0$.

$$s(t) = m(t)\cos(\omega_c t + \phi) + \hat{m}(t)\sin(\omega_c t + \phi).$$

22 Two digital signals, $b(t)$ and $d(t)$, with identical probability density functions

$$p_B(x) = p_D(x) = \frac{1}{4}[\delta(x + 2a) + \delta(x + a) + \delta(x - a) + \delta(x - 2a)]$$

modulate a QAM carrier. Compute the power of the resultant (or carrier envelope) signal $a(t)$ and the power of the modulated signal $s(t)$.

$$s(t) = b(t)\cos(\omega_c t + \phi) + d(t)\sin(\omega_c t + \phi).$$

Sketch the constellation diagram.

23 For the signal $s(t) = [1 + b(t)]\cos(\omega_c t + \phi) - [1 + d(t)]\sin(\omega_c t + \phi)$, determine the amplitude resultant $a(t)$ the modulated signal power P_S and sketch the phasor diagram. Assume now that $b(t)$ and $d(t)$ are digital signals, which use symbols from the set $\{-1, 1\}$, and draw the constellation diagram for the modulated signal.

24 Show that the sum of two distinct and uncorrelated SSB signals, that used the same sideband, produce an SSB signal. Compute the power of the composite signal.

25 A radio aficionado tries to demodulate an ISB signal using his synchronous AM equipment. The power spectral densities for the modulating signals are given as follows. Is it possible? What is the modulated signal power spectrum density? Consider $\omega_M < \omega_N$.

$$S_M(\omega) = \delta(\omega + \omega_M) + \delta(\omega - \omega_M)$$

$$S_N(\omega) = u(\omega + \omega_N) - u(\omega - \omega_N).$$

26 For a signal $s(t) = [1 + b(t)]\cos(\omega_c t + \phi) + [1 + d(t)]\sin(\omega_c t + \phi)$, determine the amplitude resultant and the modulated signal power P_S. Consider that $b(t)$ and $d(t)$ are digital signals, using symbols from the set $\{-3, -1, 1, 3\}$, sketch the constellation diagram for the modulated signal.

27 For the QAM signal, $s(t) = b(t)\cos(\omega_c t + \phi) + d(t)\sin(\omega_c t + \phi)$, whose constellation diagram is shown in Figure 5.15, determine: *a)* values for input symbols b_j, d_k, amplitude resultant a_l and phase resultant ϕ_l; *b)* average power for signal $s(t)$, assuming equiprobable symbols.

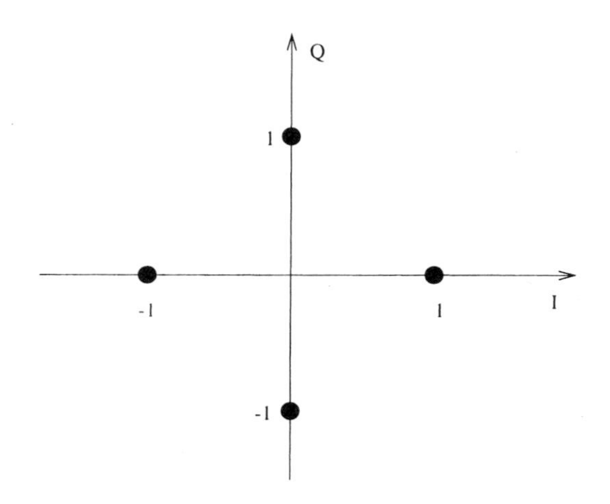

Figure 5.15. Constellation for the QAM signal.

28 Show that the correlation between $m(t)$ and $\hat{m}(t)$ does not influence the computation of the SSB modulated signal power.

29 A QUAM signal, $s(t) = b(t)\cos(\omega_c t + \phi) + d(t)\sin(\omega_c t + \phi)$, where $S_B(\omega) = S_A[1 - \frac{|\omega|}{\omega_M}]$, for $-\omega_M \leq \omega \leq \omega_M$, and $S_D(\omega) = S_D[u(\omega - \omega_D) - u(\omega - \omega_D)]$, for $\omega_D < \omega_M$, is transmitted through an additive white Gaussian channel, and $S_N(\omega) = S_0$. The demodulator filter has the same bandwidth as the transmitted signal. Compute the SNR for each of the received signals. Sketch the power spectrum density diagram.

30 An SSB signal is received through an additive white Gaussian noise channel and synchronously demodulated by a system that has a local carrier shifted in frequency. Assume that the local oscillator generates a carrier $2\cos(\omega_l t + \phi)$ and compute the SNR and demodulator gain.

31 A company uses the following AM-stereo modulation system.

$$s(t) = A[1 + \Delta_B b(t)] \cos(\omega_c t + \phi) + \Delta_D d(t) \sin(\omega_c t + \phi),$$

$$b(t) = l(t) + r(t) \text{ and } d(t) = l(t) - r(t).$$

What is the output power? Design the demodulator. Estimate the modulated signal bandwidth. What the advantage of transmitting signals $l+r$ and $l-r$, instead of l and r? Is it possible to demodulate the signal using a conventional AM receiver?

32 A company is willing to substitute its ASK modulating equipment, shown in Figure 5.16, for a QUAM system that presents the same bit error rate (BER). Compute and compare the output power for both systems. Compare both systems in terms of bandwidth transmission rate and complexity.

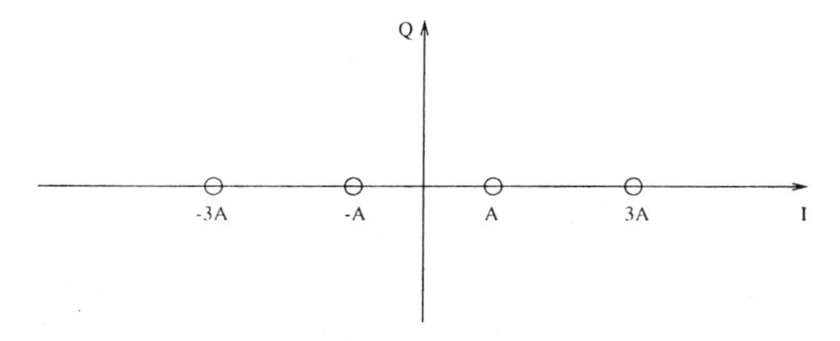

Figure 5.16. Constellation for ASK.

33 A digital signal, $a(t)$, with probability density function $p_A(a) = p\delta(a+V) + (1-p)\delta(a-V)$, modulates a QAM carrier. Design the modulator and explain the modulation process. Compute the output power. Draw the constellation diagram.

34 A company has to decide on which data communication equipment to buy, based on feedback information from costumers. The first scheme is a 4-ASK modulator and the second option is a 4QAM system. The market demands reasonable transmission rates, but minimum bit error rates. But, the cost of coherent demodulation is decisive. Explain the reasoning behind your choice.

35 The following QAM scheme uses the symbols from the sets: $b_j = \{-3A, -A, A, 3A\}$ and $d_k = \{-A, A\}$. Illustrate all phases of the modulation process and draw all waveforms. Sketch the constellation diagram and explain the advantages and disadvantages of the system.

$$s(t) = b(t) \cos(\omega_c t + \phi) + d(t) \sin(\omega_c t + \phi),$$

$$b(t) = \sum_j b_j p(t - jT_b) \text{ e } d(t) = \sum_k d_k p(t - kT_b).$$

36 Draw the constellation diagram for the 8QAM scheme and show the effects of lack of synchronism and non-linear amplification. Design the demodulator and suggest means to overcome those problems.

37 The autocorrelation functions for the uncorrelated signals $b(t)$ and $d(t)$ are given. These signals modulate a carrier in QUAM. Compute the power spectrum density and sketch the diagrams, considering $\omega_B = 2\omega_D$.

$$R_B(\tau) = S_0\omega_B[1 - \omega_B|\tau|][u(\tau + \frac{1}{\omega_B}) - u(\tau - \frac{1}{\omega_B})]$$

and

$$R_D(\tau) = \frac{\omega_D S_0}{\pi} \frac{\sin(\omega_D \tau)}{\omega_D \tau}.$$

38 Draw the constellation diagram for the QAM signal, where $b(t) = \sum_{n=-\infty}^{n=\infty} b_n p(t - nT_b)$ and $d(t) = \sum_{n=-\infty}^{n=\infty} d_n p(t - nT_b)$. Assume, for the first case, $b_n = 0, A$ and $d_n = -A, 0, A$. Consider, for the second case, $b_n = -A, 0, A$ and $d_n = \pm A$. Compute the amplitude and phase resultants for both cases. Which system is more efficient? Why?

$$s(t)_{QAM} = b(t)\cos(\omega_c t + \phi) - d(t)\sin(\omega_c t + \phi).$$

Chapter 6

ANGLE MODULATION

6.1 Introduction

It is well established nowadays that a frequency modulated signal can not be accommodated in a bandwidth which is narrower than the one occupied by the original, or modulating, signal. Ironically, though, frequency modulation was originally conceived as a means to reduce the bandwidth required for the transmission of a given signal.

The rationale behind this could be posed as follows: instead of amplitude modulation, whose bandwidth is double that of the modulating signal, one could frequency modulate a carrier using, for instance, a deviation of ± 50 Hz, which would give a transmission bandwidth of 100 Hz, regardless of the original signal bandwidth.

This argument is, of course, false. John R. Carson, was the first to recognize the fallacy of the reduction in bandwidth, and clarified the subject in a paper published in 1922 (Carson, 1922). At the time, however, he could see no advantage in frequency over amplitude modulation, and the subject was forgotten for a while, until Edwin H. Armstrong invented the first equipment to frequency modulate a carrier (Armstrong, 1936).

But, the world had to await until Carson established, in an unpublished memorandum of August 28, 1939, that the bandwidth was equal to twice the sum of the peak frequency deviation and the highest frequency of the modulating signal (Carlson, 1975).

Angle modulation is a term that encompasses both frequency modulation (FM) and phase modulation (PM). The objective of this chapter is to derive a general mathematical model to analyze stochastic angle modulated signals. Several transmission systems use either frequency or phase modulation, including analog and digital mobile cellular communication systems, satellite transmission systems, television systems and wireless telephones.

6.2 Angle Modulation Using Random Signals

This section presents a general and elegant method to compute the power spectrum density of angle modulated signals. The modulating signal is considered a stationary random process $m(t)$, that has autocorrelation $R_M(\tau)$.

6.2.1 Mathematical Model

Woodward's theorem asserts that the spectrum of a high-index frequency modulated (FM) waveform has a shape that is approximately that of the probability distribution of its instantaneous frequency (Woodward, 1952). The following method estimates the spectrum of an angle modulated signal, based on the first order probability density function (pdf) of the random modulating signal, using the previous theorem. A new proof is presented which for the first time includes the linear mean square estimator (Papoulis, 1981).

In the proposed model, the modulated signal $s(t)$ is obtained from the following equations:

$$s(t) = A\cos\left(\omega_c t + \theta(t) + \phi\right), \tag{6.1}$$

where

$$\theta(t) = \Delta_{FM} \cdot \int_{-\infty}^{t} m(t)dt, \tag{6.2}$$

for frequency modulation, and

$$\theta(t) = \Delta_{PM} \cdot m(t)dt, \tag{6.3}$$

for phase modulation, where the constant parameters A, ω_c, Δ_{FM} and Δ_{PM} represent respectively the carrier amplitude and angular frequency, and the frequency and phase deviation indices. The message signal is represented by $m(t)$, which is considered a zero mean random stationary process. The carrier phase ϕ is random, uniformly distributed in the range $(0, 2\pi]$ and statistically independent of $m(t)$.

The modulating signal, $\theta(t)$, represents the variation in the carrier phase, produced by the message. The frequency modulation index is defined as

$$\beta = \frac{\Delta_{FM}\sigma_M}{\omega_M}, \tag{6.4}$$

where $\sigma_M = \sqrt{P_M}$ and P_M represents the power of the message signal, $m(t)$, and ω_M is the maximum angular frequency of the signal.

The frequency deviation $\sigma_F = \Delta_{FM}\sigma_M$ represents the shift from the original, or spectral, carrier frequency. The modulation index gives and idea of how many times the modulating signal bandwidth fits into the frequency deviation.

The phase modulation index is defined as

$$\alpha = \Delta_{PM}\sigma_M. \tag{6.5}$$

The following steps are observed in the evaluation of the spectrum of the angle modulated signal:

1 Compute the autocorrelation function of $s(t)$ in Equation 6.1.

2 Obtain an estimate of this autocorrelation, for the case of a high modulation index, using the linear mean square estimator (Papoulis, 1981).

3 Compute the power spectrum density (PSD) of $s(t)$ as the Fourier transfor of the autocorrelation estimate.

The autocorrelation function of $s(t)$, defined by equation 6.1 is expressed as

$$R_S(\tau) = E[s(t)s(t+\tau)] \tag{6.6}$$

where $E[\cdot]$ represents the expected value operator. It follows that

$$R_S(\tau) = \frac{A^2}{2}E[\cos(w_c\tau - \theta(t) + \theta(t+\tau))]. \tag{6.7}$$

It is possible to split the problem of computing the power spectrum density of the modulated signal into three cases, which are discussed in the following, using frequency modulation as a paradigm.

6.2.2 Case I – Modulation with low index, $\beta < 0.5$

In this case, the autocorrelation function of $s(t)$ can be obtained from Equation 6.7 by expanding the cosine

$$\begin{aligned} R_S(\tau) &= \frac{A^2}{2}\cos(\omega_c\tau)E[\cos(-\theta(t) + \theta(t+\tau))] \\ &- \frac{A^2}{2}\sin(\omega_c\tau)E[\sin(-\theta(t) + \theta(t+\tau))]. \end{aligned} \tag{6.8}$$

For a low modulation index, it is possible to expand the sine and cosine of Equation 6.8 in Taylor series, neglecting the high order terms, to obtain

$$\begin{aligned} R_S(\tau) &= \frac{A^2}{2}\cos(\omega_c\tau)E\left[1-\frac{(-\theta(t)+\theta(t+\tau))^2}{2}\right] \\ &- \frac{A^2}{2}\sin(\omega_c\tau)E\left[-\theta(t)+\theta(t+\tau)\right]. \end{aligned} \quad (6.9)$$

Considering that $m(t)$ is a zero mean, stationary process, it follows that

$$R_S(\tau) = \frac{A^2}{2}\cos(\omega_c\tau)[1 - R_\Theta(0) + R_\Theta(\tau)], \quad (6.10)$$

where

$$R_\Theta(\tau) = E[\theta(t)\theta(t+\tau)] \quad (6.11)$$

and $R_\Theta(0) = P_\Theta$ is the power of signal $\theta(t)$.

The power spectral density of $s(t)$ is obtained by the use of the Wiener-Khintchine theorem, or through the Fourier transform of Equation 6.10 (Papoulis, 1983a).

$$\begin{aligned} S_S(w) &= \frac{\pi A^2(1-P_\Theta)}{2}[\delta(\omega+\omega_0)+\delta(\omega-\omega_0)] \\ &+ \frac{\Delta_{FM}^2 A^2}{4}\left[\frac{S_M(\omega+\omega_c)}{(\omega+\omega_c)^2}+\frac{S_M(\omega-\omega_c)}{(\omega-\omega_c)^2}\right] \end{aligned} \quad (6.12)$$

where $S_M(w)$ represents the PSD of the message signal $m(t)$, with has bandwidth ω_M. The modulated signal bandwidth is double the message signal bandwidth BW $= 2\omega_M$.

From Equation 6.12 one can notice that the FM spectrum has the shape of the message signal spectrum multiplied by a squared hyperbolic function. The examples which are given in the following sections consider a uniform power spectrum density for the message signal.

6.2.3 Case II – Modulation index in the interval $0.5 \leq \beta \leq 5$

Increasing the modulation index implies in considering more terms from the Taylor expansion of Equation 6.8. It can be shown that, for a Gaussian signal, for example, all terms of the sine expansion vanish, because they are joint moments of odd order (Papoulis, 1983b), therefore

$$\begin{aligned} R_S(\tau) &= \frac{A^2}{2}\cos(\omega_c\tau)E\left[1 - \frac{(-\theta(t)+\theta(t+\tau))^2}{2}\right. \\ &+ \left.\frac{(-\theta(t)+\theta(t+\tau))^4}{4!} + \dots\right] \end{aligned} \tag{6.13}$$

Considering that the bandwidth of the modulated signal does not exceed four times the bandwidth of the message signal $m(t)$, it is possible to neglect all terms of the expansion 6.13 of order four and higher, which gives

$$\begin{aligned} R_S(\tau) &= \frac{A^2}{2}\cos(\omega_c\tau)[1 - R_\Theta(0) + R_\Theta(\tau)] \\ &+ \frac{A^2}{48}\cos(\omega_c\tau)\left(2E[\theta^4(t)] + 6E[\theta^2(t)\theta^2(t+\tau)]\right. \\ &- \left.24P_\Theta R_\Theta(\tau)\right). \end{aligned} \tag{6.14}$$

It is observed that the contribution of the second term of 6.14 is less significant than the first one.

For a Gaussian signal, the previous expression can be simplified, using Price's Theorem, defined as (Price, 1958) (McMahon, 1964)

$$E[X^kY^r] = kr\int_0^c E[X^{k-1}Y^{r-1}]dc + E[X^k]E[Y^r] \tag{6.15}$$

where

$$c = E[XY] - E[X]E[Y]. \tag{6.16}$$

Applying 6.15 and 6.16 in 6.14, it follows that

$$\begin{aligned} R_S(\tau) &= \frac{A^2}{2}\cos(\omega_c\tau)[1 - R_\Theta(0) + R_\Theta(\tau)] \\ &+ \frac{A^2}{48}\cos(w_c\tau)\left(12P_\Theta^2 + 12R_\Theta^2(\tau) - 24P_\Theta R_\Theta(\tau)\right). \end{aligned} \tag{6.17}$$

Fourier transforming this expression, gives

$$\begin{aligned} S_S(w) &= \frac{\pi A^2}{2}(1 - P_\Theta + \frac{P_\Theta^2}{2})[\delta(\omega+\omega_c) + \delta(\omega-\omega_c)] \\ &+ \frac{\Delta_{FM}^2 A^2}{4}(1 - P_\Theta)\left[\frac{S_M(\omega+\omega_c)}{(\omega+\omega_c)^2} + \frac{S_M(\omega-\omega_c)}{(\omega-\omega_c)^2}\right] \\ &+ \frac{\Delta_{FM}^4 A^2}{16\pi}\left[\frac{S_M(\omega\pm\omega_c)}{(\omega\pm\omega_c)^2} * \frac{S_M(\omega\pm\omega_c)}{(\omega\pm\omega_c)^2}\right]. \end{aligned} \tag{6.18}$$

The convolution operations broaden the spectrum. The use of the previous formula permits to find the spectrum for any modulation index. The final bandwidth is obtained after the composition of several spectra.

Figure 6.1 illustrates the FM spectrum for a sinusoidal message signal, considering an intermediate modulation index.

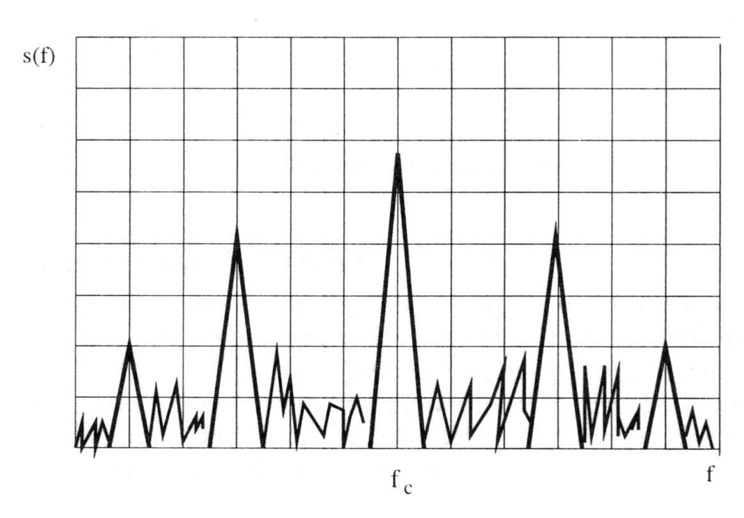

Figure 6.1. Spectrum of an FM signal for a sinusoidal message signal and intermediate modulation index.

6.2.4 Case III – High modulation index, $\beta > 5$

The power spectrum density of $s(t)$ is demonstrated to approach the probability density function of $m(t)$ in Equation 6.2, as the modulation index is increased. A high modulating index causes a spectrum broadening of the modulated signal. In addition, the modulated carrier PSD turns into the probability density function of the modulating signal. That is the result of Woodward's theorem, and will be discussed in the sequel (Woodward, 1952) (Blachman and McAlpine, 1969).

The analog mobile cellular system, the commercial FM and the wireless telephone can be considered as high modulation index systems (Lee, 1989). For the high modulation index case, it is more interesting to use Euler's formula and rewrite Equation 6.7 as

$$
\begin{aligned}
R_S(\tau) &= \frac{A^2}{4} e^{j\omega_c \tau} E[e^{j(-\theta(t)+\theta(t+\tau))}] \\
&+ \frac{A^2}{4} e^{-j\omega_c \tau} E[e^{j(\theta(t)-\theta(t+\tau))}]. \qquad (6.19)
\end{aligned}
$$

The second order linear mean square estimate of the process $\theta(t+\tau)$ includes its current value $\theta(t)$ and its derivative $\theta'(t)$

$$\theta(t+\tau) \approx \alpha\, \theta(t) + \beta\, \theta'(t). \qquad (6.20)$$

The mean square error is given by

$$e(t, \alpha, \beta) = E\left[\left(\theta(t+\tau) - \alpha\, \theta(t) - \beta\, \theta'(t)\right)^2\right]. \qquad (6.21)$$

In order to obtain the values of the optimization parameters α and β, the derivative of the error must converge to zero. The partial derivative of the error in terms of α gives

$$\frac{\partial e(t, \alpha, \beta)}{\partial \alpha} = E\left[\left(\theta(t+\tau) - \alpha\, \theta(t) - \beta\, \theta'(t)\right) \theta(t)\right] = 0, \qquad (6.22)$$

which indicates that the minimum error is orthogonal to the random process.

Recognizing the autocorrelations in the previous expression, yields

$$R_\Theta(\tau) - \alpha R_\Theta(0) - \beta R_{\Theta'\Theta}(0) = 0.$$

The last term of the equation is zero, because the autocorrelation has a maximum at the origin. Thus,

$$\alpha = \frac{R_\Theta(\tau)}{R_\Theta(0)}. \qquad (6.23)$$

The partial derivative of the error in terms of β gives

$$\frac{\partial e(t, \alpha, \beta)}{\partial \beta} = E\left[\left(\theta(t+\tau) - \alpha\, \theta(t) - \beta\, \theta'(t)\right) \theta'(t)\right] = 0 \qquad (6.24)$$

or

$$R_{\Theta'\Theta}(\tau) - \alpha R_{\Theta'\Theta}(0) - \beta R_{\Theta'}(\tau) = 0.$$

The first term of the expression is the derivative of the autocorrelation function in relation to τ. The second term is zero, as explained previously, and the third term is the negative of the second derivative of the autocorrelation function in relation to τ. Thus,

$$\beta = \frac{R'_\Theta(\tau)}{R''_\Theta(0)}. \qquad (6.25)$$

Therefore, the best approximation to the future value of the process, in the mean square sense, is

$$\theta(t+\tau) \approx \frac{R_\Theta(\tau)}{R_\Theta(0)}\theta(t) + \frac{R'_\Theta(\tau)}{R''_\Theta(0)}\theta'(t). \tag{6.26}$$

Considering that the random process is slowly varying, as compared to the spectral frequency of the modulated carrier, leads the the approximation

$$R_\Theta(\tau) \approx R_\Theta(0).$$

Expanding the derivative of the autocorrelation in a Taylor series, gives

$$R'_\Theta(\tau) \approx R'_\Theta(0) + \tau R''_\Theta(0). \tag{6.27}$$

Recalling that the autocorrelation has a maximum at the origin, simplifies the approximation to

$$R'_\Theta(\tau) \approx \tau R''_\Theta(0).$$

Finally, the future value of the random process can be approximated by

$$\theta(t+\tau) \approx \theta(t) + \tau\,\theta'(t). \tag{6.28}$$

Use of the linear mean square estimator in Equation 6.19 then gives (Alencar, 1989) (Papoulis, 1983a)

$$\begin{aligned} R_S(\tau) &= \frac{A^2}{4}e^{j\omega_c\tau}E[e^{j\tau\theta'(t)}] \\ &+ \frac{A^2}{4}e^{-j\omega_c\tau}E[e^{-j\tau\theta'(t)}]. \end{aligned} \tag{6.29}$$

But, $\theta'(t) = d\theta(t)/dt = \omega(t)$, where $\omega(t)$ is the carrier angular frequency deviation, at the instantaneous carrier angular frequency, thus

$$\begin{aligned} R_S(\tau) &= \frac{A^2}{4}e^{j\omega_c\tau}E[e^{j\tau\omega(t)}] \\ &+ \frac{A^2}{4}e^{-j\omega_c\tau}E[e^{-j\tau\omega(t)}]. \end{aligned} \tag{6.30}$$

Taking into account that

$$E[e^{j\tau\omega(t)}] = \int_{-\infty}^{\infty} p_\Omega(\omega(t))e^{j\tau\omega(t)}d\omega(t) \tag{6.31}$$

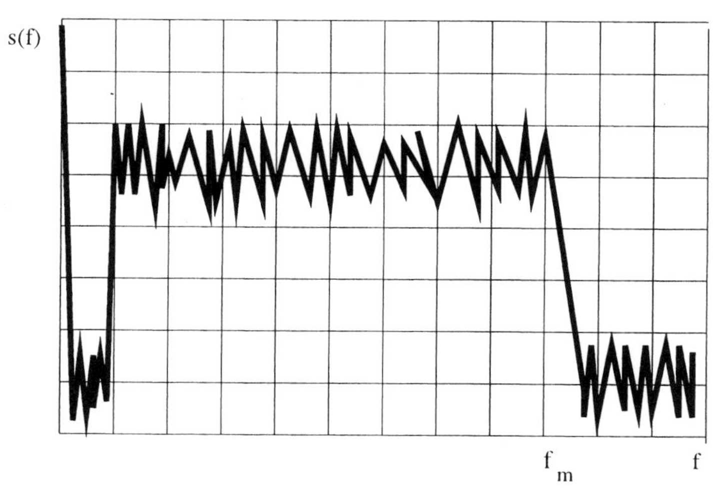

Figure 6.2. Spectrum for the Gaussian modulating signal.

represents the characteristic function of process $\omega(t) = \theta'(t)$ and $p_\Omega(\omega(t))$ is its probability density function, considered symmetrical here, without essential loss of generality.

Calculating the Fourier transform of Equation 6.30, it follows that

$$S_S(w) = \frac{\pi A^2}{2}[p_\Omega(\omega + \omega_c) + p_\Omega(\omega - \omega_c)]. \tag{6.32}$$

Considering the definition of $\omega(t)$ from Equation 6.2, it is noticed that $\omega(t) = \Delta_{FM}.m(t)$, thus

$$p_\Omega(\omega(t)) = \frac{1}{\Delta_{FM}} p_M\left(\frac{m}{\Delta_{FM}}\right) \tag{6.33}$$

where $p_M(\cdot)$ is the probability density function of $m(t)$.

Substituting 6.33 into 6.32 gives, finally, the formula for the power spectrum density for the wideband frequency modulated signal (Alencar and Neto, 1991)

$$S_S(w) = \frac{\pi A^2}{2\Delta_{FM}}\left[p_M\left(\frac{w + w_c}{\Delta_{FM}}\right) + p_M\left(\frac{w - w_c}{\Delta_{FM}}\right)\right]. \tag{6.34}$$

Following a similar line of thought it is possible to derive a formula for the spectrum of the phase modulated signal (PM). It is instructive to recall the instantaneous angular frequency, given by

$$\omega(t) = \frac{\theta(t)}{dt} = \Delta_{PM}m(t). \tag{6.35}$$

Therefore,

$$S_S(w) = \frac{\pi A^2}{2\Delta_{PM}} \left[p_{M'}\left(\frac{w + w_c}{\Delta_{PM}}\right) + p_{M'}\left(\frac{w - w_c}{\Delta_{PM}}\right) \right], \tag{6.36}$$

where $p_{M'}(\cdot)$ is the probability density function of of the derivative of the message signal $m(t)$.

For a Gaussian modulating signal

$$p_M(m) = \frac{1}{\sqrt{2\pi P_M}} e^{-\frac{m^2}{2P_M}}, \tag{6.37}$$

where $P_M = R_M(0)$ denotes the power of signal $m(t)$, which gives

$$p_M(m) = \frac{1}{\sqrt{2\pi R_M(0)}} e^{-\frac{m^2}{2R_M(0)}}. \tag{6.38}$$

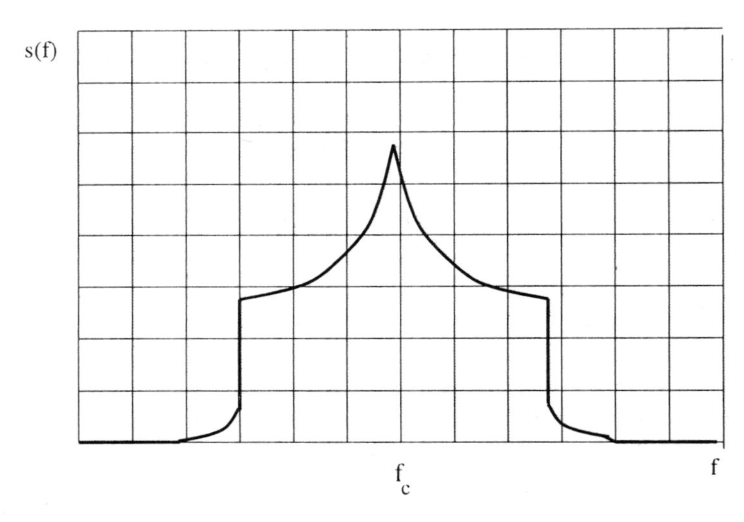

Figure 6.3. Spectrum of an FM signal with low modulation index, for a Gaussian modulating signal.

Figures 6.3, 6.4, 6.5 and 6.6, were produced from measured signals, from an experiment performed at Embratel, in Brazil, by the first author. They illustrate the changes undergone by the modulated carrier spectrum as the modulation index is increased, for a Gaussian modulating signal. The power spectrum density of the modulating signal is shown in Figure 6.2. This signal was obtained from a noise generator.

From Equation 6.39, and considering the property $R_{M'}(\tau) = -R''_M(\tau)$ (the autocorrelation of the signal derivative equals the negative of the second derivative of the autocorrelation function) and the fact that the response of a linear system to a Gaussian input is also Gaussian, one can determine the probability density function for the derivative of a Gaussian signal

$$p_{M'}(m) = \frac{1}{\sqrt{-2\pi R''_M(0)}} e^{-\frac{m^2}{(-2R''_M(0))}}. \tag{6.39}$$

That is, the spectrum of a wideband PM signal, when modulated by a Gaussian signal, presents similar characteristics to his FM counterpart. For a narrowband modulation, the PM spectrum approximates the spectrum of the modulating signal which is different from the FM spectrum for the same conditions.

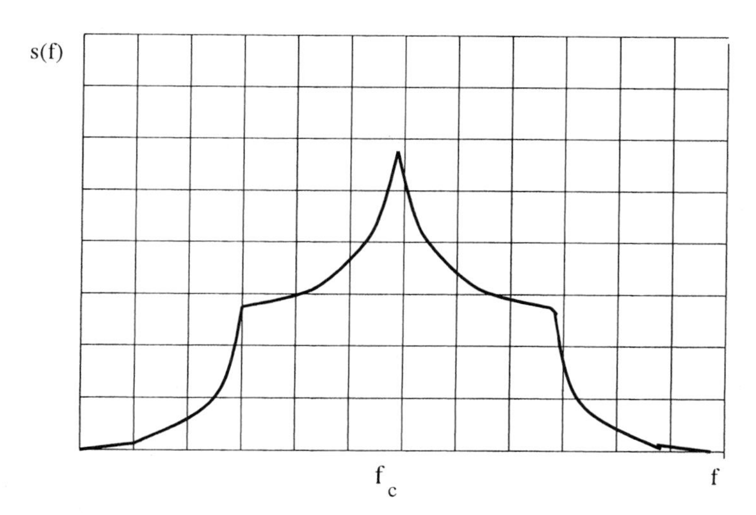

Figure 6.4. Spectrum of FM signal for a Gaussian modulating signal, and increasing modulation index.

It is noticeable that for a high modulation index the frequency deviation is given by $\Delta_{FM}\sqrt{P_M}$ and the bandwidth approximated by BW $= 2\Delta_{FM}\sqrt{P_M}$, in order to include most of the modulated carrier power. As previously derived, the bandwidth for a narrowband FM is BW $= 2\omega_M$. Interpolating between both values gives a formula that covers the whole range of modulation indices β

$$\text{BW} = 2\omega_M + 2\Delta_{FM}\sqrt{P_M} = 2\left(\frac{\Delta_{FM}\sqrt{P_M}}{\omega_M} + 1\right)\omega_M = 2(\beta + 1)\omega_M. \tag{6.40}$$

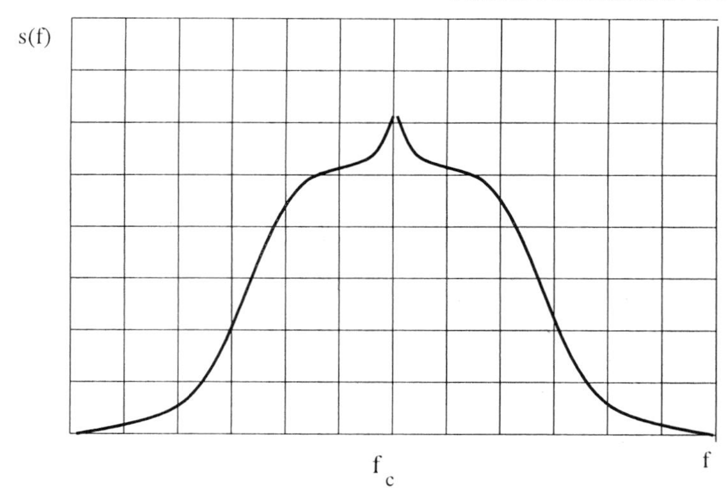

Figure 6.5. Spectrum of FM signal for a Gaussian modulating signal, and intermediate modulation index.

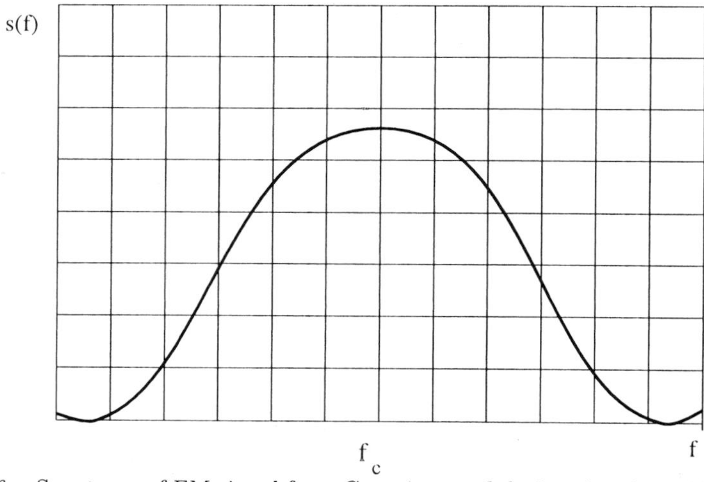

Figure 6.6. Spectrum of FM signal for a Gaussian modulating signal, and high modulation index.

This is the well known Carson's rule, whose heuristic deduction first appeared in 1922 (Carson, 1922).

A sinusoidal signal $m(t) = V \sin(\omega_M t + \varphi)$, where φ is uniformly distributed in the interval $(0, 2\pi]$, has the following probability density function

$$p_M(m) = \frac{1}{\pi\sqrt{V^2 - m^2}}, |m| < V. \tag{6.41}$$

Consequently, a carrier modulated by a sinusoidal signal, with high modulation index, has the following spectrum

$$S_S(w) = \frac{1}{2\sqrt{(V\Delta_{FM})^2 - (w \pm w_c)^2}}, |w - w_c| < \Delta_{FM}V. \tag{6.42}$$

The modulated signal, occupies, in this case, a bandwidth equivalent to $2\Delta_{FM}V$. Its PSD is shown in Figure 6.7.

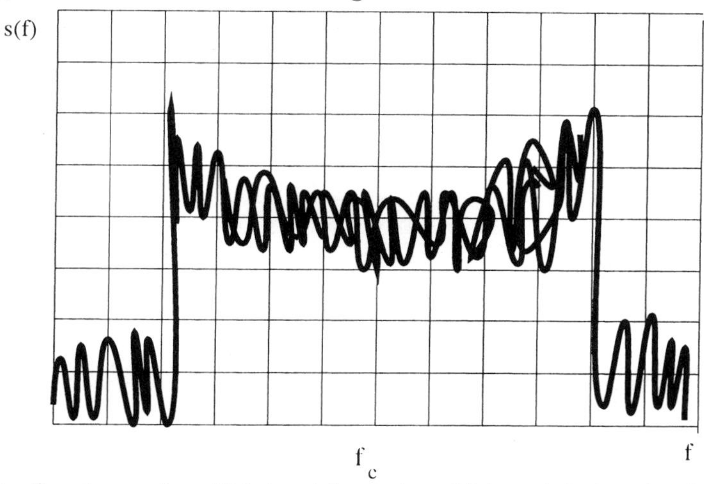

Figure 6.7. Spectrum of an FM signal for a sinusoidal modulating signal and high modulation index.

For a signal which is phase modulated, using the same sinusoid as before as the modulating signal, the derivative of the message signal is given by $m'(t) = \omega_M V \cos(\omega_M t + \varphi)$, which has the following probability density function

$$p_{M'}(m) = \frac{1}{\pi\sqrt{(\omega_M V)^2 - m^2}}, |m| < \omega_M V. \tag{6.43}$$

The spectrum can be computed, resulting in

$$S_S(w) = \frac{1}{2\sqrt{(\omega_M V\Delta_{PM})^2 - (w \pm w_c)^2}}, |w - w_c| < \omega_M\Delta_{PM}V. \tag{6.44}$$

The bandwidth is now $2\omega_M\Delta_{PM}V$ and its PSD is similar to the one shown in Figure 6.7.

The voice signal is modeled as a Gamma distribution (Paez and Glisson, 1972),

$$p_M(m) = \frac{\sqrt{k}}{2\sqrt{\pi}} \frac{e^{-k|m|}}{\sqrt{|m|}} \tag{6.45}$$

Therefore, the corresponding FM spectrum is given by

$$R_S(w) = \frac{A^2\sqrt{\pi k}}{4\sqrt{\Delta_{FM}}} \frac{e^{-\frac{k}{\Delta_{FM}}|w \pm w_c|}}{\sqrt{|w \pm w_c|}} \tag{6.46}$$

and its bandwidth is $\frac{2\Delta_{FM}\sqrt{0.75}}{k}$.

Finally, a model for a full wave rectified voice signal is the exponential distribution

$$p_M(m) = \alpha e^{-\alpha m} u(m) \tag{6.47}$$

which gives the following PSD

$$S_S(w) = \frac{\alpha A^2}{2\Delta_{FM}} \left[e^{\frac{\alpha}{\Delta_{FM}}(\omega + \omega_c)} u(-\omega + \omega_c) + e^{-\frac{\alpha}{\Delta_{FM}}(\omega - \omega_c)} u(\omega - \omega_c) \right] \tag{6.48}$$

6.3 Frequency and Phase Demodulation

The modulated carrier can be demodulated to recover the message signal, as shown in Figure 6.8. The incoming signal

$$s(t) = A\cos\left(\omega_c t + \theta(t) = \Delta_{FM} \cdot \int_{-\infty}^{t} m(t)dt + \phi\right) \tag{6.49}$$

is differentiated (passes through a discriminator), to give

$$s(t) = -A\left[\omega_c + \Delta_{FM} \cdot m(t)\right] \sin\left(\omega_c t + \theta(t) = \Delta_{FM} \cdot \int_{-\infty}^{t} m(t)dt + \phi\right). \tag{6.50}$$

The signal is envelope demodulated (non-coherent demodulation) and low-pass filtered, which results in

$$r(t) = -A\left[\omega_c + \Delta_{FM} \cdot m(t)\right] \tag{6.51}$$

The DC level then is blocked, to give the original signal $m(t)$ multiplied by a constant. Phase tracking loops, also called phase locked loops (PLL), are commonly used to demodulate FM signals.

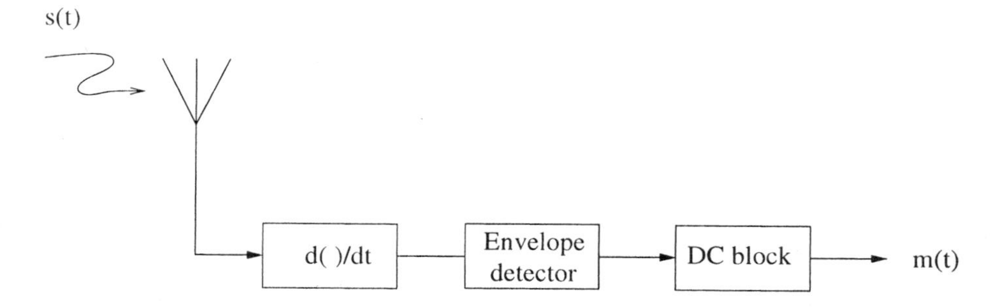

Figure 6.8. Block diagram of an FM demodulator.

Figure 6.9 illustrates the block diagram of PLL, which is composed of a mixer, followed by a low pass filter with impulse response $f(t)$ and an output filter, whose impulse response is $g(t)$. The feedback loop has a voltage controlled oscillator (VCO), which produces an output $x(t)$ whose frequency is proportional to the amplitude of $z(t)$.

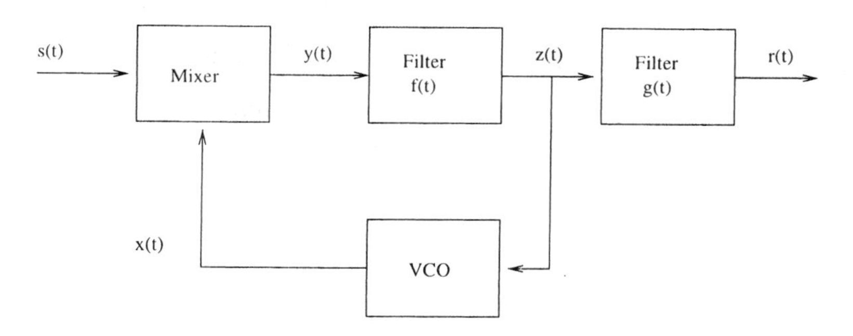

Figure 6.9. Phase locked loop.

The following analysis is performed to obtain the frequency response, or transfer function, of the PLL, in order to verify that it can actually demodulate the input signal.

The signal at the input of the PLL $s(t)$ represents the modulated carrier plus noise

$$s(t) = A\cos\left[\omega_c t + \theta(t) + \phi\right] + n(t). \tag{6.52}$$

The signal at the output of the VCO is

$$x(t) = \sin\left[\omega_c t + \gamma(t) + \phi\right]. \tag{6.53}$$

where $\gamma(t)$ is proportional to the integral of $z(t)$,

$$\gamma(t) = \Delta_{VCO} \int_{-\infty}^{t} z(t)dt, \tag{6.54}$$

which implies that the VCO instantaneous frequency $w(t)$ is proportional to the signal amplitude

$$\omega(t) = \frac{d\gamma(t)}{dt} = \Delta_{VCO} z(t). \tag{6.55}$$

At the mixer output the signal is

$$y(t) = A \sin\left[\theta(t) - \gamma(t)\right] + n_M(t). \tag{6.56}$$

where $n_M(t)$ represents the noise filtered by the mixer.

The difference between the modulated carrier phase and the VCO signal phase $\theta(t) - \gamma(t)$ must be kept to a minimum, for the PLL to work properly. Therefore, at the mixer output, the signal can be approximated by

$$y(t) = A\left[\theta(t) - \gamma(t)\right] + n_M(t). \tag{6.57}$$

Fourier transforming Equation 6.57, gives

$$Y(\omega) = A\left[\Theta(\omega) - \Gamma(\omega)\right] + N_M(\omega). \tag{6.58}$$

where

$$\Gamma(\omega) = \Delta_{VCO} \frac{Z(\omega)}{j\omega} \tag{6.59}$$

and $N_M(\omega)$ is the Fourier transform of $n_M(t)$.

At the output of the low pass filter,

$$Z(\omega) = AF(\omega)\left[\Theta(\omega) - \Gamma(\omega)\right] + F(\omega)N_M(\omega) \tag{6.60}$$

where $F(\omega)$ is the transfer function of the filter.

Substituting the Fourier transform of the VCO phase, yields

$$Z(\omega) = AF(\omega)\left[\Theta(\omega) - \Delta_{VCO}\frac{Z(\omega)}{j\omega}\right] + F(\omega)N_M(\omega) \tag{6.61}$$

Solving for $Z(\omega)$, gives

$$Z(\omega) = \left[\frac{j\omega AF(\omega)}{A\Delta_{VCO}F(\omega) + j\omega}\right]\left[\Theta(\omega) + \frac{N_M(\omega)}{A}\right]. \tag{6.62}$$

The Fourier transform of the signal at the output of the PLL is

$$R(\omega) = G(\omega)Z(\omega). \tag{6.63}$$

For an FM signal,

$$\Theta(\omega) = \Delta_{FM} \frac{M(\omega)}{j\omega} \tag{6.64}$$

where $M\omega)$ is the Fourier transform of the modulating signal $m(t)$.

Then, the transfer function of the output filter, to correctly demodulate the signal, is

$$G(\omega) = \frac{A\Delta_{VCO}F(\omega) + j\omega}{AF(\omega)}, \tag{6.65}$$

which is valid for $|\omega| \leq \omega_M$, and null outside this interval, where ω_M is the maximum frequency of the modulating signal.

This results in

$$R(\omega) = \Delta_{FM}M(\omega) + \frac{j\omega N_M(\omega)}{A}. \tag{6.66}$$

Thus, the transfer function of the PLL for the signal is just Δ_{FM}. For the noise, the transfer function is $\frac{j\omega}{A}$. Recalling that, at the output of a linear filter, the power spectral density if given by

$$S_Y(\omega) = |H(\omega)|^2 S_X(\omega)$$

and that the input noise has a uniform power spectral density N_0, implies that the noise at the output of the PLL has a quadratic power spectral density.

$$S_N(\omega) = \frac{\omega^2 N_0}{A^2}. \tag{6.67}$$

For a PM signal,

$$G(\omega) = \frac{A\Delta_{VCO}F(\omega) + j\omega}{AF(\omega)/j\omega}, \tag{6.68}$$

valid in the interval $|\omega| \leq \omega_M$ and null outside this interval.

The output signal transfer function is

$$R(\omega) = \Delta_{PM}M(\omega) + \frac{N_M(\omega)}{A}. \tag{6.69}$$

The noise power spectral density for the PM demodulator is then

$$S_N(\omega) = \frac{N_0}{A^2}. \tag{6.70}$$

6.4 Performance Evaluation of Angle Modulation

The preceding analysis has shown a bandwidth equivalent to $BW = 2\omega_M$, for narrowband angle modulation and a modulating signal of bandwidth ω_M.

For wideband frequency modulation, considering the upper limit for the modulation index, the bandwidth is $BW = 2\sigma_F = 2\Delta_{FM}\sigma_M$, where σ_F represents the root mean square (RMS) frequency deviation from the carrier spectral frequency, and σ_M is the RMS value of the Gaussian modulating signal.

The modulation index, for a random modulating signal, has been defined as

$$\beta = \frac{\sigma_F}{\omega_M} = \frac{\Delta_{FM}\sigma_M}{\omega_M} = \frac{\Delta_{FM}\sqrt{P_M}}{\omega_M}, \qquad (6.71)$$

and recall that Δ_{FM} is the frequency deviation index and P_M is the message signal average power.

The approximate formula for the bandwidth, which can be used for both narrowband and wideband case, is

$$B.P. = 2\omega_M + 2\Delta_{FM}\sigma_M = 2\omega_M\left(1 + \frac{\Delta_{FM}\sigma_M}{\omega_M}\right) = 2(\beta+1)\omega_M. \qquad (6.72)$$

As mentioned earlier, Formula 6.72 is known as Carson's rule to determine the bandwidth of an FM signal. This formula is used to compute the demodulator, or front end filter, bandwidth, as well as the intermediate filter (IF) bandwidth.

The received signal power is $P_S = \frac{A^2}{2}$. Noise is also received by the demodulator, and it is considered to have a uniform power spectrum density N_0. The noise is filtered by the front end filter, which gives a net received power of

$$P_N = \frac{N_0(\beta+1)\omega_M}{\pi}. \qquad (6.73)$$

The signal to noise ratio at the input of the demodulator is given by

$$SNR_I = \frac{A^2/2}{N_0(\beta+1)\omega_M/\pi} = \frac{\pi A^2}{2N_0(\beta+1)\omega_M}. \qquad (6.74)$$

The demodulator consists of a discriminator, which is equivalent to a circuit to compute the derivative of the input signal, operating at the intermediate frequency, and produces at its output a signal whose power is $P_{\hat{M}} = \Delta_{FM}^2 P_M$.

Noise is, as well, differentiated by the discriminator, and its output power spectrum density is given by the formula $S_{N'}(\omega) = |H(\omega)|^2 S_N(\omega)$, where $H(\omega) = \frac{j\omega}{A}$ represents the discriminator, or the PLL, transfer function and $S_N(\omega) = N_0$.

At the output of the demodulator, after being filtered by a bandpass filter of bandwidth ω_M, the noise power is given by

$$P_{N'} = \frac{1}{2\pi} \int_{-\omega_M}^{\omega_M} S_{N'}(\omega) d\omega = \frac{1}{2\pi} \int_{-\omega_M}^{\omega_M} |H(\omega|^2 S_N(\omega) d\omega. \tag{6.75}$$

Substituting the previous results into 6.75, it follows that

$$P_{N'} = \frac{1}{2\pi} \int_{-\omega_M}^{\omega_M} \frac{N_0 \omega^2}{A^2} d\omega = \frac{N_0 \omega^3}{3\pi A^2}. \tag{6.76}$$

The signal to noise ratio, at the demodulator output is then

$$SNR_O = \frac{3\pi A^2 \Delta_{FM}^2 P_M}{N_0 \omega_M^3}. \tag{6.77}$$

Finally, the demodulation gain, for the frequency demodulator is given by

$$\eta = \frac{SNR_O}{SNR_I} = 6\beta^2(\beta + 1). \tag{6.78}$$

A similar derivation for the PM demodulation scheme, leads to

$$\eta = \frac{SNR_O}{SNR_I} = 2\alpha^2(\alpha + 1), \tag{6.79}$$

where $\alpha = \Delta_{PM}\sqrt{P_M}$ represents the PM modulation index.

Formulas 6.78 and 6.79 show that the demodulation gain increases with the square of the modulation index, for narrowband FM and PM, and with the third power of the index, for wideband FM and PM.

This demonstrates the importance of the modulation index to improve the quality of the angle modulated signal reception. It is worth to mention that the bandwidth increases in the same proportion as the modulation index, which imposes a tradeoff on the design of FM and PM transmitters. In order to improve reception the bandwidth has to be increased, but the telecommunication regulators limit the amount of bandwidth allowed to each operator.

Phase modulation presents a similar result for the demodulation gain, as well as, the same considerations regarding the tradeoff between power and bandwidth. But, wideband PM is not so popular as wideband FM, because it requires absolute phase deviation measurement, instead of detection modulo 2π, and this limits its usefulness for practical purposes (Gagliardi, 1988).

6.5 Angle Modulation with Digital Signal

The angle modulation with a digital signal, for the phase shift keying (PSK) case, can be described by the following equations

$$s(t) = A\cos(\omega_c t + \Delta_{PM} m(t) + \phi) \tag{6.80}$$

$$m(t) = \sum_{j=-\infty}^{\infty} m_j p(t - jT_b). \tag{6.81}$$

The constellation diagram of a PSK modulation scheme, with parameters $\Delta_{PM} = \pi/4$, $\phi = 0$ and $m_j = \{0, 1, 2, 3, 4, 5, 6, 7\}$, is illustrated in Figura 6.10.

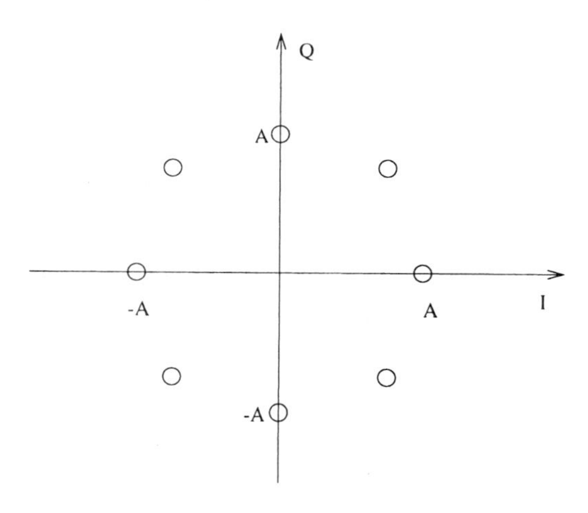

Figure 6.10. Constellation for a PSK signal.

The main feature of the PSK modulation is its carrier constant envelope, with the constellation symbols positioned on a circle. This gives a certain advantage for the scheme regarding multiplicative noise fading channels. The power of the modulated carrier is $P_S = A^2/2$.

An increase in the transmission rate, which implies the addition of new symbols, makes the PSK scheme more susceptible to noise, increasing the probability of error, because the symbols get closer in signal space. This modulation technique is not as efficient as the quadrature modulation to allocate symbols on the constellation diagram.

The probability of error for the coherent binary PSK is (Haykin, 1988)

$$P_e = \frac{1}{2}\text{erfc}\left(\sqrt{\frac{E_b}{N_0}}\right), \tag{6.82}$$

where, E_b is the pulse energy and N_0 represents the noise power spectral density.

For the coherent FSK, the probability of error has 3 dB penalty, compared to the PSK result (Haykin, 1988)

$$P_e = \frac{1}{2}\text{erfc}\left(\sqrt{\frac{E_b}{2N_0}}\right). \tag{6.83}$$

On the other hand, the noncoherent FSK has probability of error (Haykin, 1988)

$$P_e = \frac{1}{2}\exp\left(\frac{E_b}{2N_0}\right). \tag{6.84}$$

6.6 Problems

1 A signal, whose probability density function is given, frequency modulates a carrier. Consider a high modulation index and compute the power spectrum density of the modulated carrier, as well as its bandwidth.

$$p_M(m) = \frac{1}{2D}[u(m+D) - u(m-D)]$$

2 A PSK modulator has the following parameters: $\Delta_{PM} = \pi/2$, $\phi = \pi/4$, A $= 1$ and $m_j = \{0, 1, 2, 3\}$. Draw the constellation diagram for the modulation scheme. Compute the phase and amplitude resultants for the modulated carrier. What happens if synchronism is lost? Sketch a diagram that shows the result. Consider

$$s(t) = A\cos(\omega_c t + \Delta_{PM} m(t) + \phi)$$

and

$$m(t) = \sum_{j=-\infty}^{\infty} m_j p(t - jT_b).$$

3 Using the formula for the spectrum of the wideband frequency modulated carrier, compute the carrier power for the following power spectral density

$$S_S(\omega) = \frac{\pi A^2}{2\Delta_{FM}} \left[p_M \left(\frac{\omega + \omega_c}{\Delta_{FM}} \right) + p_M \left(\frac{\omega - \omega_c}{\Delta_{FM}} \right) \right]$$

4 Signal $m(t) = V \cos(\omega_M t + \varphi)$ modulates an FM and a PM carrier, respectively. Determine the relation between the respective frequency deviation indices to guarantee that the bandwidth is the same in both cases. Consider

$$S_S(\omega) = \frac{\pi A^2}{2\Delta_{PM}} \left[p_{M'} \left(\frac{\omega + \omega_c}{\Delta_{PM}} \right) + p_{M'} \left(\frac{\omega - \omega_c}{\Delta_{PM}} \right) \right]$$

where $p_{M'}(\cdot)$ is the probability density function of the modulating signal derivative.

5 A signal $m(t)$, whose power spectrum density is given, frequency modulates a carrier. Consider a low modulation index and compute the spectrum and bandwidth for the modulated signal.

$$S_M(\omega) = [u(\omega + \omega_M) - u(\omega - \omega_M)]$$

6 A signal, whose autocorrelation is given, modulates a carrier in phase. Assume narrowband modulation and compute the power spectrum density and bandwidth for the modulated signal.

$$R_M(\tau) = A^2[1 + e^{-|\tau|}].$$

7 A message signal, whose probability density function is given, is applied to the input of an FM modulator. Determine the output spectrum. Compute the bandwidth of the output signal in terms of the parameter V.

$$p_M(m) = \frac{1}{\pi\sqrt{V^2 - m^2}}$$

8 Using Carson's rule, compute the lower and upper limit for the modulated signal bandwidth, considering frequency and phase modulation and a message signal $m(t) = V \sin(\omega_M t + \varphi)$.

9 A Gaussian signal, whose autocorrelation function is given, modulates a carrier in frequency. Consider both the narrowband and the wideband cases to compute the modulated carrier bandwidth.

$$R_M(\tau) = \frac{\omega_M S_0}{\pi} \frac{\sin(\omega_M \tau)}{\omega_M \tau}.$$

10 Using Carson's rule, for a wideband frequency modulated carrier, determine the signal bandwidth, for a message signal input whose probability density function is given by

$$p_M(m) = r(m+1) - 2r(m) + r(m-1).$$

11 Signal $m(t)$ phase modulates a signal. Consider wideband modulation, equiprobable levels, and compute the spectrum of the modulated carrier. Determine the bandwidth of the message and modulated signal.

$$m(t) = \sum_{j=-\infty}^{\infty} m_j q(t - jT), \text{ for } q(t) = (1 - \frac{2|t|}{T}) \text{ e } m_j = \{A, -A\}.$$

12 Show that the following 4PSK signal is equivalent to a 4-QAM, comparing the amplitude and phase resultants, using $b(t)$ e $d(t)$. Draw the constellation diagram and discuss the advantages of each scheme if the number of symbols is increased.

$$s_{PSK}(t) = A\cos(\omega_c t + \theta(t) + \phi)$$

and

$$s_{QAM}(t) = b(t)\cos(\omega_c t + \phi) + d(t)\sin(\omega_c t + \phi),$$

for

$$\theta(t) = \sum_{k=-\infty}^{\infty} \theta_k p(t - kT_b),$$

where $\theta_k = \{-\pi/4, \pi/4, 3\pi/4, -3\pi/4\}$ and T_b represents the bit interval.

13 An FM signal $s(t)$ is transmitted through an additive channel, whose noise signal $n(t)$ has a power spectrum density $S_N(\omega) = N_0$. The received signal $r(t) = s(t) + n(t)$, is demodulated. What is the ratio between the modulating $m(t)$ signal power and the noise power resulting from the demodulation process?

$$s(t) = A\cos[\omega_c t + \theta(t) + \phi], \; \theta(t) = \Delta_{FM} \int_{-\infty}^{t} m(t)dt,$$

$$S_M(\omega) = M_0, \text{for } |\omega| \leq \omega_M.$$

14 How many voice channels can be inserted in a 6 MHz video channel using FM with modulation index $\beta = 4$ and a message signal whose bandwidth is 15 kHz? How many channels would be allocated if one uses the SSB or AM schemes?

15 Draw the constellation diagram for the following PSK signal, where $m(t) = \sum_{k=-\infty}^{\infty} m_k p(t - kT_b)$. Consider $m_k = \{0, 1, 2, 3\}$. Compute the phase and amplitude resultants.

$$s(t) = A\cos[\omega_c t + \frac{\pi}{2}m(t) + \frac{\pi}{3}]$$

Compare this system to the 4-ASK scheme.

16 Show, mathematically, that the 4-QAM and 4PSK signals, given as follows, can be expressed by the same formula and produce the same constellation diagram. Determine $b(t)$ and $d(t)$ as a function of A and $\theta(t)$.

$$s(t)_{QAM} = b(t)\cos(\omega_c t + \phi) + d(t)\sin(\omega_c t + \phi),$$

$$s(t)_{PSK} = A\cos(\omega_c t + \theta(t) + \phi).$$

17 Compute the mean value and the total power for the signal $X(t) = f(t)[\sin(\omega_o t + \theta) - \cos(\omega_o t + \theta)]$, where $R_F(\tau) = Be^{-\alpha\tau^2}$ and θ is a random variable, uniformly distributed in the interval $(0, 2\pi]$.

18 Considering that the random signals $X(t)$ and $Y(t)$, which modulate a QUAM carrier, are wide sense stationary and that $R_{XY}(\tau) = R_{YX}(\tau)$, show that the carrier autocorrelation, $R_S(\tau)$, is independent from the correlation between $X(t)$ and $Y(t)$.

19 The constellation diagrams shown in Figure 6.11 represent QAM and ASK carriers. Determine the average power for the transmitted signals, write the equations for the modulated carriers and compare both schemes in terms of probability of error, demodulator simplicity and symbol rate.

20 For the following modulation scheme, compute the autocorrelation, the power spectrum density and determine the condition to eliminate the carrier. Consider that $b(t)$ and $d(t)$ are uncorrelated, zero mean, stationary signals. Draw the corresponding diagrams.

$$s(t) = A\cos(\omega_c t + \Delta_{PM} m(t) + \phi),$$

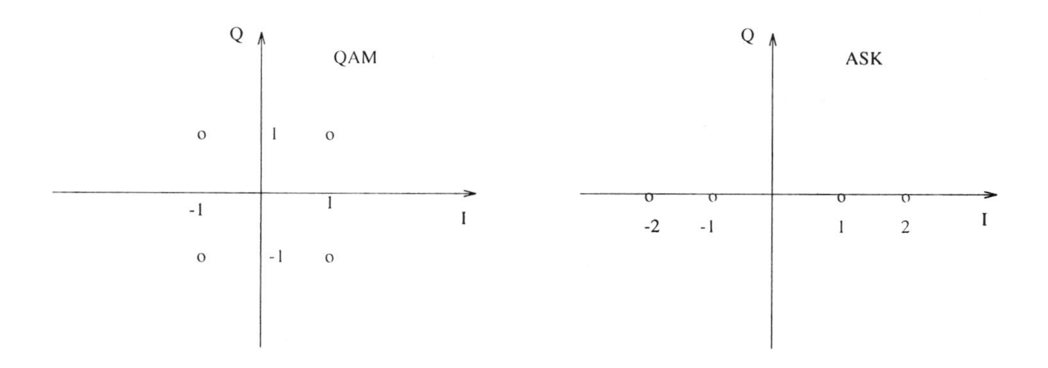

Figure 6.11. Constellation diagrams.

$$m(t) = b(t) + d(t)\cos(\omega_P t + \theta)$$

21 Given the following probability distribution for a signal which modulates a carrier in frequency, compute the power spectrum density, power and autocorrelation of the modulated carrier. Consider wideband modulation.

$$p_M(m) = \frac{1}{b}[1 - \frac{|m|}{b}], \ |m| \leq b.$$

22 A signal, whose probability density function is given as follows, phase modulates a carrier. Consider a high index of modulation and compute the power spectrum density of the modulated signal. Sketch the spectrum and explain the result.

$$p_M(m) = \frac{1}{\sigma_M\sqrt{2\pi}}e^{-\frac{(m-\eta)^2}{2\sigma_M^2}},$$

where η represents the modulating signal mean value and σ_M is its RMS value.

23 For the following scheme, compute the power spectrum density and the bandwidth of the modulated signal. Consider that $b(t)$ and $d(t)$ are uncorrelated, zero mean stationary signals. The modulation index is low. Also consider that $S_B(\omega) = S_D(\omega) = S_o[u(\omega + \omega_M) - u(\omega - \omega_M)]$, $\omega_P > 2\omega_M$ and P_Θ is a given parameter. Draw the corresponding diagrams.

$$s(t) = A\cos(\omega_c t + \Delta_{FM}\int_{-\infty}^{t} m(t)dt + \phi),$$

$$m(t) = b(t) + d(t)\cos(\omega_P t + \theta)$$

24 A signal, whose autocorrelation function is given as follows, modulates a carrier in phase. Consider narrowband modulation in this case and compute the autocorrelation, power spectrum density and bandwidth for the modulated carrier.

$$R_M(\tau) = [1 - \frac{|\tau|}{T}]\cos\omega_M\tau,\ |\tau| \leq T,\ \omega_M \geq \frac{2\pi}{T}.$$

25 A signal, whose characteristic function is given in the following, frequency modulates a carrier. Consider a high modulation index and compute the autocorrelation and power spectrum density and bandwidth of the transmitted signal. Sketch the spectrum and autocorrelation. What happens to the spectrum if the parameter a is changed?

$$P_M(\omega) = \frac{\sin\omega a}{\omega a}.$$

26 A given signal has the following probability density function and modulates a carrier in frequency, with low modulation index. Calculate the autocorrelation and power spectrum density and bandwidth of the modulated carrier. Draw the associated diagrams.

$$p_M(m) = \frac{\beta}{4}e^{-\beta|m+\alpha|} + \frac{\beta}{4}e^{-\beta|m-\alpha|}.$$

27 Compute the power spectrum density and bandwidth for a carrier which is modulated by a signal with the following probability density function. Draw the spectrum and analyze what happens if the mean and RMS values of the signal are changed.

$$p_M(m) = \frac{1}{\sigma_M\sqrt{2\pi}}e^{-\frac{(m-\eta)^2}{2\sigma_M^2}},$$

where η represents the modulating signal mean value and σ_M is its RMS value.

28 A signal has power spectrum density $S_M(\omega) = S_M\omega^2$, for $-\omega_M \leq \omega \leq \omega_M$, and modulates a carrier in frequency, using low modulation index. The following autocorrelation is obtained

$$R_S(\tau) = \frac{A^2}{2}\cos(\omega_c\tau)[1 - R_\Theta(0) + R_\Theta(\tau)].$$

Determine the power which is contained in the carrier spectral line. Compute the power of the remaining spectrum. Draw the diagrams for the modulating signal and modulated carrier spectra.

29 Two PSK signals are transmitted. The first one is modulated by a signal whose symbols are $m_j \in \{0, 1\}$, the phase deviation index is π the carrier amplitude is A. The second signal has symbols $m_k \in \{0, 1, 2, 3\}$, phase deviation index $\pi/2$ and amplitude B. Determine the power for each modulated signal. Sketch the constellation diagrams for both signals. Which values of A and B give the same bit error probability for both signals? Which signal transmits more information per Hz, if both have similar bit interval T_b?

30 The formula for the autocorrelation of an angle modulated signal is given as follows

$$R_S(\tau) = \frac{A^2}{2} P_\Omega(\tau) \cos \omega_c \tau.$$

For a characteristic function $P_\Omega(\tau) = \cos a\tau$, compute the power spectrum density for the modulated signal. Sketch the diagram. Determine the signal bandwidth.

31 Two signals are transmitted, using ASK and PSK modulation. The ASK signal has symbols $a_j \in \{0, 1, 2, 3\}$ and amplitude A. The PSK signal, has symbols $m_k \in \{0, 1, 2, 3\}$, phase deviation index $\pi/2$ and amplitude B. Determine the power for each modulated signal. Draw the constellation diagrams for both signals. Which values of A and B give the same bit error probability for both signals? Which signal transmits more information per Hz, if both have similar bit interval T_b? Which one is more efficient in terms of transmitted energy?

32 A PM signal, $s(t) = A\cos(\omega_c t + \Delta_{PM} m(t) + \phi)$, where $m(t)$ has maximum frequency ω_M, is modulated in narrowband. Show that this signal can be approximated by the following

$$s(t) = C\cos(\omega_c t + \phi) + Dm(t)\sin(\omega_c t + \phi).$$

Determine parameters C e D. The signal is demodulated by a receiver which presents a frequency shift $\Delta\omega$ at the local oscillator. Analyze what happens if: *a)* $\Delta\omega = \omega_M$; *b)* $\Delta\omega = 2\omega_M$. After the mathematical analysis, use diagrams to explain the results.

33 Write the expression in time for the modulated signal shown in Figure 6.12. Explain which type of modulation is used, its advantages and disadvantages. Compute the modulated signal power.

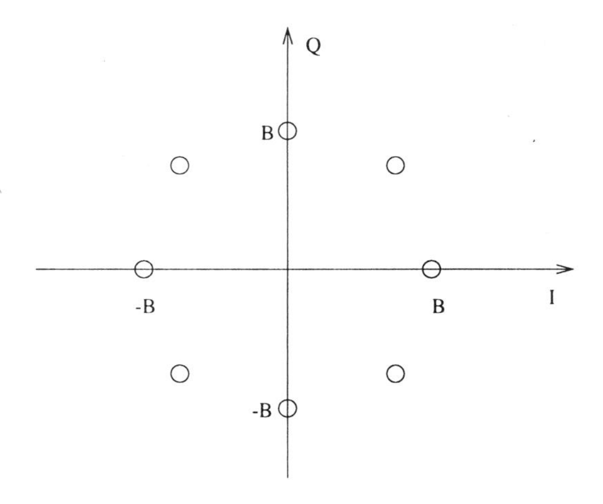

Figure 6.12. Signal PSK.

34 For a mono TV transmission the maximum frequency deviation for the audio signal is 25 kHz. The modulating signal occupies a bandwidth of 15 kHz. What is the modulation index? Compute the demodulation gain.

35 The maximum frequency deviation for an FM signal is 75 kHz. The audio program has bandwidth 15 kHz. What is the modulation index? Compute the signal to noise ratios and the demodulation gain. Consider a carrier peak value of 2 V and a noise power spectrum density $S_N(f) = 1$ nW/Hz.

36 A digital signal $x(t)$, with levels A and $-A$, phase modulates a carrier and is transmitted through an additive white Gaussian channel. The noise $n(t)$, has variance σ_N^2. Assume equal probabilities for the symbols $P\{A\} = P\{-A\} = 0.5$ and determine the received signal symbol error probability.

Chapter 7

PROPAGATION CHANNELS

The objective of this chapter is to compare the main types of wireless communication channel models in respect to capacity, transmission speed, operation characteristics, cost and operability.

The non-guided propagation channel is the free-space between two antennas, where one antenna is used to transmit a signal and the other is used to receive it. These channels can be classified as terrestrial wave channels, space wave channels, sky wave channels and outer space channels(Gagliardi, 1988).

7.1 Basic Concepts

When comparing communication channels two fundamental concepts are those of bandwidth and frequency. In practice the rate at which a source generates information is related to both concepts. The channel bandwidth is directly related to the rate at which information can be sent through the channel, and the channel operating frequency range is related to the carrier modulating frequency (Schwartz et al., 1966). The choice for the channel that best adapts to these basic concepts is of fundamental importance in the design of a communication system. The choice must be made such that it takes into account the noise introduced by various system components, trying to minimize this noise. It must also be taken into consideration the use of repeaters placed along the signal path in order to amplify the signal level, keeping the signal magnitude at an adequate level so as to compensate for transmission losses (Schwartz, 1970).

The design of a communication system and the choice of channel must take into consideration the fact that the transmission of a large amount of information in a short time interval requires a wide bandwidth system to accommodate the signals. The width of a frequency band appears thus as a fundamental limitation. When signal transmission occurs in real time the design must allocate an adequate bandwidth for the system. If the system bandwidth is insufficient it may be necessary to reduce the rate of information transmission, thus causing an increase in transmission time.

It must also be taken into account that the design of equipment design not only considers bandwidth required but also the ratio (fractional band) between bandwidth and the center frequency in the frequency band. The modulation of a wideband signal by a high frequency carrier reduces the fractional band and consequently helps to simplify equipment design. Analogously, for a given fractional band defined by equipment consideration, the system bandwidth can be increased almost indefinitely by raising the carrier frequency. A microwave system with a 5 GHz carrier can accommodate 10,000 times more information in a given time interval than a system with a 500 kHz carrier frequency. On the other hand, a laser with frequency $5 \cdot 10^{14}$ Hz has a theoretical information carrying capacity exceeding that of the microwave system by a factor of 10^5 or, equivalently, can accommodate 10 million TV channels. For this reason communication engineers are continually searching for new sources of high-frequency carriers, as well as for channels that are best matched to such carriers in order to provide wider bandwidths.

As a final comment in this section, for a given communication system and a fixed signal to noise ratio, there is an upper limit defined for the rate of information transmission with a prescribed measure of quality. This upper limit, called the channel capacity, is one of the fundamental concepts of information theory. Being capacity a finite quantity, it is possible to say that the design of a communication system is in a certain way a compromise solution involving transmission time, transmitted power, bandwidth and signal to noise ratio. All those depend also on the choice of the proper channel, a fact that imposes even more strict constraints due to technological issues (Schwartz, 1970).

7.2 Non-guided Channels

The main characteristics of non-guided channels are the following.

- Propagation of an electromagnetic wave from the transmitting antenna through free space until it reaches the receiving antenna.
- Irradiated power loss with the square of distance in free space.
- Propagation in the terrestrial atmosphere, causing greater losses due to particles, water vapor and gases, of dimension comparable to the

wavelength of the transmitted radiation; attenuation with rain, snow and other weather factors and attenuation due to obstacles.

The free space non-guided channel can still be subdivided into four categories, according to the wave type, antenna elevation angle, propagation characteristics of the tropospheric and ionospheric layers and the type of link employed, as follows.

- Terrestrial wave (free space and earth surface);
- Tropospheric wave;
- Sky wave;
- Outer space.

Terrestrial Wave

The terrestrial wave can be decomposed into a surface wave and a space wave.

The surface wave propagates close to the Earth surface in a manner similar to the wave that propagates in a transmission line. The surface wave follows the Earth surface contours, causing the electric field to bend in respect to the Earth surface, due to signal power losses on the Earth soil due to a relative permittivity and a finite surface conductivity. Mobile communication systems, for example, benefit from the surface wave for signal transmission.

The space terrestrial wave is due to the diffraction of the wave in the propagation media, consisting of a direct wave and a refracted wave. This type of wave is characteristic of simple system links or radio links, as illustrated in Figure 7.2.

The generation of the space terrestrial wave can also be due to wave diffraction in obstacles called "knife's edge" the dimension of which is close to the transmitted wavelength, creating the so called "silence zone", responsible for signal loss within this zone.

The reflected (diffracted) wave can cause problems in reception because of the phase difference introduced by the lengthening of the propagation path. For certain links it is desirable to avoid obstacles that otherwise would preclude a line-of-sight link. For that purpose passive reflectors are introduced along the signal path.

Tropospheric Wave

Figure 7.3 illustrates a tropospheric wave propagation, where the wave is reflected or refracted in the troposphere, according to the variation of the refraction index of the layers in the latter. It turns out that the

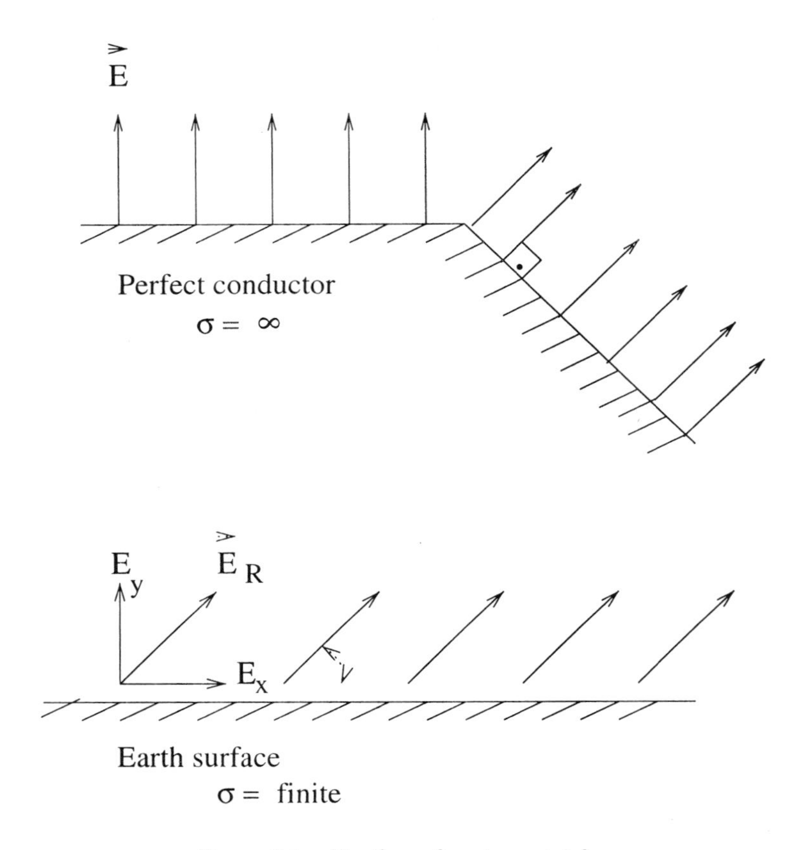

Figure 7.1. Earth surface terrestrial wave.

signal path is bent due to the refraction index non-homogeneity. When a statistical model is applied to the tropospheric channel the Rayleigh amplitude distribution results, as seen in Section 7.6. Transmission through a tropospheric channel has been discussed in Chapter 6.

Sky Wave

The sky wave uses the ionosphere as the natural ionized media for the reflection of radio waves within certain a frequency range and a certain antenna elevation angle, as illustrated in Figures 7.5 and 7.6.

Consider the behavior of an electromagnetic wave penetrating an ionized gas. The ionosphere is the region which extends approximately between 70 and 450 km above sea level and in which the constituent gases are ionized due to the Sun ultraviolet radiation. At such altitudes air pressure is so low that free electrons and ions can exist for relatively

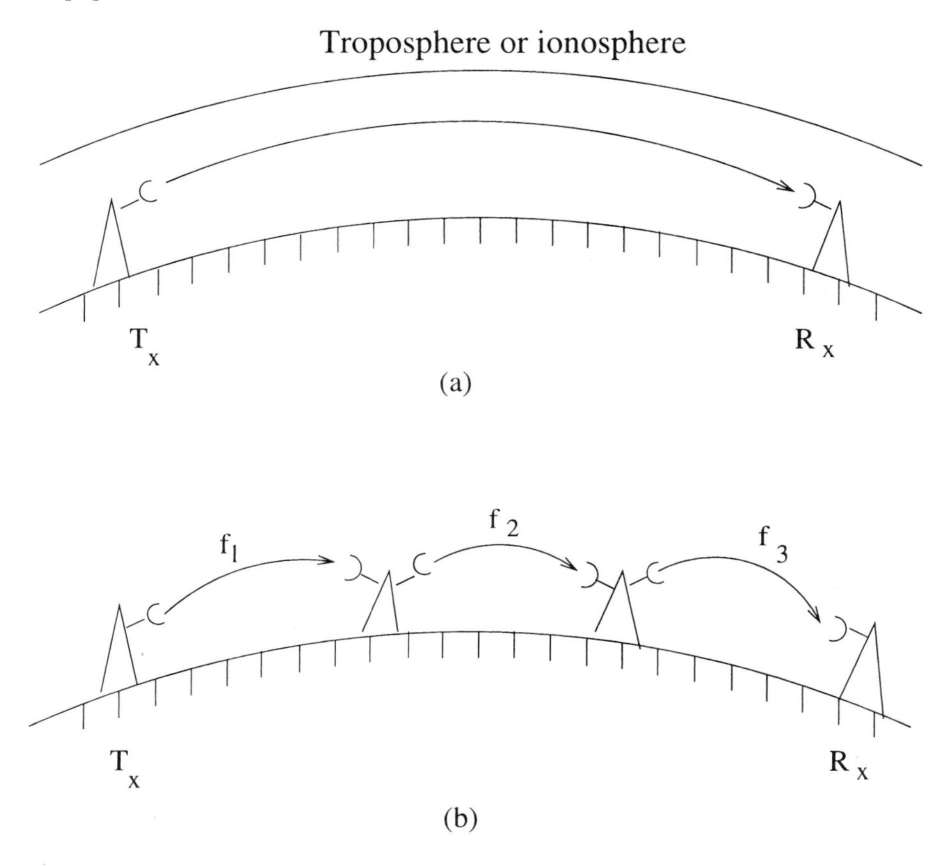

Figure 7.2. Space terrestrial wave. a) Simple system, tropospheric channel. b) Radio link.

long periods of time without recombining to form neutral atoms. The ionization density however does not vary uniformly with altitude. On the contrary, there are four layers designated as layers D, E, F1 and F2, with distinct vertical ionization densities, which occur in this order at well defined altitudes. In each of these layers the ionization varies with the time of the day, with the season of the year, with the geographical region, with the eleven year cycle of the Sun spots and with meteor trails, thus causing irregularities in the ionosphere. Overall, the ionosphere tends to reflect back to Earth waves that originate at ground level and hit the ionosphere, according to some incidence angle.

The propagation of signals by tropospheric wave or by sky wave characterizes communication systems using for wave scatter the ionosphere or the troposphere.

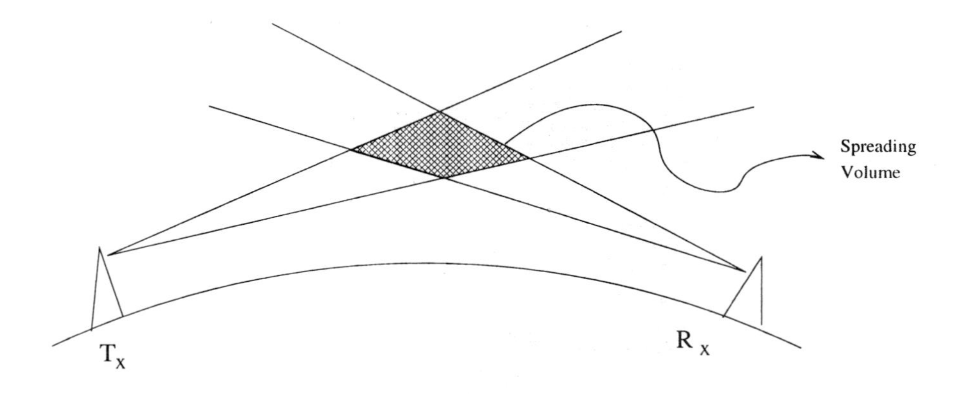

Figure 7.3. Tropospheric wave.

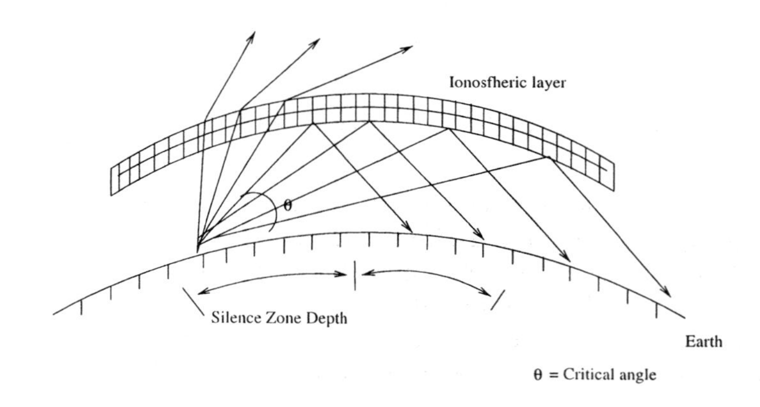

Figure 7.4. Ionospheric reflection and refraction.

The noise received by the antenna comes from many distinct sources. The antenna noise is usually characterized by the antenna temperature according to the frequency of operation, since the latter is an important factor to be considered in this type of propagation. It is also possible to obtain curves of minimum and maximum atmospheric noise, also known as static precipitation, originated by the natural occurrence of electrical discharges such as lightening. Figure 7.7 shows the curve of absolute temperature for galactic noise. de

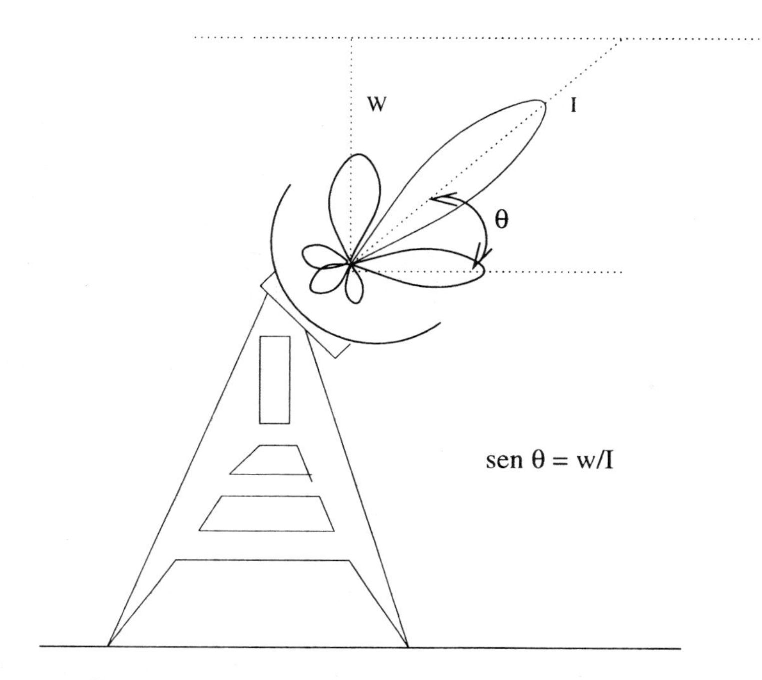

Figure 7.5. Definition of antenna elevation angle θ.

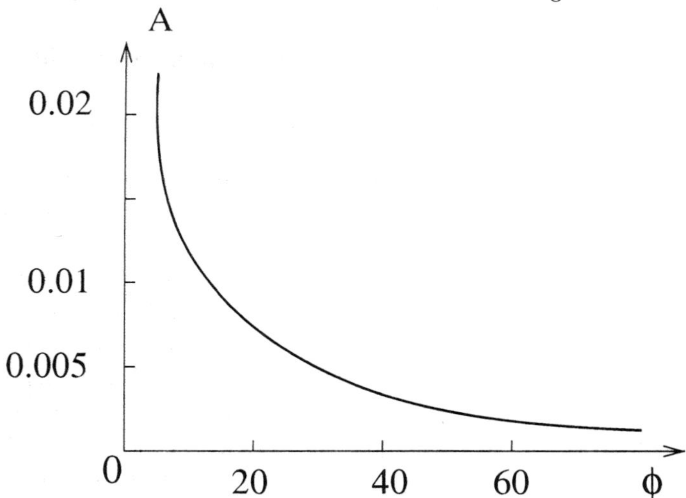

Figure 7.6. Atmospheric absorption as a function of elevation angle.

Between 50 and 500 MHz, the atmospheric noise becomes negligible, and consequently the cosmic noise, or galactic noise, begins to predominate. This noise exists due to the emission of cosmic sources belonging

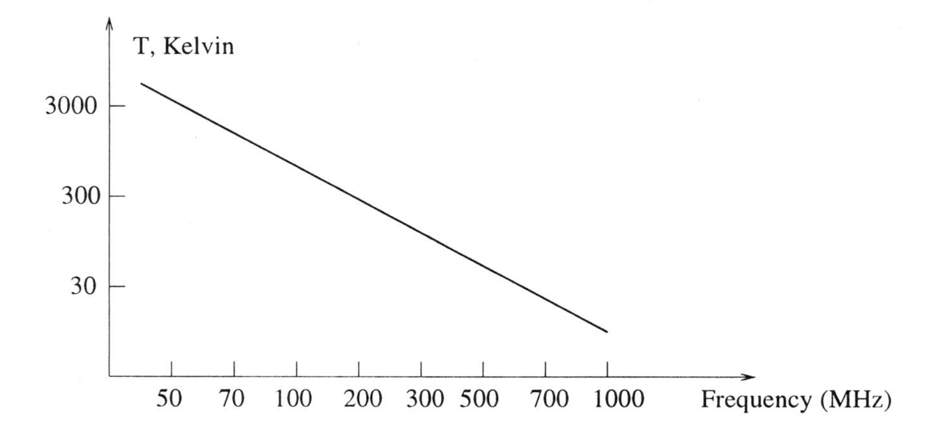

Figure 7.7. Galactic noise temperature, in degrees Kelvin.

to our galaxy. Above 1 GHz, the noise contribution due to the phenomenon of atmospheric absorption of atmospheric oxygen becomes important. Heating of the lower atmosphere, where molecular oxygen is found, provides conditions for the occurrence of this type of noise, the absorption peaks of which occur between 22 and 60 GHz.

The atmosphere absorption curve (due to oxygen and other gases) is shown in Figure 7.8 as a function of frequency. The combination of the two types of noise mentioned earlier constitutes what is known as sky noise. An important detail in the curve shown in Figure 7.9 is in the region situated between 1 and 10 GHz, in which the sky oriented antenna receives a noise power smaller than it would receive if it was working over higher or lower frequencies. This lower power noise level is particularly advantageous in satellite communication systems due to the low power levels of the processed signals (Gagliardi, 1988).

Outer Space

In this type of propagation the radiation elevation angle with respect to the earth surface is such that the signal manages to go through the ionospheric layer, penetrating the outer space. This propagation mode is usually employed for satellite transmission.

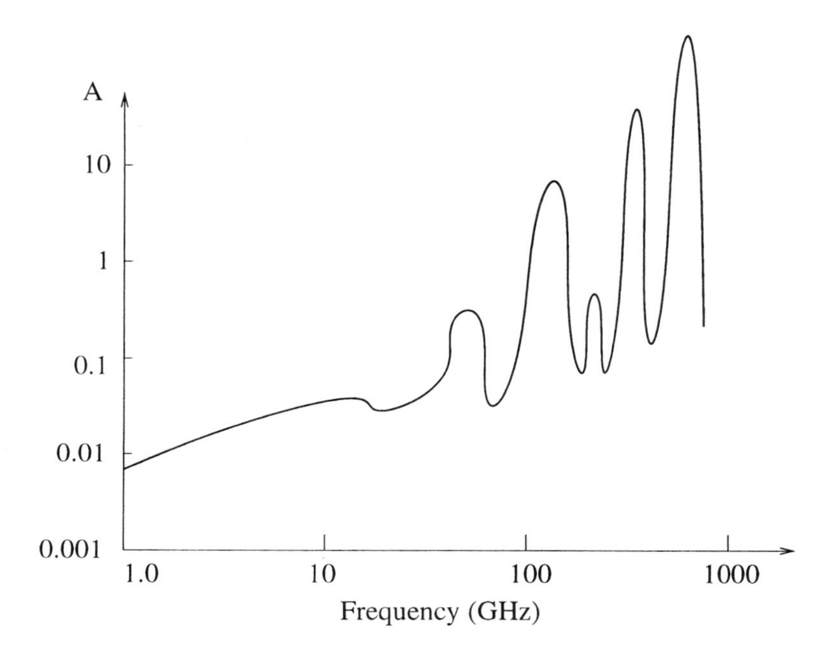

Figure 7.8. Atmosphere absorption curve, as a function of frequency.

The adequate use of each channel is a function of the system design for that particular end, as a function of its characteristics; therefore, each communication channel has its importance due to some specific application, each one having advantages and disadvantages. The system design must take into account all those aspects. Usually propagation requirements, bandwidth and cost dictate the solution.

7.3 Effects on the Transmitted Signal

The transmitted signal suffers various effects when traversing the channel. The main effects are the following.

- **Filtering** – This type of effect tends to reduce the available bandwidth of the modulated carrier, since filtering affects the shape of the modulated waveform causing also phase distortion;

- **Doppler** – This effect makes the carrier frequency at the reception to differ from that at the transmitter. This effect is due to the frequency

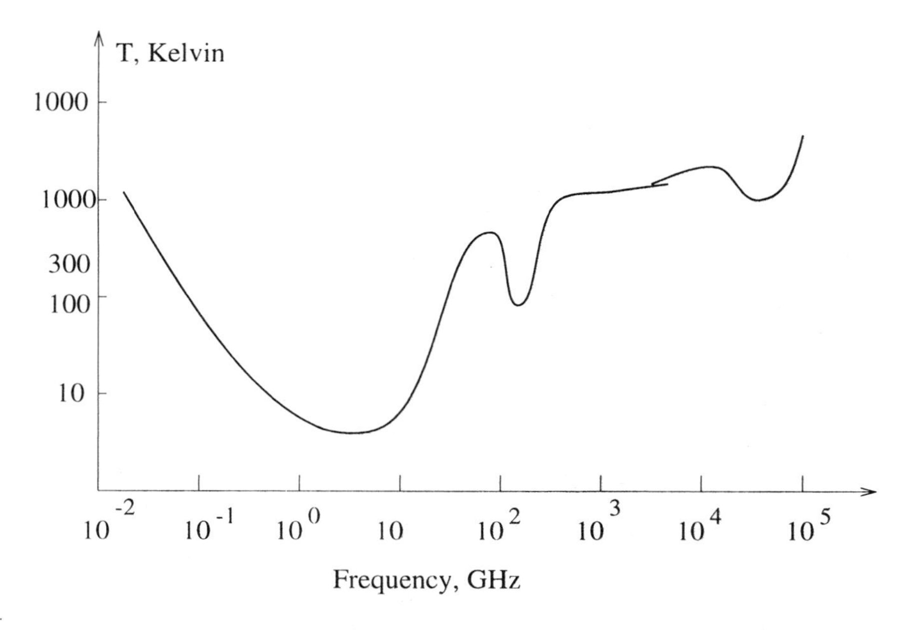

Figure 7.9. Typical sky temperature curve.

deviation provoked by the relative movement between transmitter and receiver. It affects reception of the synchronization signal;

- **Fading** – This is the name given to the phenomenon causing random variations in the received signal magnitude, in time. This variation has as a reference the value of the received electric field in free space. The causes for fading are found in the propagation media or, in other words, such phenomenon would not be present in the communication links if it were not for the existence between antennas of a media subject to changes in its characteristics;

- **Multipath** – The received signal is the resultant of the sum of a direct ray between antennas and other rays which follow paths distinct from that of the direct ray. These distinct paths, characterized by multiple paths, originate from refractions and reflections (even if of small intensity) resulting from irregularities in the atmosphere dielectric constant with altitude. The energy transported by means of such multiple paths is, in general, much less than that associated with the main beam. However, when for some reason (partial obstruction, interference caused by reflection in the terrain, etc.) the main beam suffers a considerable attenuation, the energy received by means of multiple paths then plays an important role, giving origin to non-negligible interference phenomena.

7.4 The Mobile Communication Channel

Radio propagation in a mobile communication environment, outdoors or indoors (buildings) is certainly a complex phenomenon. Multipath dispersion caused by the various structures in a building causes a series of echoes for each transmitted pulse. The performance of mobile communication systems is affected by the effect of multipaths of short duration and by large scale path losses. These problems tend to introduce errors in transmission and to reduce the area of signal coverage (Hashemi, 1991).

Channel characterization is an important topic in the investigation of the mobile channel. The usual way to address this theme is as follows. Modeling attenuation as a function of frequency, distance and location; measurement of the time variation of amplitude and phase; use of correlation to assess potential improvements with the the technique of diversity and to establish the channel impulse response, so as to determine the delay spread.

On the other hand, a channel probabilistic analysis is necessary to provide the decoder with essential information about the state of the channel, so as to make communication more effective. Typical methods to combat fading and interference in a channel include the use of burst error correcting codes and interleaving. The choice of the best technique will depend on the channel characteristics.

7.5 Multipath Effects

Fading as caused by multipath, due to reflections on buildings and natural obstacles, for outdoor mobile communication, or on walls, roof and floor, for indoor mobile communication, provokes a series of dips in the received signal spectrum. The pattern thus received represents the signature of selective fading, which can be fairly well predicted by means of an accurate channel analysis.

Fading can also be caused by the movement of people, equipment or cars, such that the resulting interference is the combined effect of multiple access interference (MAI), thermal and impulse noise provoked by electrical equipment. Fading due to transmitter or receiver movement can predominate, in the case of indoor communication, when the indoor objects move relatively slow (Rappaport, 1989).

Many experiments have been performed in the UHF band, between 300 MHz and 3 GHz (Bultitude, 1987), (Lafortune and Lecours, 1990a).

This is due in part to the authorization granted in USA, by the Federal Communications Commission (FCC), for operation with spread spectrum in the 900 MHz, 2,400 MHZ and 5,725 MHz bands. This is also due to Japan for having chosen the 400 MHz and 2,450 MHz bands for indoor mobile systems (Rappaport, 1989), (Newman Jr., 1986). These frequency bands are usually used in low capacity voice communication systems, since this application presents less stringent transmission requirements. However, in order to provide enough capacity to cope with the needs of digital transmission systems, further research is needed into the EHF band.

In fact, both the microwave and millimeter wavebands are adequate for data transmission at rates above 10 Mbits/s and coverage of micro or picocells. These frequencies present a series of advantages, usually related to their dual behavior in terms of propagation. They combine properties of UHF frequencies and those of infra-red light (IR), i.e., diffusion, reflection, refraction, occupation of spaces and are blocked by large objects. This characteristic allows the signal to fill in the cell space almost completely, with a minimum of signal spillage over neighboring cells, in a manner to provide an adequate company environment, with micro or picocells defined by natural indoor barriers, i.e., floor and walls. Mutual interference with existing equipment is normally kept at a minimum level, due to the low power levels employed or due to the high frequencies in use. Although interference in indoor mobile communication due to industrial equipment is significant in HF and VHF, it drops quickly above 1 GHz (Rappaport, 1989). Operations at high frequencies allow also the use of much smaller antennas.

Nowadays it is accepted that the indoor mobile channel varies in time and space and is characterized by high path losses and sudden changes in average signal level. The time variations in the signal statistics are produced mainly by the movement of people and equipment inside a building.

Propagation via multipath is a very often used expression to describe the propagation characteristics in the radio channel. In particular, a signal propagates through different paths in the channel, where each path presents an associated gain, phase and delay. Multipath signals recombine in the receiver in a manner that makes the received signal to be a distorted version of that transmitted.

There are three different effects: reflection, diffraction and scattering. Diffraction occurs when the signal rays bend around obstacles; reflection occurs when the rays collide with hard and even surfaces, and scattering occurs when a ray splits into various rays, after an impact with a hard and uneven surface.

The manner by which multipath signals combine at the receiver leads to the particular type of distortion in the received signal. Two types of recombination may occur, i.e., paths with identical delay will be amplified or attenuated, depending whether the respective signals are in phase or not. This type of distortion is known as narrowband fading and as a consequence causes bursts of errors. Recombination which do not occur simultaneously or paths with different delays will produce echoes. This type of distortion, called wideband fading, causes pulse spreading which leads to intersymbol interference.

The effects of fading observed in a mobile channel can be grouped into three different categories: spatial fading, time fading and frequency selective fading. Spatial fading is characterized by a variation of signal intensity as a function of of distance and occurs in two ways: slow fading and fast fading. Slow fading is characterized as the mean value of the set of signal fluctuations whereas fast fading is characterized by the variation of the signal values around its average value.

The following equation describes slow fading as a function of distance

$$P = Ad^{-n} \tag{7.1}$$

where P is the received signal power, A is a constant gain, d is the distance between the transmitter and the receiver and n is the attenuation factor.

For a perfect outdoor channel (free space) the attenuation factor is $n = 2$, however for an indoor channel (building interior) a strong correlation exists between n and the channel topography. For example, in roomy environments and corridors $n < 2$, indicating that signal intensity is greater than that obtained in free space. This occurs due to a phenomenon called channelling, which produces a kind of waveguide containing the signal. Inside rooms or offices $n > 2$ is observed. The signal intensity is smaller than that in free space. This is due to absorption caused by obstacles such as furniture and half walls.

Besides channelling and absorption it was demonstrated that the parameter n depends on three other parameters (Lafortune and Lecours, 1990b): the distance between the transmitter and the nearest wall, the distance between the transmitter and any door, and whether the doors are open or closed. Due to the gradual drop in signal level with distance, the characteristics of selective fading are used to calculate coverage areas and to determine the location of sites for placing cells.

Similar to slow fading, fast fading depends also on the terrain topography. Statistical analysis shows that the variations in fast fading are usually of the Rayleigh, Rice, or log-normal type, depending on the terrain topography (Rappaport, 1989). These findings are consistent with what would be expected. That is, if a direct signal is dominant then fading tends to be of the Rice type. On the other hand, if there is no direct signal ray, or whether the signal is dominated by random multipath, then fading will be of the Rayleigh type.

The second type of fading is time fading, which is characterized by the presence of a signal intensity variation, measured at a particular fixed frequency, as a function of time. Other names often employed are narrowband fading, flat fading and frequency non-selective fading. The time fluctuations in an office environment are in general bursty, while variations in a factory environment are more continuous.

One of the causes for time fading is the physical movement inside the channel, i.e., mobile entities can block signal paths and absorb part of the signal energy temporarily, in a manner so as to create momentary fading. In an outdoor channel rain, water vapor, aerosols and other substances are energy absorption agents. In the indoor channel people are the primary absorption agents.

The other cause of time fading is the time-varying nature of the channel. Contrasting with the primary source of time fading, physical movement of entities in the channel, the second source can be seen as movement of the channel itself. This movement is caused by changes in propagation characteristics, resulting from changes in room temperature, changes in relative humidity, doors opening and doors closing. Movement inside the channel has a dominant effect on time fading. However, channel movement can affect long term statistics of the channel fluctuations.

The third type of fading, called frequency selective fading, is characterized by a variation on signal intensity as a function of frequency, also known as wideband fading. Frequency selective fading is the dual of time fading. In time fading the time-varying signal is measured at a fixed frequency while in frequency selective fading the frequency-varying signal is measured in a fixed time instant.

The main cause of frequency selective fading is multipath propagation. At certain frequencies, a combination of multipath signals provokes a reduction in the received signal level, for out of phase signals, and an increase in signal level for in-phase signals. A received signal resulting from a combination of a few components in general will show deep fades, while a signal resulting from a combination of many components in general will show shallow fades.

7.5.1 Statistical Modeling of the Mobile Channel

A complete description of the mobile communication channel would be too complex and beyond hope. Models for the mobile channel must take into account unwanted effects like fading, multipath as well as interference. Many models have been proposed to represent the behavior of the envelope of the received signal.

A very welcome model to extract statistics of the amplitude of signals in an environment subject to fading is provided by the Rayleigh distribution (Kennedy, 1969). This distribution represents the effect of the amplitude of many signals, reflected or refracted, reaching a receiver, in a situation where there is no prevailing component or direction (Lecours et al., 1988). The Rayleigh probability density function is given by (Proakis, 1990)

$$p_X(x) = \frac{x}{\sigma^2} e^{-\frac{x^2}{2\sigma^2}} u(x), \tag{7.2}$$

with average $E[X] = \sigma\sqrt{\pi/2}$ and variance $V[X] = (2 - \pi/2)\sigma^2$. The associated phase distribution, in this case, is considered uniform in the interval $(0, 2\pi)$. It is possible to approximate with a Rayleigh distribution the amplitude distribution of a set of only six waves with independently distributed phases (Schwartz et al., 1966).

Considering the existence of a strong direct component in the received signal, in addition to multipath components, the Rice distribution is used in this case to describe fast envelope variations in this signal. This line-of-sight component reduces the variance of the signal amplitude distribution, as its intensity grows in relation to the multipath components (Lecours et al., 1988) (Rappaport, 1989). The Rice probability density is given by

$$p_X(x) = \frac{x}{\sigma^2} e^{-\frac{x^2+A^2}{2\sigma^2}} I_o\left(\frac{xA}{\sigma^2}\right) u(x), \tag{7.3}$$

where $I_o(\cdot)$ is the modified Bessel function of order zero and A denotes the signal amplitude. The corresponding variance, for a unit average value, is $V[X] = A^2 + 2\sigma^2 + 1$. The average or expected value is given by

$$E[X] = e^{-\frac{A^2}{4\sigma^2}} \sqrt{\frac{\pi}{2}}\sigma \left[\left(1 + \frac{A^2}{2\sigma^2}\right) I_o\left(\frac{A^2}{4\sigma^2}\right) + \frac{A^2}{2\sigma^2} I_1\left(\frac{A^2}{4\sigma^2}\right)\right], \tag{7.4}$$

where $I_1(\cdot)$ is the modified Bessel function of order one.

The term $A^2/2\sigma^2$ is a measure of the fading statistics. As the term $A^2/2\sigma^2$ increases, the effect of multiplicative noise or fading becomes less important. The signal pdf becomes more concentrated around the main component. The remaining disturbances show up as phase fluctuations.

On the other hand, signal weakening can cause the main component not to be noticed among the multipath components, originating thus the Rayleigh model. An increment of A, with respect to the standard deviation σ makes the statistics to converge to a Gaussian distribution with average value A (Schwartz, 1970).

Under the conditions giving origin to the Rice distribution, it makes no sense to assume a uniform phase probability distribution. The joint amplitude and phase distribution, which originate the Rice model for amplitude variation, is given by

$$p_{X\Theta}(x,\theta) = \frac{xe^{-A^2/2}}{2\pi\sigma^2} e^{-(x^2-2xA\cos\theta)/2\sigma^2} u(x). \tag{7.5}$$

Integrating the above expression for all values of x, the marginal probability distribution θ for the phase results

$$p_\Theta(\theta) = \frac{e^{-s^2}}{2\pi} + \frac{1}{2}\sqrt{\frac{s^2}{\pi}} \cos\theta \; e^{-s^2\sin^2\theta}[1 + 2(1 - Q(s/\sqrt{2}))\cos\theta], \tag{7.6}$$

where the function $Q(x)$ is defined in the usual manner

$$Q(x) = \frac{1}{\sqrt{2\pi}} \int_x^\infty e^{\frac{-y^2}{2}} dy. \tag{7.7}$$

The signal to noise ratio is an auxiliary parameter given by $s^2 = A^2/2\sigma^2$. Expression (7.6) generates a curve with a bell shape for high values of the signal to noise ratio. For $A = 0$ this probability distribution converges to a uniform distribution (Schwartz, 1970).

Another distribution that has found application for modeling fading in multipath environments is the Nakagami pdf. This distribution can be applied in the situation where there is a random superposition of random vector components (Neal H. Shepherd, Editor, 1988). The Nakagami probability density function is given by

$$p_X(x) = \frac{2m^m x^{2m-1}}{\Gamma(m)\Omega^m} e^{-\frac{mx^2}{\Omega}} u(x) \tag{7.8}$$

where $\Omega = P_X$ denotes the received signal average power and $m = \Omega^2/E[(X^2 - E^2[X])^2]$ represents the inverse of the normalized variance of X^2. The parameter m is known as the distribution modeling factor and can not be less than $1/2$. It is easy to show that Nakagami's distribution contains other distributions as particular cases. For example, for $m = 1$ the Rayleigh distribution is obtained.

Finally, the log-normal distribution is used to model certain topographic patterns which appear due to non-homogeneities in the channel, or due to transmission in densely packed spaces or spaces with obstacles (Hashemi, 1991) (Rappaport, 1989).

The log-normal distribution is represented by the expression

$$p_R(r) = \frac{1}{\sigma r\sqrt{2\pi}} e^{-\frac{(\log r - m)^2}{2\sigma^2}} u(r) \tag{7.9}$$

has average value $E[X] = e^{\sigma^4/2+m}$ and variance $V[X] = e^{\sigma^4+m}(e^{\sigma^4} - 1)$, and can be obtained directly from the Gaussian distribution by means of an appropriate transformation of variables.

The sequence of arrival times for the various paths $\{\sigma_k\}$ assumes in general a Poisson distribution, given that such sequences are completely random. The Poisson probability distribution function is given by

$$P_k(t) = \frac{(\lambda t)^k}{k!} e^{-\lambda t} \ , \ k = 0, 1, 2, \ldots \tag{7.10}$$

where k represents the number of events counted in the given time interval and λ can be interpreted as the mean rate of occurrence of events (Blake, 1987). The expected value of the Poisson distribution and its variance are both equal to λt.

Exhaustive measurements performed in factory environments led to the conclusion that the time interval between consecutive arrivals $\{t_i = \sigma_i - \sigma_{i-1}\}$ obeys a Weibull distribution with pdf given by (Hashemi, 1991) (Rappaport, 1989)

$$p_T(t) = \beta\alpha t^{\beta-1} e^{-\alpha t^\beta} u(t), \tag{7.11}$$

where $\alpha, \beta > 0$ are parameters of the distribution. For $\beta = 1$ the distribution converges to the Exponential distribution (Blake, 1987).

The Weibull distribution appears when results of radio propagation measurements are plotted on a flat surface with scales adjusted in a manner that the plot of the Rayleigh distribution appears as a straight line with slope equal to -1 (Neal H. Shepherd, Editor, 1988). Propagation measurements in medium size buildings, however, indicate a tendency of echoes arriving in groups (Saleh and Valenzuela, 1987). This effect is attributed to the building structure, whereby multipath components arriving in groups are due to reflection from objects in the space next to the transmitter or next to the receiver.

The average $E(T)$ and the variance $V(T)$ of the Weibull distribution are given by (Leon-Garcia, 1989)

$$E[T] = \alpha^{-1/\beta}\Gamma(1 + \frac{1}{\beta}) \tag{7.12}$$

$$V[T] = \alpha^{-2/\beta}\{\Gamma(1 + \frac{2}{\beta}) - [\Gamma(1 + \frac{1}{\beta})]^2\}. \tag{7.13}$$

7.5.2 The Two-Ray Model of the Mobile Channel

The two-path model of the mobile channel is based on the diagram shown in Figure 7.10, in which a base station, with an antenna height h_1, transmits to the mobile terminal the antenna height of which is h_2. The distance between the two stations is d (Lee, 1989).

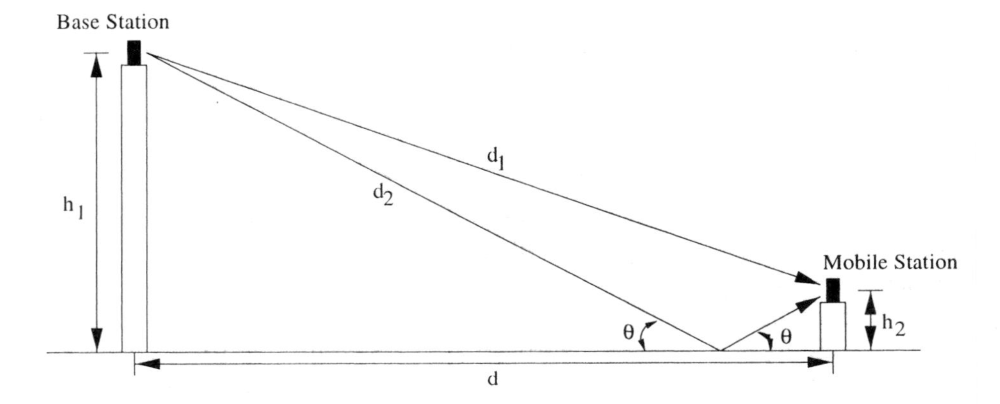

Figure 7.10. Two-path model.

In Figure 7.10, the distances traveled by the direct and by the reflected rays are respectively

$$d_1 = \sqrt{(h_1 - h_2)^2 + d^2}$$

and

$$d_2 = \sqrt{(h_1 + h_2)^2 + d^2}$$

and it follows that the difference between traveled distances is given

$$\Delta d = d_1 - d_2 = d\left[\sqrt{1 + \left(\frac{h_1 + h_2}{d}\right)^2} - \sqrt{1 + \left(\frac{h_1 - h_2}{d}\right)^2}\right]. \quad (7.14)$$

This formula can be simplified by considering that the mobile unit is far from the base station, i.e., that

$$\frac{h_1 + h_2}{d} \ll 1$$

which leads to

$$\Delta d \approx \frac{2h_1 h_2}{d}. \quad (7.15)$$

The received power is given by

$$P_r = P_0 \left(\frac{\lambda}{4\pi d}\right)^2 |1 + a_\nu e^{j\Delta\phi}|^2 \tag{7.16}$$

where P_0 represents the transmitted power, $a_\nu = -1$ is the ground reflection coefficient and $\Delta\phi$, the phase difference, is given by

$$\Delta\phi = \beta\Delta d = \frac{2\pi\Delta d}{\lambda} \approx \frac{4\pi h_1 h_2}{\lambda d}.$$

The expression for the received power can then be written as

$$P_r = P_0 \left(\frac{\lambda}{4\pi d}\right)^2 |1 - e^{j\Delta\phi}|^2$$

which can be simplified even further for $\Delta\phi \ll 1$, resulting in

$$P_r = P_0 \left(\frac{\lambda}{4\pi d}\right)^2 (\Delta\phi)^2 = P_0 \left(\frac{h_1 h_2}{d^2}\right)^2. \tag{7.17}$$

From Equation 7.17 it can be concluded that the received power, under the model conditions, is proportional to the square of the antenna heights – and decreases with the fourth power of the distance between them.

7.5.3 Two-Ray Model with Frequency Selectivity

Channel characterization is an important research topic in the design of communication systems. In fact, the choice of the best technique for multiple access depends on the channel characteristics. This section presents a model for the mobile communications environment, which characterizes the channel properties in terms of its frequency response. This model is not intended to be complete but will include also some spectral scattering characteristics which can be useful in the design of communication systems.

A fading model will now be presented, as a function of frequency, which considers the effect of the specular component and the environmental noise. The model is illustrated in Figure 7.11.

It is assumed that the attenuation functions, α and β, have a multiplicative effect on the direct and on the reflect signal, respectively. The net effect of multipath is the introduction of frequency selective fading in the transmitted signal. The relevance of the analysis of this model comes from the fact that frequency selectivity is the main cause of error bursts in digital communication systems. In this model $s(t)$ represents

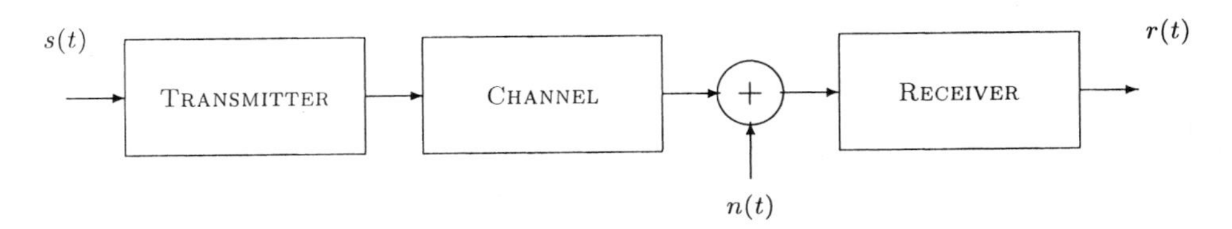

Figure 7.11. Channel model including frequency selectivity.

the transmitted signal, $r(t)$ denotes the received signal, selected among M distinct sequences, and $n(t)$ denotes the environmental noise. All stochastic processes considered in this model are assumed to be stationary, at least wide sense stationary. As usual, the noise is assumed to be independent of the transmitted signal, and having two components. The first component $n_A(t)$ takes into account equipment noise and environmental noise, considered as an AWGN process with zero mean. The second component $n_I(t)$ represents the interference due to the remaining $M - 1$ signals.

$$n(t) = n_A(t) + n_I(t). \tag{7.18}$$

Considering the assumed independence between the noise processes and the transmitted signals, in the worst case, it follows that

$$S_N(w) = S_A(w) + S_I(w) \tag{7.19}$$

or

$$S_N(w) = S_o + (M - 1)S_S(w) \tag{7.20}$$

where S_o represents the AWGN noise power spectral density – PSD, and $S_S(w)$ is the transmitted signal PSD. For simplicity it is assumed that the various transmitted signals are statistically similar.

The time delay σ for the reflected signal is assumed to be constant, even if that may appear to be a rather restrictive assumption for a mobile system. This consideration is based on the fact that the distance traveled by the user is usually much less than the average distance between the user and the base station transmitting antenna. The consequence of a time-varying delay is a change in the separation between fading valleys in the spectrum of the received signal. This effect can be significant for short distances. The received signal is given by

$$r(t) = \alpha(t)s(t) + \beta(t)s(t-\sigma) + n(t). \tag{7.21}$$

Three important cases are considered in this section. In the first case the attenuation functions are considered time independent. In the second case, these functions are considered to be time-varying but deterministic. Finally, in the third case it is considered that the two attenuation functions are of a random nature. The autocorrelation function and the power spectral density of the received signal are computed in order to provide information about the channel behavior. The autocorrelation and the power spectral density for the preceding equation are given, respectively, by

$$R_R(\tau) = E[r(t)r(t+\tau)] \tag{7.22}$$

and

$$S_R(w) = \int_{-\infty}^{\infty} R_R(\tau)e^{jw\tau}dw. \tag{7.23}$$

Consider constant attenuation functions $\alpha(t) = \alpha$ and $\beta(t) = \beta$ for the first case. By substituting these values in the previous two equations it follows that

$$R_R(\tau) = (\alpha^2 + \beta^2)R_S(\tau) + \alpha\beta R_S(\tau - \sigma) + \alpha\beta R_S(\tau + \sigma) + R_N(\tau) \tag{7.24}$$

and

$$S_R(w) = [\alpha^2 + \beta^2 + 2\alpha\beta\cos w\sigma]S_S(w) + S_N(w). \tag{7.25}$$

The square of the channel transfer function modulus can be obtained directly from the preceding equation, i.e.,

$$|H(w)|^2 = [\alpha^2 + \beta^2 + 2\alpha\beta\cos w\sigma]. \tag{7.26}$$

¿From Equation 7.26 it is easy to grasp the role of the parameters α, β and σ with respect to channel behavior as a function of attenuation and frequency selectivity. For example, σ controls the separation between valleys in the channel transfer function, while α and β control the depth of these valleys. Figure 7.12 illustrates the calculated transfer function.

As a second example, consider that the attenuation functions are time-varying, however of a deterministic nature. Again, introducing the respective parameters in Equations 7.22 and 7.23 leads to

$$\begin{aligned} R_R(\tau) &= \alpha(t)\alpha(t+\tau)R_S(\tau) + \beta(t)\beta(t+\tau)R_S(\tau) + R_N(\tau) \\ &+ \alpha(t)\beta(t+\tau)R_S(\tau-\sigma) + \alpha(t+\tau)\beta(t)R_S(\tau+\sigma) \end{aligned} \tag{7.27}$$

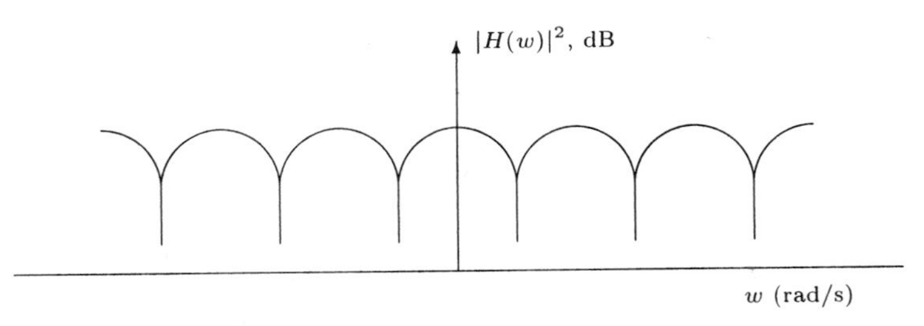

Figure 7.12. Transfer function for a channel with multipath.

and

$$\begin{aligned} S_R(w) &= [\alpha(t)A(w)e^{jwt} \\ &+ \beta(t)B(w)e^{jwt}] * S_S(w) + [\alpha(t)B(w)e^{jwt}] * [S_S(w)e^{-jw\sigma}] \\ &+ [\beta(t)A(w)e^{jwt}] * [S_S(w)e^{jw\sigma}] + S_N(w). \end{aligned} \quad (7.28)$$

Finally, assuming random parameters $\alpha(t)$ and $\beta(t)$, leads to

$$\begin{aligned} R_R(\tau) &= [R_A(\tau) + R_B(\tau)]R_S(\tau) + R_{AB}(\tau)R_S(\tau - \sigma) \\ &+ R_{BA}(\tau)R_S(\tau + \sigma) + R_N(\tau) \end{aligned} \quad (7.29)$$

and

$$\begin{aligned} S_R(w) &= [S_A(w) + S_B(w)] * S_S(w) + S_{AB}(w) * [S_S(w)e^{-jw\sigma}] \\ &+ S_{BA}(w) * [S_S(w)e^{jw\sigma}] + S_N(W) \end{aligned} \quad (7.30)$$

where $R_A(\tau)$ and $R_B(\tau)$ represent, respectively, the autocorrelation of $\alpha(t)$ and $\beta(t)$. An analogous notation follows for the spectrum.
freqüência

7.5.4 Effect of Multiple Rays

A more accurate model can be devised, by considering a large number of rays for the transmitted signal. The general expression for the signal received by one of the users is given by

$$r(t) = \sum_{k=1}^{K} \alpha_k s(t - \sigma_k). \quad (7.31)$$

Here, the attenuation parameters α_k are considered dependent only on the path followed by the signal, l_k, besides dependency on the time instant. The delays, σ_k, are also functions of l_k. It is important to notice that a distinction must be made between models that describe the signal intensity as a function of time and those that describe the signal intensity as a function of the distance traveled. The first is useful for computing the error probability and the margin of operation, while the second is appropriate to determine the coverage area and co-channel interference. A great simplification results when independence between the effects of distance and time is assumed. That implies writing the probability density function for the attenuation parameters as $p(\alpha(t, l_k)) = p(\alpha(t))p(\alpha(l_k))$.

The channel transfer can be obtained directly from the previous equation as

$$h(t) = \sum_{k=1}^{K} \alpha_k \delta(t - \sigma_k). \tag{7.32}$$

The Fourier transform of the previous expression leads to the channel transfer function

$$H(w) = \sum_{k=1}^{K} A_k * e^{jw\sigma_k} \tag{7.33}$$

where A_k represents the Fourier transform of α_k.

In order to consider the effects of phase displacement of the reflected signals, it is more convenient to use the complex notation to represent the received signals. This leads to the low-pass model for the transfer function

$$h(t) = \sum_{k=1}^{K} \alpha_k \delta(t - \sigma_k) e^{j\theta_k} \tag{7.34}$$

where $\{\theta_k = w_c \sigma_k\}$ represent the phase rotations and w_c denotes the carrier frequency (Schwartz et al., 1966). All parameters involved are considered to be random variables obeying their respective distributions, discussed earlier.

Obtaining the $\{ \alpha_k \}$ parameters is of great importance when modeling the mobile channel. However, two of the derived parameters are useful when describing global characteristics of the propagation profile in multipath (Saleh and Valenzuela, 1987) (Pahlavan et al., 1989). These are the multipath power gain

$$G_M = \sum_{k=1}^{K} \alpha_k^2. \tag{7.35}$$

and the RMS delay spread

$$\sigma_{RMS} = \sqrt{E[\sigma^2] - E^2[\sigma]} \tag{7.36}$$

where

$$E[\sigma^n] = \frac{\sum_{k=1}^{K} \alpha_k^n \sigma_k^n}{\sum_{k=1}^{K} \alpha_2^k}, \quad n = 1, 2 \tag{7.37}$$

or

$$E[\sigma^n] = \frac{1}{G_M} \sum_{k=1}^{K} \alpha_k^n \sigma_k^n. \tag{7.38}$$

The power gain is useful to estimate the system signal to noise ratio. The RMS delay spread is a good measure of multipath scattering, besides being related to the performance degradation caused by intersymbol interference. As a consequence, the RMS delay spread limits the maximum signaling rate allowed for a given error rate. Values found in the literature situate the RMS delay spread in the 20 ns to 50 ns range, for small office buildings of small to medium size, (Saleh and Valenzuela, 1987), between 30 ns and 300 ns for factory environments (Rappaport, 1989) (Pahlavan et al., 1989), and up to 250 ns for large office buildings (Devasirvatham, 1984). Based on the previous values, it is estimated that digital transmission rates above 400 kbit/s may not be feasible for an error probability of 10^{-3} or smaller in the above referred building (Devasirvatham, 1984). It should be observed that measurements were performed in the 800 MHz to 1.5 GHz frequency band.

7.5.5 Time Varying Channels

An appropriate channel model must take into account three fading effects: spatial, temporal and frequency selective. The spatial fading strongly depends on the topography, indicating that it is unique in each case, i.e., a model designed for a building may not be adequate for use in another building. Therefore, the dependency with distance will not be included in the channel models to be considered in the sequel. The variables considered will be time and frequency (Thom, 1991).

The frequency response of a time-varying channel is denoted by $h(\tau, t)$. The time-varying frequency response is obtained by means of the Fourier transform of the impulse response with respect to the variable τ

$$H(\omega, t) = \int_{-\infty}^{\infty} h(\tau, t) e^{-j\omega\tau} \, d\tau. \tag{7.39}$$

Both $h(\tau, t)$ and $H(\omega, t)$ are random processes due to variations in t, representing time. From the variables τ and t other functions can be derived, by means of the autocorrelation function or the Fourier transform of $h(\tau, t)$, i.e., $H(\omega, t)$.

The autocorrelation function of the impulse response $h(\tau, t)$ is defined as

$$R_H(\tau_1, \tau_2; t_1, t_2) = E\left[h^*(\tau_1, t_1)h(\tau_2, t_2)\right]. \tag{7.40}$$

Assuming that the random variables are wide sense stationary, which is valid for symmetric time variations, it follows that the autocorrelation will depend on the difference $\sigma = t_2 - t_1$ only. This model is known as the wide sense stationary (WSS) model. Equation 7.40 then becomes

$$R_h(\tau_1, \tau_2; \sigma) = E\left[h^*(\tau_1, t_1)h(\tau_2, t_1 + \sigma)\right] \tag{7.41}$$

The autocorrelation of the frequency response $H(\omega, t)$ is given by

$$R_H(\omega_1, \omega_2; t_1, t_2) = E\left[H^*(\omega_1, t_1)H(\omega_2, t_2)\right] \tag{7.42}$$

The dual of the wide sense stationary model is the uncorrelated scattering (US) model (Bello, 1963). For the WSS model the autocorrelation depends on the time difference $\sigma = t_2 - t_1$ only. A situation analogous to this occurs when the autocorrelation in Equation 7.42 for the US model depends on the frequency difference $\nu = f_2 - f_1$ only. Thus, Equation 7.42 becomes

$$R_H(\nu; t_1, t_2) = E\left[H^*(\omega_1, t_1)H(\omega_1 + \nu, t_2)\right]. \tag{7.43}$$

A third possible model, which is the simplest of them all, is that model which is wide sense stationary uncorrelated scattering (WSSUS). This type of channel model has the properties of channels already described. In other words, the autocorrelation in Equations 7.40 and 7.42 depend both of σ and ν. Equation 7.42 then becomes

$$R_H(\nu; \sigma) = E\left[H^*(\omega_1, t_1)H(\omega_1 + \nu, t_1 + \sigma)\right]. \tag{7.44}$$

In this case, $R_H(\nu; \sigma)$ is called autocorrelation function spaced in time and frequency. By making $\sigma = 0$, the following function results

$$R_H(\nu; \sigma)\,|_{\sigma=0} = R_H(\nu), \tag{7.45}$$

which is the frequency correlation function. $R_H(\nu)$ measures the degree of correlation between different frequencies. A measure of uncorrelatedness is obtained when $\nu = \omega_c$ such that $R_H(\omega_c) = 0$. Frequencies separated by ω_c will be uncorrelated. The resulting parameter ω_c is called correlation bandwidth.

Thus, if a transmitted signal bandwidth W is narrower than the correlation bandwidth, i.e., if $W < \omega_c$, then that signal frequency content is highly correlated and all its spectrum will suffer similar phase and amplitude variations. This condition is referred to as *non-selective fading.*

On the other hand, if a transmitted signal bandwidth W is wider than the correlation bandwidth, i.e., if $W > \omega_c$, then the time variation for some frequencies will be independent of the time variation of other frequencies, and that characterizes what is called *frequency selective fading.*

Other functions can be derived from the relationship between $h(\tau, t)$ and $H(\omega, t)$. Since $h(\tau, t)$ and $H(\omega, t)$ are a Fourier transform pair, it follows that $R_h(\tau_1, \tau_2; t_1, t_2)$ and $R_H(\omega_1, \omega_2; t_1, t_2)$ are also a Fourier transform pair, i.e.,

$$R_H(\omega_1, \omega_2; t_1, t_2) = \int_{-\infty}^{\infty}\int_{-\infty}^{\infty} R_h(\tau_1, \tau_2; t_1, t_2) e^{j(\omega_1\tau_1 - \omega_2\tau_2)}\, d\tau_1\, d\tau_2 \quad (7.46)$$

Thus, for a WSSUS channel, the replacement of $\sigma = t_2 - t_1$ and $\nu = w_2 - w_1$ in Equation 7.46, after some manipulation leads to

$$\begin{aligned} R_H(\tau_1, \tau_2; \sigma) &= \frac{1}{2\pi}\int_{-\infty}^{\infty} R_H(\nu; \sigma) e^{j\tau_1\nu}\, d\nu\, \delta(\tau_2 - \tau_1) \\ &= R(\tau_1, \sigma)\delta(\tau_2 - \tau_1), \end{aligned} \quad (7.47)$$

where

$$R(\tau_1, \sigma) = \frac{1}{2\pi}\int_{-\infty}^{\infty} R_H(\nu; \sigma) e^{j\tau_1\nu}\, d\nu \quad (7.48)$$

$R(\tau_1, \sigma)$ is the Fourier transform of the autocorrelation of the channel frequency response. This function is called the *delay cross-power spectral density function.* For $\sigma = 0$ it follows that

$$R(\tau, \sigma)\,|_{\sigma=0} = R(\tau) \quad (7.49)$$

and Equation 7.47 becomes

$$R_H(\tau_1, \tau_2) = R(\tau_1)\delta(\tau_2 - \tau_1) \quad (7.50)$$

$R(\tau)$ in Equation 7.49 is the power impulse response, which expresses a measure of the channel wideband fading characteristics. From Equations 7.45, 7.48, and 7.49, it can be seen that $R(\tau)$ is the inverse Fourier transform of the frequency correlation function $R_H(\nu)$, i.e.,

$$R(\tau) = \frac{1}{2\pi}\int_{-\infty}^{\infty} R_H(\nu) e^{j\tau\nu}\, d\nu \quad (7.51)$$

An important parameter which can be extracted from $R(\tau)$ is the delay spread σ_{RMS}. This result is obtained by making $\tau = \sigma_{RMS}$, in a manner that $R(\sigma_{RMS}) = 0$. Considering that $R(\tau)$ and $R_H(\nu)$ are Fourier transform pairs, then the delay spread and the correlation bandwidth are related as

$$\sigma_{RMS} = \frac{1}{\omega_c}. \tag{7.52}$$

Two other functions can still be obtained from the WSSUS channel. Differing from the previous functions, which focus on the channel frequency variations, the following functions focus on the channel time variations. The first function is obtained by making $\nu = 0$ in the correlation function spaced in frequency and time $R_H(\nu; \sigma)$. The result is

$$R_H(\nu; \sigma)\,|_{\nu=0} = R_H(\sigma). \tag{7.53}$$

Thus, $R_H(\sigma)$ is called the *correlation function spaced in time* and describes how fast the channel fades. A useful parameter which can be derived from $R_H(\sigma)$ is the coherence time t_c, which results by making $\sigma = t_c$, such that $R_H(t_c) = 0$. High values of t_c indicate a channel which is slow in terms of time variations, while small values of t_c indicate a channel with fast variations with respect to time.

The second function is obtained by means of the Fourier transform of $R_H(\sigma)$ with respect to σ, i.e.,

$$S(\lambda) = \int_{-\infty}^{\infty} R_H(\sigma) e^{-j\lambda\sigma}\, d\sigma. \tag{7.54}$$

This function is called the Doppler power spectrum. A useful parameter derived from $S(\lambda)$ is the Doppler bandwidth W_D, which measures the amount of Doppler scattering or movement within the channel. Since the echo power spectrum and the correlation function spaced in time constitute a Fourier transform, then the Doppler bandwidth and the coherence time are related as

$$W_D = \frac{1}{t_c}. \tag{7.55}$$

Consequently, a large Doppler bandwidth indicates a channel with fast time variations, while a narrow Doppler bandwidth indicates a channel with slow time variations.

One last function will now be derived from the WSSUS channel. This function results by considering the two-dimensional Fourier transform of the correlation function spaced in time and frequency $R_H(\nu;\sigma)$, i.e.,

$$S(\tau;\lambda) = \int_{-\infty}^{\infty}\int_{-\infty}^{\infty} R_H(\nu;\sigma)e^{-j\lambda\sigma}e^{j\tau\nu}\,d\sigma\,d\nu. \tag{7.56}$$

In this case $S(\tau;\lambda)$ is called the channel scattering function. If the channel transfer function is not known in advance, some form of combating its effects must be found. Equalization is the usual procedure for obtaining a flat amplitude response and a linear phase response, which characterize the ideal channel. An adaptive equalizer is designed from observations of the channel output, for a given input known as training sequence. Alternately, the output sequence can be the channel response to a carrier modulated by a random data sequence. The former approach is used in point-to-point communication systems while the latter is more adequate for broadcast (Blahut, 1990).

One of the first studies about the convergence of equalizers showed that the use of isolated test pulses as the training signal is sub-optimum when compared with the use of pseudo-random sequences, or the data signal itself (Gersho, 1969) as training signals. More complete essays on equalization can be found in (Macchi et al., 1975) and (Qureshi, 1985).

As discussed in this chapter, propagation of signals in mobile channels is subject to severe frequency fading. In an indoor channel presenting a high delay spread, performance will seriously suffer with intersymbol interference (ISI). Adaptive equalization is efficient against this type of degradation. The final result,as a consequence of improvement in performance, is an increase in the channel signaling rate (Valenzuela, 1989).

A few improvements can be done to the previous model, in order to make it suitable for indoor use. The proposed model is based on the assumption that the indoor channel can be modeled as a WSSUS channel (Yegani and Mcgillen, 1991). This model concentrates on the channel impulse response $h(\tau,t)$, where the parameters of the model are related to the propagation parameters. It follows that all channel functions derived from Equations 7.44, 7.45, 7.51, 7.53, 7.54, and 7.56 will follow from the knowledge of $h(\tau,t)$.

The time varying channel impulse response is given by

$$h(\tau,t) = \sum_{k=0}^{K} \alpha_k(t)\delta(\tau-\sigma_k(t))e^{-j\theta_k(t)}, \tag{7.57}$$

where t denotes the sampling instant and $\alpha_k(t)$, $\theta_k(t)$ and $\sigma_k(t)$ are random variables representing the time varying attenuation, phase and delay in the k-th channel path, respectively.

The statistical distributions of multipath parameters are the following.

- The variable $\alpha_k(t)$ has a Rayleigh, Rice, or log-normal distribution, depending on the terrain topography. For indoor channels usually the Rice distribution is considered while the Rayleigh distribution is frequently used for outdoor channels.

- The variable $\theta_k(t)$ is uniformly distributed between 0 and 2π radians if $\alpha_k(t)$ is Rayleigh or Rice distributed.

Most of the recent research has concentrated on the $\alpha_k(t)$ distribution and a little on the $\theta_k(t)$ distribution, where a uniform distribution is usually assumed. Far less attention has been given to the $\sigma_k(t)$ distribution.

7.5.6 Propagation Model in Urban Area

An estimate of signal attenuation, as seen in previous sections, is done either by stochastic analysis or by an empirical formulation based on experimental data. Okumura's model and the multiple ray models are employed for the analysis of propagation in urban areas (Yacoub, 1993).

Propagation in Mobile Systems

In order to obtain an empirical model for propagation prediction for mobile communications, Y. Okumura performed various field intensity measurements in Tokyo's urban area and also in a suburban profile area (Lee, 1989). The measurements were performed in the 150 MHz to 20 GHz frequency range with an effective transmitting antenna height varying between 30 and 1000 m, over distances varying from 1 a 100 km. The height of the antenna in the mobile unit varied from 1 to 10 m.

The following equations, listed here for reference only, compare various signal propagation models, for the case where frequency selectivity is not considered, i.e., flat fading is assumed. P_T e P_R represent, respectively, the transmitted and the received power, d is the distance between antennas, G_T and G_R are the gains respectively for the transmitting antenna with height h_B, and receiving antenna with height h_M.

- Propagation in Free Space

$$\frac{P_R}{P_T} = \left(\frac{\lambda}{4\pi d}\right)^2 (G_T G_R) \tag{7.58}$$

- Propagation with path including ground reflection

$$\frac{P_R}{P_T} = \left(\frac{h_B h_M}{d^2}\right)^2 (G_T G_R) \tag{7.59}$$

- Propagation in an Urban Area

$$P_R = P_T - A_M(f,d) + H_B(h_B,d) + H_M(h_M,f) \tag{7.60}$$

- Propagation in Non-Urban Environment

$$P_R = P_T + K_{SO} + K_{SP} + K_{Terr} \tag{7.61}$$

The factors $A_M(f,d)$, $H_B(h_B,d)$ e $H_M(h_M,f)$ represent the effect of attenuation or gain in an urban environment as a function of distance, frequency of operation and antenna height. The factors K_{SO}, K_{SP} e K_{Terr} represent the effect of the type of terrain in an area with no buildings.

7.6 Problems

1 Explain how the Doppler effect can disturb the reception of a transmitted signal in the case of frequency modulation and in the case of amplitude modulation.

2 The following signal is received by a radar equipment.

$$r(t) = s(t) + n(t).$$

Assume that the noise $n(t)$, with power spectral density (PSD) $S_N(\omega) = S_0$, is independent of the signal $s(t) = a(t)\cos(\omega_R t + \phi)$, where $a(t)$ represents the modulating signal, appropriate for radar detection, with power spectral density $S_A(\omega)$, in the range $-\omega_M \le \omega \le \omega_M$, having zero mean value. Calculate the resulting autocorrelation and PSD, and explain how to recover the modulating signal. Plot the corresponding graphs.

3 In the two-ray model (Lee) the transmitted power is 100 W and the antenna gain at the base station and at the mobile terminal are identical, $G_T = G_R = 1,4$. The base station antenna is 100 m high, while the antenna in the mobile unit is 1,5 m. high. For an average distance of 5 km, calculate the received power in dBm.

4 The amplitude of a received signal is Rayleigh distributed, as indicated below. Calculate the most likely signal amplitude.

$$p_R(r) = \frac{r}{\sigma^2} e^{-\frac{r^2}{2\sigma^2}} u(r).$$

5 What causes multipath fading and what is its effect on the signal?

6 Which are the effects that cause multipath to occur? Explain the occurrence of each one of them.

7 Discuss the types of fading for a mobile channel: spatial fading, time fading and frequency selective fading. Indicate their basic characteristics and their respective causes.

8 Slow fading, as discussed, is described by the equation

$$P = Ad^{-n} \tag{7.62}$$

where P is the received signal power, A is a constant gain, d is the distance between the transmitter and the receiver and n is the attenuation factor. Explain the variation of n with the type of environment.

9 Analyze the statistical effects present in the mobile channel and draw a relationship with the various distributions used in their characterization.

10 The Rice distribution is given by

$$p_X(x) = \frac{x}{\sigma^2} e^{-\frac{x^2+A^2}{2\sigma^2}} I_o\left(\frac{xA}{\sigma^2}\right) u(x) \tag{7.63}$$

where $I_o(\cdot)$ is the modified Bessel function of order zero and A is the signal amplitude. Show that this distribution converges to the Rayleigh distribution when the signal amplitude approaches zero.

11 Show that the following Nakagami distribution represents the Rayleigh distribution for $m = 1$.

$$p_X(x) = \frac{2m^m x^{2m-1}}{\Gamma(m)\Omega^m} e^{-\frac{mx^2}{\Omega}} u(x) \tag{7.64}$$

where $\Omega = P_X$ is the received signal average power and $m = \Omega^2/E[(X^2 - E^2[X])^2]$ represents the inverse of the normalized variance of X^2. The parameter m is the modeling factor for the distribution.

12 Which transformation in the Gaussian distribution leads to the following log-normal distribution?

$$p_R(r) = \frac{1}{\sigma r \sqrt{2\pi}} e^{-\frac{(\log r - m)^2}{2\sigma^2}} u(r) \tag{7.65}$$

with mean value $E[X] = e^{\sigma^4/2+m}$, and variance $V[X] = e^{\sigma^4+m}(e^{\sigma^4} - 1)$.

13 Measurements made in factory environments led to the conclusion that the time interval between consecutive arrivals $\{t_i = \sigma_i - \sigma_{i-1}\}$ obeys a Weibull distribution with pdf given by

$$p_T(t) = \beta \alpha t^{\beta-1} e^{-\alpha t^{\beta}} u(t) \tag{7.66}$$

where $\alpha, \beta > 0$ are parameters of the distribution. Find out for which value of β this distribution converges to the Exponential distribution?

14 What is the effect of the delay σ, for the model with frequency selectivity, in the mobile channel transfer function? What is the meaning of the channel signature?

15 Why RMS delay spread limits the transmission rate in a mobile channel?

16 Analyze the effect of fading in a signal with bandwidth W in respect to the correlation bandwidth parameter.

17 Classify and analyze the types of purposeful interference.

18 Define outage for both time-varying and time invariant channels.

19 Explain how the effect of frequency selective fading can disturb reception of a transmitted signal, in the case of angle modulation. Show how to combat this effect.

Chapter 8

CARRIER TRANSMISSION

In classic frequency division multiplexing (FDM), now revived with the emergence of orthogonal frequency division multiplexing (OFDM), the signals are modulated in SSB (Single Sideband) and allocated in sets of twelve channels to form a group. Five groups, duly transferred to specific frequencies, compose the master group, which contains sixty channels, and so on. In the multiplexing in time process, the signals are digitized and grouped in frames and multiframes, which form tributaries, with a transfer rate of 2 Mbits/s.

These hierarchies are commonly known as Plesyochronous Digital Hierarchies (PDH). To establish standardized structures, the unification of the hierarchies became necessary, and the establishment of a new and unique digital hierarchy the Synchronous Digital Hierarchy (SDH). The synchronous network was conceived to solve several problems found in the plesyochronous network.

For any system, after the multiplexing, the allocated signals form a transmission baseband. The signal baseband to be transmitted through space modules a carrier, usually using an intermediate frequency carrier, which is amplified and propagated in a transmission channel. This chapter presents some characteristics of that process.

8.1 Carrier amplification

The classic transmission process of a baseband signal is formed by a modulator which associates the baseband signal to the carrier, and a power amplifier which amplifies the carrier signal to produce a power gain. That amplification of the carrier should be in accordance with the transmission support, whether it is from an antenna or a guided line.

This amplifier is very common in the systems which possess a high channel capacity, and because of that, it should be a broadband amplifier. The amplification should be extended to the full length of the modulated carrier band, in such way as to not cause a distortion in the phase and amplitude

In some cases there is a need to transmit several carriers through a single power amplifier. In those cases the amplifier should have a sufficient gain and bandwidth to simultaneously amplify all the carriers.

An aspect to be analyzed is the influence of the non-linearity of the amplifier on the output signal's spectral power density. It is common to model the power amplifier by an equation which involves first and second order factors.

Considering the signal $Y(t)$ at the amplifier's output

$$Y\left(t\right)=\alpha X\left(t\right)-\beta X^{2}\left(t\right) \tag{8.1}$$

and that $X(t)$ is a zero mean stationary Gaussian signal, one can determine the output signal autocorrelation in the following way

$$\begin{aligned} R_Y\left(\tau\right) &= E\left[Y\left(t\right)Y\left(t+\tau\right)\right] \qquad (8.2)\\ &= E\left[\left(\alpha X\left(t\right)-\beta X^{2}\left(t\right)\right)\left(\alpha X\left(t+\tau\right)-\beta X^{2}\left(t+\tau\right)\right)\right]\\ &= \alpha^{2}R_X\left(\tau\right)-\alpha\beta E[X\left(t\right)X^{2}\left(t+\tau\right)+X^{2}\left(t\right)X\left(t+\tau\right)]\\ &+ \beta^{2}E[X^{2}\left(t\right)X^{2}\left(t+\tau\right)]. \end{aligned}$$

Using Price's theorem, which is given by equation

$$E[X^kY^l]=kl\int_0^{C(X,Y)}E[X^{k-1}Y^{l-1}]dC(X,Y)+E[X^k].E[Y^l], \tag{8.3}$$

one obtains

$$R_Y\left(\tau\right)=\alpha^{2}R_X\left(\tau\right)+2\beta^{2}R_X^{2}\left(\tau\right)+\beta^{2}P_X^{2}. \tag{8.4}$$

Applying the Wiener-Khintchin theorem, one obtains

$$S_Y\left(\omega\right)=\alpha^{2}S_X\left(\omega\right)+\frac{2\beta^{2}}{2\pi}S_X\left(\omega\right)*S_X\left(\omega\right)+\beta^{2}P_X^{2}\delta\left(\omega\right). \tag{8.5}$$

Thus it is concluded that, in function of the appearance of the convolution in the above obtained expression, that the signal spreads throughout a band twice as big as the original. The Equation 8.1 describes particularly well the behavior of a power amplifier, of the FET type, operating in the non-linear region, but is typical of most power amplifiers.

The main characteristic of the signal amplifiers is the low figure of noise, meaning that the noise generated by the amplifier is extremely small, permitting its use in the amplifying of very weak signals, obtaining an acceptable signal to noise ratio.

The figure of noise can be defined as the ratio between the total noise of the amplifier's output and its input. This can be given by

$$F = \frac{P_{int} + KTBG}{KTBG} = 1 + \frac{P_{int}}{KTBG} \tag{8.6}$$

where P_{int} represents the noise due to the amplifier internal sources and $KTBG$ represents the input noise power.

From this last expression, it is possible to obtain the equation for the noise figure for a system composed of several amplifiers. The equations for two and three amplifiers are shown in the following.

$$F = F_1 + \frac{F_2 - 1}{G_1}, \tag{8.7}$$

$$F = F_1 + \frac{F_2 - 1}{G_1} + \frac{F_3 - 1}{G_1 G_2}. \tag{8.8}$$

Another important parameter in any transmitter is the quantity of power at the output of the amplifier which is coupled to the antenna or the guided line. This depends not only on the amplifier's gain, but also on the circuit connection between the amplifier output and the antenna or cable's input terminals. To prevent losses by coupling, it is necessary to have a match of impedances between the amplifier and the signal's irradiation system.

If there is an impedance mismatch, part of the signal's power P_T is reflected to the amplifier, which defines the reflection coefficient Γ, given by

$$\frac{P_T}{P_{AMP}} = 1 - |\Gamma|^2 \tag{8.9}$$

When the coupling is done by a wave guide, cable, or transmission line, the losses can be obtained directly by determining the Voltage Standing Wave Ratio (VSWR), which is the ratio between the maximum and minimum tension present at the guide.

The reflection coefficient is related to the VSWR by the expression

$$|\Gamma| = \frac{VSWR - 1}{VSWR + 1} \tag{8.10}$$

since the VSWR is easy to be measured because it needs only one voltmeter. The losses by coupling can be easily determined without the need to know the internal impedance values.

The power amplifiers are widely used in radio telescope, artificial satellites, controlling and tracking satellites and in tropodiffusion receptors, in which the signals received are of a very low level (Gagliardi, 1988). Figure 8.1 shows the carrier transmission in block diagram.

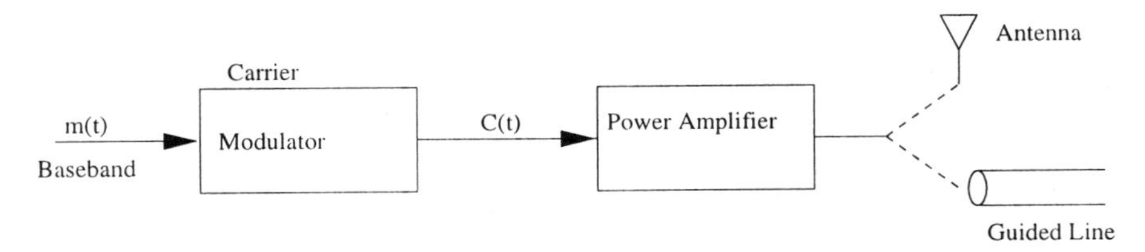

Figure 8.1. Carrier transmission in block diagram.

8.2 Features of the Cavity and Solid State Amplifiers

The conventional electronic amplifiers are limited by parasitic capacitance and by the transition time of the electrons, and as a consequence there is a reduction of the carrier gain when in a high frequency. For this reason the cavity and solid state amplifiers are used, since in high frequencies they possess a satisfactory gain and a good relative efficiency.

Basically, there are four types of amplifiers at the microwave frequency, two cavity and two solid state. They are respectively: *Klystron* and *TWT*, *TDA* and *MASER*.

The Klystronis used when there is a need for bandwidth and a reasonable gain, with an excellent noise figure. The noise figure, depending on the type of refrigeration used, can achieve values around 0.3 dB. The need for an RF source, generally in a far superior frequency than the signal's frequency, and of a DC source, together is some cases associating refrigeration to low temperatures, makes the Klystron voluminous and greatly power consuming.

The traveling wave tube (TWT), or progressive wave valve, is a power amplifier applied to the output stages of microwave transmitters. However, some types of TWT can show a good noise figure (2.25 a 5.5 dB) maintaining an output power tens to hundreds of times superior to the tunnel diode amplifier (TDA), used in receivers. The disadvantages of this kind of low noise amplifier are the extremely high source voltage and its weakness, limiting its use in many applications.

The tunnel diode amplifier (TDA) presents a slightly bigger noise figure when compared to the other low noise amplifiers (typically around 5 dB), although for moderate gain values, like, for example, 10 dB, a high bandwidth can be obtained (around 4 GHz) around 10 GHz, or a high gain (around 25 dB) and a bandwidth of 1 GHz. Noise figures around 2.8 dB can be reached in only some frequency ranges. The main advantage of the TDA is to be compact and only need one battery for operation. The TDA's are already found commercially in frequencies up to 20 GHz, with gains superior to 50 dB, when projected with several amplification stages.

The microwave amplifier that used stimulated emission of radiation (MASER) is a low noise amplifier which is based on atomic effects to produce amplification. The MASER is without a doubt the amplifier which presents the best noise figure (around 0.1dB) and a gain from 10 to 20 dB with a moderate bandwidth, although its refrigeration based on liquid helium is essential, which makes it complex and voluminous (Gagliardi, 1988).

8.3 Communication Channels

Communication channels can take various shapes and have various properties. They differ basically on the transmission capacity, interconnection distance, transmission speed and the fidelity of the signals they transmit. The channels are usually noisy and limited both in transmission capacity and speed.

The current technological developments try to naturally explore channels which transport large rates of information at ever larger speeds, which imply on the reduction of cost and permit a larger compacting of the system. The main and most used channels can be divided in:

- Guided channels;
- Non-guided channels.

Guided waves are those which carry the energy through the transmission lines or similar structures. Non-guided waves are those which conduct energy through space. The guided wave's path is fixed by the transmission structure, and the non-guided wave's path is determined by the characteristics of the medium of propagation. Thus being, the signals irradiated by an antenna are non-guided waves, because even if the antenna supplies a certain referential direction of irradiation, it does not influence the wave's path through space.

As an example, the physical lines (twisted pair and coaxial) and the wave guides are guided wave systems, while the radio systems are non-guided wave systems. The guided waves, subjects of this section, include the physical lines, which exist in several types, such as: pair of wires, cable of pairs, open line, high voltage line, coaxial cable, wave guides and optical fibers.

The non-guided waves, studied previously, include the radio systems, which are classified in: terrestrial wave, spatial wave, celestial wave and outer space.

8.3.1 Guided Channels

The main characteristics if guided channels are:

- Confinement of the transmitter's electromagnetic wave to the receptor by means of the guide;
- Adequate for areas of intense electromagnetic fields;
- Applications in privates links of communication;
- Possibility of skirting obstacles;
- Attenuation, depending on the frequency, type and size of the guide material;
- The need for an impedance match, to avoid losses by reflection;
- High cost of implementation, in the case of microwave guides;
- Short-ranged links;

Below there is a brief description of the peculiarities of some guided channels.

Twisted Pair

It is the simplest means of transmission. It is composed of a set of two conductor wires (usually copper), and each wire is covered in an electric insulator. It is used in telephone lines, to connect one telephone to a terminal block located near the buildings, where pairs corresponding to several users arrive.

Cable of Pairs

The cable of pairs are constituted of sets of pairs of wires united, normally using paper or polyethylene for isolation of the conductors, containing up to 2800 pairs. They can be classified as three types, depending on their installation:

- Aerial: fixed to light posts and supported by messenger cables. Usually up to 300 pairs of aerial cables are used.

- Underground: are installed in subterranean ducts.

- Buried: buried directly under the ground without the use of ducts. The underground and buried cables are usually pressurized to avoid the penetration of moisture.

When necessary the usage of pairs of cables can be optimized with use of multiplexing equipments (normally with the capacity of up to 30 voice channels per pair of wires).

The losses by attenuation can be reduced in the pairs of cables, and can be extended to their use distance, by means of a technique called pupinization, which consists of the insertion of inductances in series in the conductors, at regular intervals, which exert a compensating effect in relation to the capacitance that exists between the conductors of the pair.

The open lines are aerial lines (installed in posts) in which naked conductors are used (without electric isolation) and the distance is far wider than the diameter.

Open Line

The open lines are aerial (installed in posts) and use naked conductors (no electrical isolation). The separation between the lines in far wider then their diameter.

The attenuation produced by the open lines is a lot smaller than the corresponding systems with pairs of cables, especially because of the wider diameter of the conductors used in the lines and the space between them. As a setback, due to the larger use of raw material (usually copper), the cost (per channel) is higher in the open lines than in the pair of cables.

The open lines can be used for bidirectional connections, covering distances of up to 100 km without the use of repeaters, or larger distances (usually up to 400 km) with repeaters typically spaced by 100 km.

Normally they are used in mountainous and unleveled regions where a radio call, for example, would demand lots of repeaters due to the obstruction of the radio wave by the terrain.

The open line systems are subjected to bad weather conditions, their characteristics are exposed to temperature variations and humidity. Usually, the capacity of these systems, voice channel wise, is of the order of dozens of channels.

High Voltage Line

The high voltage lines are extensively used for the transmission of various types of communication, including telephony, telegraphy, data transmission and telemetry. The transmission of those signals can occur simultaneously with the transmission of electric energy – without mutual interference – or by means of a special optical cable, called Optical Ground Wire (OPGW), which usually substitutes one of the ground cables in the electric energy transmission line.

The high voltage communication systems were widely used to supply the internal communications needs of the electric energy generation and distribution companies. Nowadays in the electric energy companies there is a tendency to distribute means of communications for commercial purposes. This is due to the utilization of optical cables in the lines, which has considerably increased the system's transmission capacity and has left unused a considerable bandwidth in the channeling. This remaining capacity needs to be sold.

Theoretically, the electric energy transmission lines operating at any voltage can be used for communications services, or Power Line Communication (PLC), however the higher voltage ones (above 33 kV) are used more frequently. This occurs because the lower voltages are interrupted or derived in a great number of power stations, greatly increasing the costs of installation and maintenance of the communications services, due to the necessity of a larger number of additional equipments to be installed (communications and protection).

An important consideration in the connection is the isolation of the communications equipments in the electric energy network, as well as the protection against discharges and transitory effects, This way it is necessary to have at least three types of additional equipments for the connection to the high voltage lines: high voltage coupling capacitor, protection device and tuning and matching devices.

Coaxial Cable

To compensate for certain deficiencies, such as: reception of natural noise, interference and the irradiation phenomenon when transmitting higher frequencies, the coaxial cables were created, which work like shielded lines, avoiding the radiation of energy, and the reception of external signals. The diameter of the internal and external conductors (determining the space between them) as well as the isolating material's dielectric constant, determine the characteristics of the transmission via coaxial cables. The coaxial cables, besides permitting the transmission of a great number of telephonic channels, offer the possibility for the transmission of television signals.

As for the installation, they can be aerial (only for low capacity), underground or buried, like the pair of cables, and as a special case the submarine cable.

Waveguides

Although there are various types of waveguides, the system analyzed here refers to a special case of the utilization of these structures as a means of transmission. In the guided waves systems, special hollow conductors are used, with a straight circular section (the straight circular guarantees the least attenuation), which internally guide very high frequency radio waves. These systems present a very superior capacity when compared to the systems mentioned before, reaching up to 200.000 voice channels. They require a very precise and costly technology.

Optical Fibers

The optical fiber transmission systems are based on the propagation of modulated light, with wavelength in the range of 850 to 1550 nm.

The optical fiber is a silica or glass high purity compound, conveniently doped with phosphor ions and germanium, which, among other advantages, increases the light conductivity, lessens the losses by attenuation, and reduces the group retard. Nowadays, the loss by attenuation in optical fiber is around 0.2 a 1 dB/km and their diameters vary between 5 e 50 μm, without the cover, and reach 125 μm, with the cover.

With such loss, the tendency of this system is to have a great distance between repeaters, reaching around 1000 km today, reducing the number of these, and consequently, the costs of equipment. A typical configuration can be observed in the figure. The modulated current (in frequency, amplitude, pulse width and pulse position) is applied to the LED or LASER diode, which transforms it into light, and using lens, the new signal is sent to the fiber. These lens can also be used to compatabilize the sizes of the useful area of emission and the detection of diodes and the straight section of the fiber.

On the reception's side, the light from the fiber is directed by the lens to the photodiode which retransforms it into an electric signal, which is amplified and delivered to the demodulator, in the case of a terminal station, or retransformed into light and delivered to the fiber, in the case of a repeating station.

As typical applications, there are: connections between telephone stations in big cities, interconnection between computers, and between computers and interfaces, substations supervision systems, substitution of conventional cables in military planes (lessening the weight) and in satellites.

Some of the greatest advantages of the fiber optics systems are the available bandwidth and the immunity to interferences, due to the frequencies they run on.

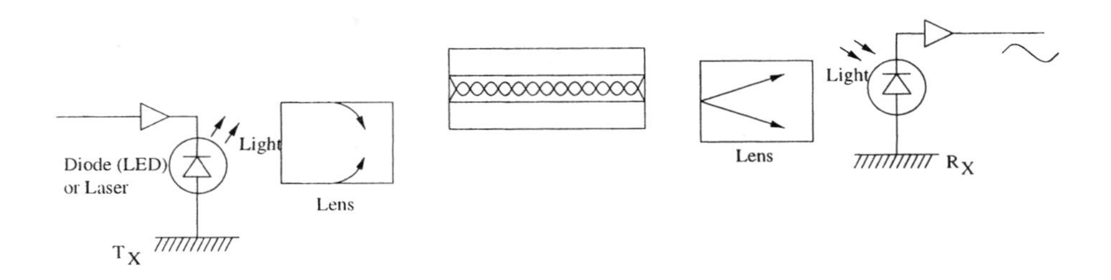

Figure 8.2. Optical transmission system.

8.4 Microwave Systems

The word microwave does not specify clearly a band of the radiofrequency spectrum, corresponding to the superior band of UHF grouped with the SHF. Therefore, a well defined frequency which classifies the microwave systems does not exist. However, it is usually considered the start of the microwave band at 900 MHz. And can be divided into radiovisibility and tropodiffusion systems.

Line of Sight Systems

In these systems the signal irradiated by an antenna is received by another, which should be visible by the first one, there should be a clear direct view between the transmitting and the receiving antenna, as can be seen in Figure Figure 8.3.

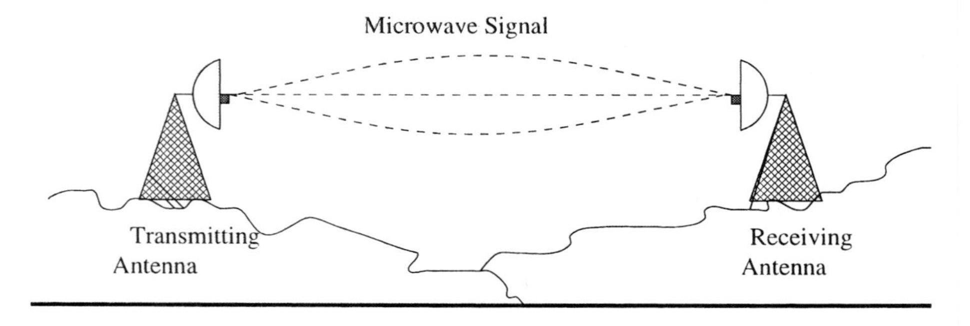

Figure 8.3. Microwave line of sight transmission system.

Tropodiffusion Systems

The troposphere is the layer adjacent to the earth's surface and extends to an altitude of approximately 11 km. In that layer there is the presence of many types of gases, like oxygen and carbon dioxide, besides water vapor and eventual precipitations, like rain and snow.

Consequently the physical behavior of that layer can be described by three parameters: atmospheric pressure, temperature and water vapor pressure.

Concerning the radio waves, the main phenomena to be analyzed when propagating through that layer are the following:

- Wave refraction (specially the resulting effects of the variations of the refraction index values)
- Wave energy absorption by the oxygen and the water vapor, besides other types of absorption.
- Influence of the precipitations (attenuation caused by rain or snow)

These systems operate on the superior band of the UHF and the inferior band of the SHF (900 MHz a 2 GHz), by means of propagation by diffusion in the troposphere.

As they do not require a direct view between the antennas, they are normally utilized for long distance communications, in regions where the installation of repeaters with spacing of 50 km would be non practical.

Not requiring a connection in visibility, a tropodiffusion connection without the use of repeaters can have a range of 450 km, depending on the intermediary terrain's configuration.

They utilize transmitters with power between 100 W and 1 kW and present a regular capacity of 180 voice channels (very inferior to the capacity reached by visibility systems).

In a tropodiffusion network, the antennas concentrate the energy in one given direction, like a flashlight, in conic and thin beams, tangent to the ground, the beams are directed to a certain region of the tropospheric layer as can be seen in Figure Figure 8.4.

The lower the height of the ground compared to the common volume and the angle between the beams, the better will be the transmission.

There are two theories which are more adequate to justify the considerable reception beyond the horizon, as occurs in the tropodiffusion connections: the theory of diffusion in turbulent regions and the reflection in stratified layers. The transmission by tropodiffusion has been used very little nowadays and can be characterized statistically by the Rayleigh distribution.

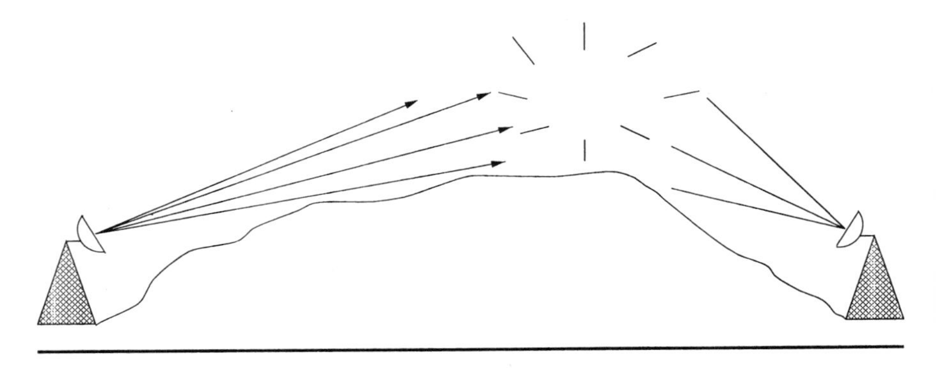

Figure 8.4. Tropodiffusion transmission.

8.5 Antennas

In communications systems using non-guided wave transmission, the carrier is coupled to space, using a transmission antenna. An antenna is simply a transducer which converts electric signals into electromagnetic fields, or vice-versa.

A transmission antenna converts the amplified carrier into a propagating electromagnetic field. Any transmission antenna is composed of a metallic frame which illuminates the reflective surface, in which the field is radiated. This field is transmitted by the antenna as a propagating plane wave with a specific polarization and spatial distribution of its field power density.

The spatial distribution of a transmission antenna's power is described by its irradiation pattern, which is defined as the power density irradiated by unit of solid angle in the direction of the elevation angle and the azimuth angle (ϕ, θ). The angles are shown in Figure 8.5.

- The elevation angle has a vertical span from $-\pi/2$ to $\pi/2$;
- The azimuth angle has a horizontal span from $-\pi$ to π;
- The solid angle is defined as the ratio between the limited spheric area and the square of the radius.

So the antenna pattern indicates the quantity of field power which goes through a unit of solid angle in a given direction, leaving from the center of the antenna. Besides the power content, an electromagnetic field has a polarization (orientation in space) which is associated to it, which is determined by the way the electromagnetic field is excited close to the antenna terminals before its propagation. The polarization can be linear (horizontal or vertical) or circular (left or right).

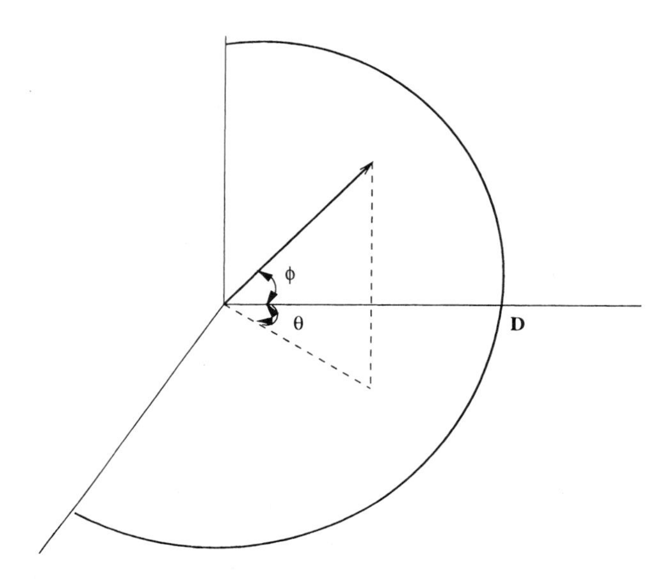

Figure 8.5. Definition of elevation and azimuth angles.

The antenna pattern is normalized in relation to the isotropic antenna, which gives the gain function of the antenna

$$g(\phi, \theta) = \frac{\omega(\phi, \theta)}{\omega_i(\phi, \theta)}. \tag{8.11}$$

The isotropic antenna radiates equally in all directions. The gain is usually given in dBi.

The total power P_T transmitted by an antenna is the integral of $\omega(\phi, \theta)$ over the spherical hull

$$P_T = \int_{-\pi}^{\pi} \int_{-\pi/2}^{\pi/2} \omega(\phi, \theta) \cos \phi \, d\phi \, d\theta, \tag{8.12}$$

where $\cos \phi \, d\phi \, d\theta$ is the differential solid angle. The power per unit of solid angle of an isotropic antenna, with same total power, is given by $P_T/4\pi$. Therefore,

$$g = \frac{\omega(\phi, \theta)}{P_T/4\pi}. \tag{8.13}$$

So the gain function of a transmission antenna is simply a normalized form of the antenna standard.The calculation of the integral above the spheric surface with a uniform field distribution is 4π. This means that if the gain function increases in a determined direction, it will be diminished in the other directions, in a way that the result of the calculation remains the same constant. Usually a transmission antenna is projected to transmit its radiated field in a specific direction, called the antenna's main lobe.

Instead of working with the gain function, the gain and angle of view are usually specified. The gain of a transmission antenna g can be defined as the maximum value of its gain function expressed in decibels,

$$g = \max_{\phi,\theta} g(\phi, \theta). \tag{8.14}$$

The field of view of an antenna is a measure of the solid angle in which there is a larger concentration of the transmitted field power, the angle of view is a measure of the antenna's directional properties. Usually the angle of view is defined in terms of planar angles, For a symmetric standard antenna, the antenna's beamwidth Φ_b in radians, in any plane, is related to the angle of view Ω_{fv} in spheroradians by

$$\begin{aligned} \Omega_{fv} &= 2\pi \left(1 - \cos \frac{\Phi_b}{2}\right) \\ &\cong \frac{\pi}{4}\Phi_b^2, \qquad \Phi_b \ll 1. \end{aligned} \tag{8.15}$$

.

This way, using the expression above, the beamwidth and view angles can be easily converted.

An intrinsic property to any antenna is its maximum gain g and its planar angle of half power in radians, combined with a wavelength λ, which relates to the physical area of the antenna A by

$$g = \left(\frac{4\pi}{\lambda^2}\right) \rho_a A, \tag{8.16}$$

where A represents the structural sectional area of the antenna.

The planar angle is giver by

$$\phi_b = c\frac{\lambda}{d\sqrt{\rho_a}}, \tag{8.17}$$

where d is the antenna diameter, c is a proportionality constant which depends on the antenna type and ρ_a is the aperture loss factor.

For a parabolic antenna, for example, these parameters are defined by the following expressions

$$g = \left(\frac{\pi}{3}\right)^2 \left(\frac{df_c}{10^8}\right)^2, \tag{8.18}$$

$$\phi_b = 3,06\left(\frac{10^8}{f_c d}\right), \tag{8.19}$$

where the angle is expressed in radians, d is given in meters, $\rho_a = 1$ and f_c the carrier frequency.

The most common types of antennas are the linear dipole, the helix and the parabolic reflector, the last one is the most well-known of them.

The Effective Isotropic Radiated Power (EIRP) is used by Engineers to quantify the effective power which is radiated by an isotropic antenna in the direction (ϕ_0, θ_0).

$$EIRP = P_T g(\phi_0, \theta_0) \tag{8.20}$$

The satellite antennas can be classified as global or sectorial:

- **Global antennas**: cover all of Earth's surface, from a satellite;
- **Sector antennas**: cover only a predetermined region of the Earth.

The global antenna has an opening of approximately 18°, enough to cover almost a whole terrestrial hemisphere. The sectorial antenna has an opening of about 4°, which permits the illumination of a country with Brazil's dimensions, for example.

8.6 Link Budget

In the following, the basic theory needed to design a line-of-sight link is described.

8.6.1 Influence of the Earth's Curvature and the Equivalent Radius

The energy trajectory for a transmitted signal is curved, and this makes the calculations of the radioelectric connections difficult. To facilitate the calculations, the trajectory is considered straight and the Earth's curvature is modified using a correction factor for its radius.

In the classic method, proposed by Schelleng, Burrous and Ferrell, taking in consideration the effects of atmospheric refraction, an equivalent radius is used for the Earth r. For this, the vertical distances between the path and the Earth's surface and the horizontal distances between the antennas are maintained. The equivalent radius, so calculated, is given by $r = K.h$ where h=6357 km is the Earth's radius and K is given by the expression

$$K = \frac{1}{1 + \frac{a}{n} \cdot \frac{dn}{dh} \cdot \cos\theta}. \tag{8.21}$$

Typical values for K are: $K = \infty$ for a flat Earth, $K = 4/3$ usual value, and $K = 1/3$ for regions which present thermal inversion zones. After the computation of the equivalent radius, one draws the terrains profile wherė the link is supposed to be deployed.

8.6.2 Profile Design

The terrain profile is a sketch between two points which shows the terrain altitude in relation to the fictional curve located at the sea level and which connects those two points.

To obtain a terrain profile an aerophotogrametric or a radioaltimetric process of the region where the intersect points are located is done. If a process of this type can not be done, the region's terrain can be obtained by means of a topographic process or even existing cards to cover with sufficient precision the region of interest.

Next a graph is sketched where the ordinate corresponds to the terrain elevation above the sea level and the abscissa corresponds to the distance between the projection of each point of the sketch over the fictional curve located at sea level and the origin of the coordinates. To consider the effect of the wave refraction in the troposphere, the sketch must be done on a curved reticulated paper, calculated in accordance with the equivalent radius r of the region.

8.6.3 Fresnel Zones

The concept of the Fresnel's first zone radius is very important in the radiovisibility network dimensioning. This theory is summarized below.

In its simplest form, the Huygens principle establishes that each point of a wave front (produced by a primary source) is the source of a new secondary spheric wave. Using the Huygens principle, it is possible to calculate the field for any point in space produced by a known wave front.

Based on the Huygens principle it is possible to determine which regions in space contribute significantly to radiopropagation. To determine those regions, it is useful to analyze Figure 8.6, in which the field in B is resulting of the composition of the fields on the S surface.

In 1818, Fresnel demonstrated that the Figure 8.6, could be interpreted in a very simple way, using optical principles.

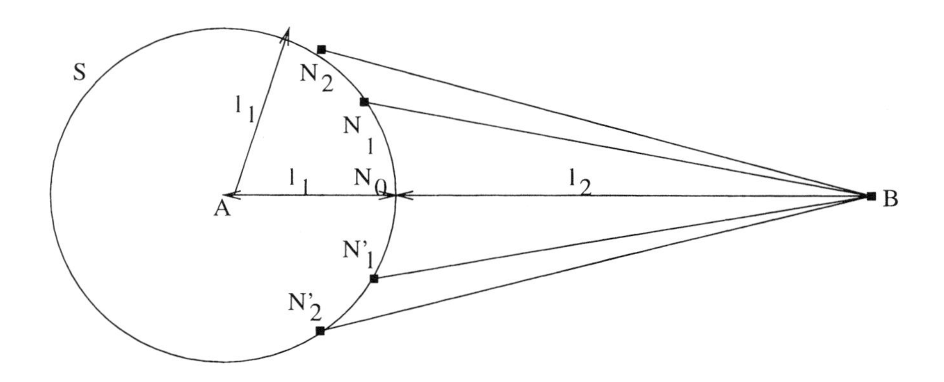

Figure 8.6. Field on a point B due to surface S.

1 On the surface S the field presents constant phase;

2 The phase difference introduced by the difference in path lengths is responsible for the additive or subtractive effect in the field at B.

To reach a perfect definition of the Fresnel zones, it is first necessary to introduce the Fresnel ellipsoids. Fresnel ellipsoids are the geometric places of the points, where the sum of the distances to the transmission antenna and the receptor is constant and exceeds the distance in free view, of an integer number of half lengths of a wave. Figure 8.7 illustrates the section of the Fresnel ellipsoid in the plane which contains the transmitter and the receptor, showing the several parameters of interest.

The several Fresnel zones correspond to the regions limited by two consecutive ellipsoids. The opening radius of any Fresnel zone can be calculated considering the existence of an obstacle of the knife edge type (for example a hill) at a distance d_1 from the transmitter and d_2 from the receptor, in which $d_1 + d_2 = d$, and a sum as shown in Figure 8.7.

Regions of space corresponding to the path lengths $\lambda/2, 2\lambda/2, 3\lambda/2, \ldots$, in relation to the line of sight of the ellipsoids, with focuses in T and R. The nth Fresnel surface is the one in which the sum of the distances between the transmitter and the receiver and a point on the surface of the revolution ellipsoid exceeds in $n\lambda/2$ the distance between the transmitter and the receiver.

Therefore,

$$(r_1 + r_2) - (d_1 + d_2) = \frac{n\lambda}{2}, \tag{8.22}$$

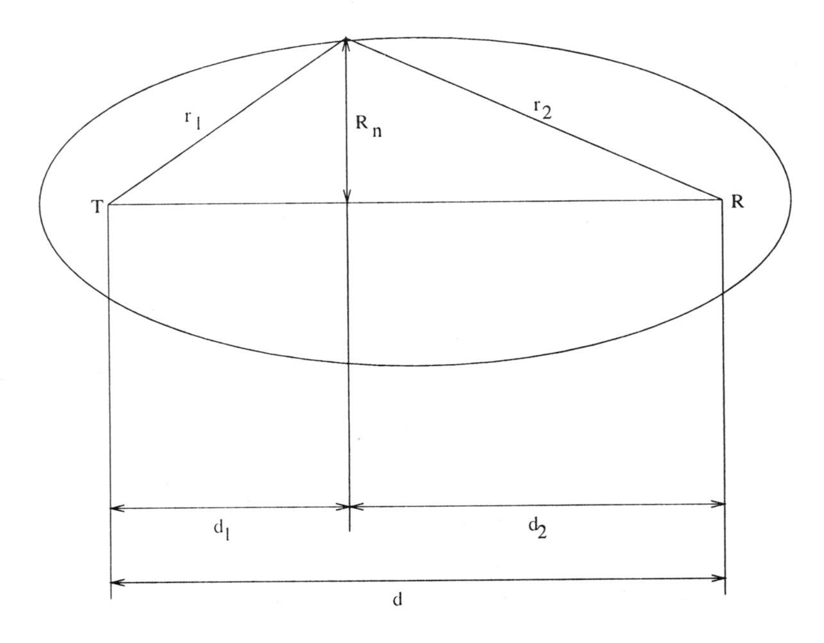

Figure 8.7. Section of the Fresnel ellipsoid.

where

$$r_1 = \sqrt{d_1^2 + R_n^2} = d_1\sqrt{1 + \left(\frac{R_n}{d_1}\right)^2} \tag{8.23}$$

and

$$r_2 = \sqrt{d_2^2 + R_n^2} = d_2\sqrt{1 + \left(\frac{R_n}{d_2}\right)^2}. \tag{8.24}$$

Considering that $R_n/d \ll 1$, the previous formulas can be simplified to

$$r_1 = d_1\left[1 + \frac{1}{2}\left(\frac{R_n}{d_1}\right)^2\right] \tag{8.25}$$

and

$$r_2 = d_2\left[1 + \frac{1}{2}\left(\frac{R_n}{d_2}\right)^2\right]. \tag{8.26}$$

Substituting the approximations in the original equation, one obtains

$$\frac{R_n^2}{2}\left[\frac{1}{d_1} + \frac{1}{d_2}\right] = \frac{n\lambda}{2}. \tag{8.27}$$

Finally, the opening radius of the nth Fresnel zone is given by the relation (Panter, 1972)

$$R_n = \sqrt{\frac{n\lambda d_1 d_2}{d}} \tag{8.28}$$

in which n is th order of the Fresnel zone considered and the other parameters are shown in the Figure 8.7.

Finally, the opening radius of the nth Fresnel zone is given by (Panter, 1972)

$$R_n = \sqrt{\frac{n\lambda d_1 d_2}{d}} \tag{8.29}$$

where n is the order of the considered Fresnel zone and the remaining parameters are shown in Figure 8.7.

8.7 Problems

1 Describe the transmission subsystems.

2 What are the main channel types? Give examples.

3 Explain the channel effects on the transmitted signal.

4 The transmitted signal $s(t)$ is attenuated in free space and phase shifted in its path to the receiver, producing at the antenna a signal $r(t) = \alpha m(t)\cos(\omega_C t + \theta)$, where θ represents a random phase shift uniformly distributed in the interval $[0, 2\pi]$. The signal passes through an ideal passband filter $H_C(\omega)$, tuned to the carrier frequency, with bandwidth $2\omega_M$. A local mixer, operating at frequency $\omega_L = \omega_C + \omega_I$, translates the signal, which is then filtered by an ideal filter, $H_I(\omega)$, with bandwidth $2\omega_M$, in the intermediate frequency (IF).

(a) What is the signal autocorrelation at the $H_I(\omega)$ filter's output?

(b) Determine and sketch the power spectrum density of that signal.

5 The noise figure can be computed as a function of the input and output signal to noise ration. Considering that the the signal source produces a noise of power spectrum density KT_0B, that the amplifier gain is G and that the internal amplifier noise power is P_N, determine the formula for the noise figure.

6 Define fading and cite the main types of fading usually considered in the design of communication systems.

7 What defines a wave-front? What is a plane wave? Define vertical polarization. What characterized the circular polarization?

8 What is the power received by an antenna of diameter $d = 4$ m, which is located in a satellite at an altitude of 50.000 km, if the isotropic effective radiated power is given by EIRP $= 10$ kW?

9 Compute the gain for an antenna which operates at a frequency of 5 GHz, presents a physical area $A = 10$ m^2 and a loss factor $\rho_{ap} = 0,7$.

10 What is the coupling loss, in dB, of a signal which is coupled to an antenna by means of a coaxial cable, if the antenna impedance is $Z_T = 377\ \Omega$, the cable impedance is $Z_G = 360\ \Omega$ and amplifier impedance is $Z_A = 300\ \Omega$? The amplifier output power and the cable loss are, respectively, $P_{AMP} = 1$ kW and $L_G = -20$ dB.

11 Compute the power received from a satellite by a 5 m diameter antenna located in Manaus, Brazil, if itis known that the signal has been transmitted to the satellite by a base station in New York, United States. What is the power amplification needed on board the satellite to obtain a received power of 1 W, considering that the low noise amplifier (LNA) at the Manaus station is 50 dB and the cable loss is 2 dB? Data:

(a) Uplink frequency: $f_U = 6$ GHz;

(b) Downlink frequency: $f_D = 4$ GHz;

(c) Diameter of the transmitting antenna: $d_T = 10$ m;

(d) Diameter of the antenna on board the satellite: $d_S = 2$ m;

(e) Transponder gain at the satellite: $G_S = 80$ dB;

(f) Satellite altitude in relation to Manaus: $h_S = 35.800$ km;

(g) Potência do transmissor: $P_T = 100$ W.

12 Two amplifiers presenting noise figures $F_1 = F_2 = 2\ dB$ are mounted in series. Compute the total noise figure of the assembly for the following gains:

(a) $G_1 = G_2 = 30$ dB;

(b) $G_1 = 40\ dB$, $G_2 = 20$ dB.

13 A geostationary satellite at 35.800 km of altitude illuminates the United States, using a parabolic antenna of diameter 2 m. Consider the downlink frequency 4 GHz and compute the illuminated area. What is the antenna gain in dB?

14 An Earth station of a satellite network has has a GaAs FET pre-amplifier of gain 27 dB and a TWT amplifier of gain 69 dB. The antenna diameter is 10 m, the uplink frequency is 6 GHz. What is the gain of the link?

15 A line-of-sight link of 50 km transmits at a frequency of 7 GHz. Both ends of the link use parabolic antennas, whose diameter is 2 m. The atmospheric losses area given by

$$L = 10^{-0,9\alpha d}$$

where d is the distance in km and α represents the attenuation in dB/km. The coupling losses are negligible, as well as the radiation losses. For an atmospheric attenuation of 0.2 dB, what the the received power for a transmitted power of 10 W.

16 A FET amplifier response is given by the equation $Y = \alpha X - \beta X^2$. If it is known that $R_{X^2}(\tau) = 2R_X^2(\tau) + P_X^2$ and that all the odd moments are null, compute and sketch the power spectrum density at the output of the amplifier, for a white Gaussian input signal.

17 A line-of-sight link of 50 km, operates at 7.5 GHz and has two towers of 80 m located on elevations of 250 m and 300 m, respectively. Consider the flat Earth model and the existence of a hill of height 338 m, at half distance between the two towers. Determine the opening radius of Fresnel's first zone ($h = 1$). Compute the attenuation caused by the obstruction. Assume that the parabolic antennas' diameter is 1.5 m.

18 Three amplifiers, with gain $g(X) = \alpha - \beta X$, are amounted in series. Compute the gain of the assembly. Determine the value of the input voltage for which the assembly no longer works as am amplifier. What is the output voltage in this case?

Chapter 9

MOBILE CELLULAR TELEPHONY

9.1 Introduction

A mobile telephone system is defined as a communications network via radio which allows continuous mobility by dividing its coverage area into cells. Wireless communications, on the other hand, implies radio communications without necessarily requiring handover from one cell to another during an ongoing call (Nanda and Goodman, 1992).

The cellular communications network was originally proposed by the Bell Laboratories, in USA, in 1947. However, experiments only started in 1978. The first country to offer a cellular service was Switzerland, in 1981, although Motorola had patented the cellular telephone in 1973. In spite of its release in 1979, in Chicago, the system only started full installation in 1984 in USA.

The effort of the industry to provide ever more efficient mobile communications via radio to the population, has demanded intense research and development over the last few years (Hashemi, 1991, Dhir, 2004). One of the aims of this research is to avoid high costs associated with the installation and relocation in places interconnected by wires (Freeburg, 1991).

Various cellular telephone systems have been proposed and some are already operational to handle the control and the information flow in mobile systems. The main proposed systems are the following.

1) Code division multiple access (CDMA);

2) Time division multiple access (TDMA);

3) Frequency division multiple access (FDMA).

A conventional mobile telephone system selects one or more RF channels for use in specific geographic areas. The area of coverage is planned to be as large as possible, which may require a rather large transmitting power.

In a cellular mobile telephone system, the area of coverage is divided into regions called cells, in a manner that the transmitted power is low and the available frequencies for RF channels may be reused.

9.2 Systems in Commercial Operation

The majority of the mobile cellular analog telephone systems in commercial operation are very similar. All of them, in one way or another, are based on the *Advanced Mobile Phone Service* (AMPS), developed in USA in the seventies, and are currently being replaced by digital systems. However, the procedures for completing and maintaining a call are similar to those of the digital systems but are easier to understand. For this reason the analog system will be used as a reference in this chapter.

In the AMPS system, the voice signals modulate the channel carriers in frequency (FM). Signaling, which operates at a 10 kbit/s rate, uses frequency shift keying (FSK), i.e., digital FM. Voice channels as well as signaling channels occupy, individually, a bandwidth of 30 kHz.

The use of frequency modulation, i.e., of a constant envelope carrier, in the AMPS standard has the purpose of fighting multipath effects. The multipath effect, common in the mobile channel, produces a multiplicative noise which acts on the carrier as an amplitude modulating signal. For a transmitted signal $s(t)$ the received signal is given by (9.1)

$$r(t) = \alpha(t)s(t) + n(t), \tag{9.1}$$

where $n(t)$ represents additive noise, usually with a Gaussian probability distribution, and $\alpha(t)$ represents the multiplicative noise.

If the signal $s(t)$ were amplitude modulated, the multiplicative noise would be incorporated to the modulating signal, making it almost impossible to recover the original modulating signal. Angle modulation (frequency or phase) provides for the elimination of a considerable part of the noise by limiting (clamping) the carrier amplitude before demodulation. This will not affect signal recovery, since the desired information is contained in the frequency, or phase, of the carrier.

An important characteristic of these systems is their concern with the protection of transmitted messages. The analog systems still in operation in the world are: NTT (Japan), AMPS (USA and Brazil), TACS (UK), NMT (Scandinavia) and C-450 (Germany), which make use of the following protection methods:

Characteristics	Japan	USA	UK	Germany
System	NTT	AMPS	TACS	C-450
Frequency band (MHz)				
base → mobile	870 - 885	869 - 894	935 - 960	461,3 - 465,74
mobile → base	925 - 940	824 - 849	890 - 915	451.1 - 455.74
Frequency spacing between Tx and Rx (MHz)	55	45	45	10
Channel spacing (kHz)	25	30	25	20
Number of channels	600	832 †	1000 †	122
Coverage radius (km)	5 - 10	2 - 20	2 - 20	5 - 30
Audio signal				
Modulation type	FM	FM	FM	FM
Frequency deviation	±5	±12	±9, 5	±4
Control signal				
Modulation type	FSK	FSK	FSK	FSK
Frequency deviation	±4.5	±8	±6.4	±2.5
Data Transmission Rate de Dados (kbit/s)	0.3	10	8	5.28
Message Protection	ARQ	PMD	PMD	ARQ

†42 control channels

Table 9.1. Main technical characteristics of mobile communications standards.

- Principle of Majority Decision (PMD), in which various copies of the signal are transmitted;

- Automatic Repeat Request (ARQ), in which whenever an error is detected, then a signal retransmission is requested.

In environments with strong fading, the PMD is a good option whereas in environments with slow fading ARQ is the best option. The AMPs system uses PMD. Table 9.1 summarizes the main technical characteristics of the analog standards.

Considering that the analog system is used in many countries for interconnecting different digital standards, it still constitutes an important basis for mobile cellular telephony, with frequency modulation being used in all implemented systems. The basic principles of digital telephony are independent of the specific type of system employed.

9.3 Description of the Cellular System

The design of a cellular system consists in dividing into smaller areas the area to be covered by mobile telephony, thus allowing the use of smaller power transmitters and efficient use of the spectrum by means of frequency reuse.

In this context it is said that cellular systems are limited by interference while conventional mobile systems, using higher power, which are limited by thermal noise.

9.3.1 Cellular Structure

In principle a given region or coverage geographic area to be served by the mobile cellular service is divided into subregions, called cells. The cell is a geographic area *illuminated* by a base station, within which signal reception conforms to the system specification.

Extension of the coverage area of a base station depends on the following.

- radio transmitter output power;
- frequency band employed;
- antenna height and location;
- antenna type;
- terrain topography;
- receiver sensitivity.

In conventional systems the approach used to cover a large area is to radiate a relatively high power level. This technique however is not appropriate for cellular systems, except in areas with a very low traffic density, or with coverage by large cells.

Radio waves propagate from a base station in a straight line, called line of sight. This means that a user connected to a base station located behind a large obstacle as for example a hill or a tunnel, will be in an area without radio coverage called an area of shade.

However, the presence of buildings in large cities is not a critical impediment to propagation of radio waves thanks to their reflection and refraction properties, which minimize the shade effect. On the other hand, areas with a high traffic density are usually divided into a large quantity of small cells.

By adjusting the output power at the base station transmitter, the coverage area can be increased or reduced to the appropriate dimension. Currently there are cells with a coverage radius of less than 500 m for outdoor environments and micro-cells, used in shopping centers, viaducts and convention centers.

In its basic conception, the operator of the cellular system divides an area into equal size cells of a hexagonal shape, in a manner that the center of each cell contains a base station (CNTr, 1992). The most common types of cells are the following.

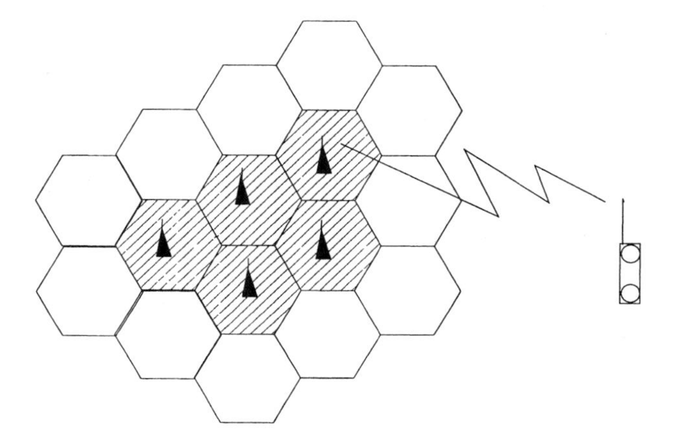

Figure 9.1. Description of the cellular system.

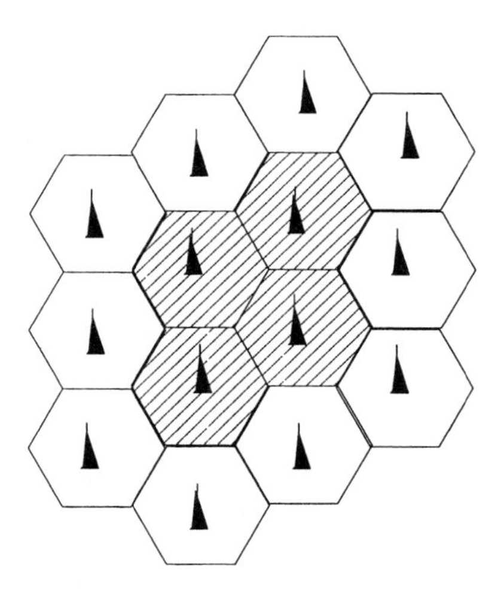

Figure 9.2. Sets of omnidirectional cells – four cells ($i = 2; j = 0$).

- **Omnidirectional cell** – in omnidirectional cells, the base station is equipped with an omnidirectional antenna, i.e., an antenna which transmits the same power in all directions, in the azimuthal plane, forming thus a coverage area approximately circular, the center of which is the base station itself. This type of cell is graphically represented by a hexagon.

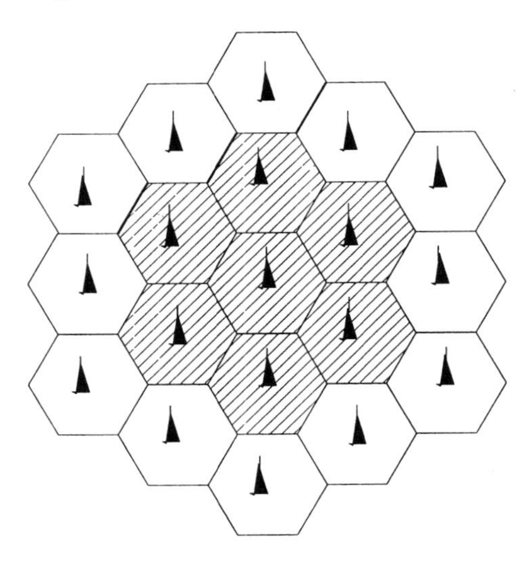

Figure 9.3. Sets of omnidirectional cells – seven cells ($i = 2; j = 1$).

- **Sector cell** – in sector cells the base station in each cell is equipped a directive antenna, so as to cover a specific region.

9.3.2 Cellular Structures

A group of neighbor cells which use the whole group of 395 voice channels and 21 control channels available in the Advanced Mobile Phone System (AMPS) is called a *cluster*, or set, of cells. In other words, within a cluster there is no frequency reuse.

The number of cells forming cluster differs according to the cell structure employed. The most common cell structures are the following.

- Four cell standard, all omnidirectional cells, as shown in Figure 9.2;
- Seven cell standard, all omnidirectional cells, illustrated in Figure 9.3;
- 12 Cell standard, all omnidirectional cells. Figure 9.4 illustrates the cell distribution;
- 21 Cell standard, with seven base stations, each base station associated to three sector cells, as shown in Figure 9.5;
- 24 Cell standard, with four base stations, each base station associated to three sector cells, as illustrated in Figure 9.6.

In a cluster with seven cells there seven different sets of frequencies, which can receive for example the designation A, B, C, D, E, F, G. However, in a cluster with 21 sector cells a more convenient designation is A1, A2, A3, B1, B2, B3, C1, C2, C3, D1, D2, D3, E1, E2, E3, F1, F2, F3, G1, G2 e G3.

In a cluster with 21 cells, each pair of channels among those used by a specific cell are spaced by 21 other channels. This characteristic, besides facilitating transmitter use, eliminates also the risk of possible adjacent channel interference within a cell. Channel allocation takes into consideration the traffic density between various cells.

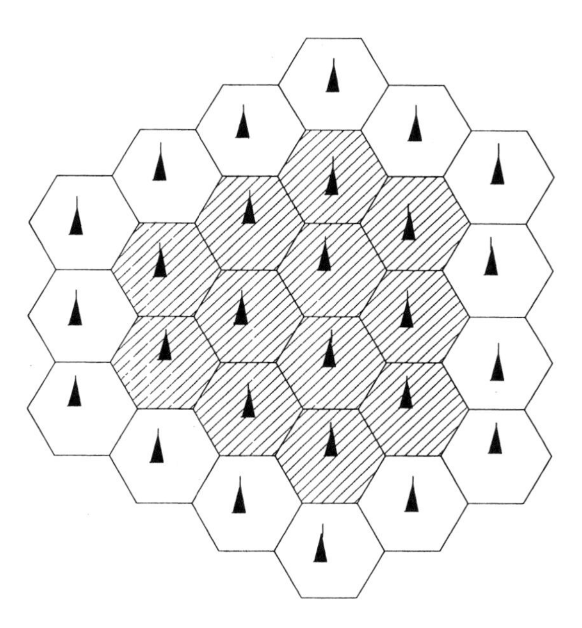

Figure 9.4. Sets of omnidirectional cells – 12 cells ($i = 2; j = 2$).

9.4 Frequency Reuse

Frequency reuse means the simultaneous use of the same frequency in two distinct sets of cells. The distance between cells for frequency reuse (reuse distance) is limited by the maximum co-channel interference allowed in the system.

In the case of a homogeneous system, formed be sets of $N = i^2 + ij + j^2$ cells, where i and j are positive integers, the reuse distance D is given

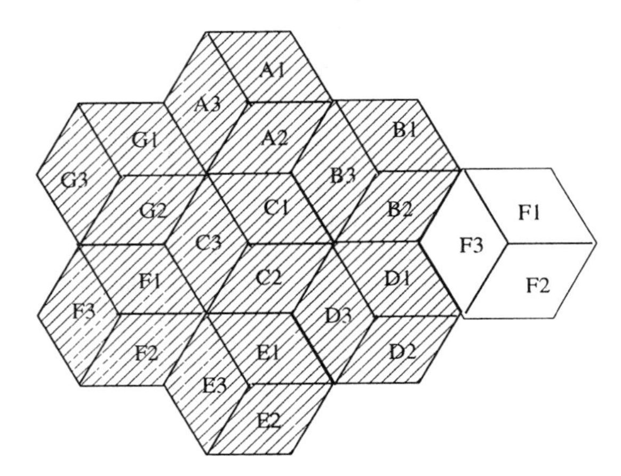

Figure 9.5. Sets of sector cells – Set of seven cells with three sectors.

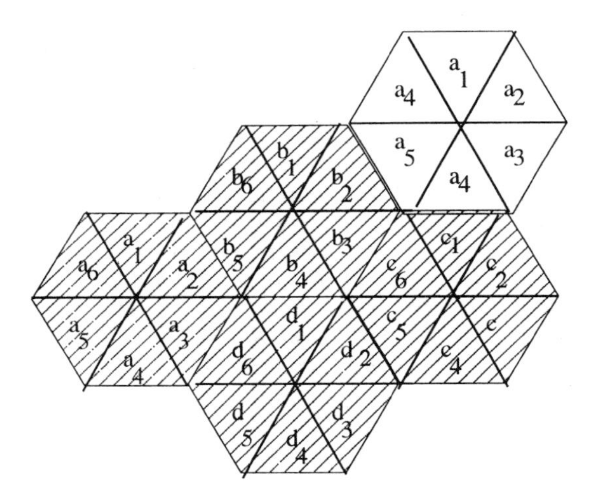

Figure 9.6. Sets of sector cells – Set of four cells with six sectors.

by

$$\frac{D}{R} = \sqrt{3N},$$

where R is the radius.

Evidently, the quality of the transmissions is reduced as the size of the clusters becomes smaller, but the traffic capacity increases due to the possibility of distributing all channels among a few cells. Table 9.2 shows the relationship existing between traffic capacity and co-channel interference, for various hypothetical sizes of cell sets (Yacoub, 1993).

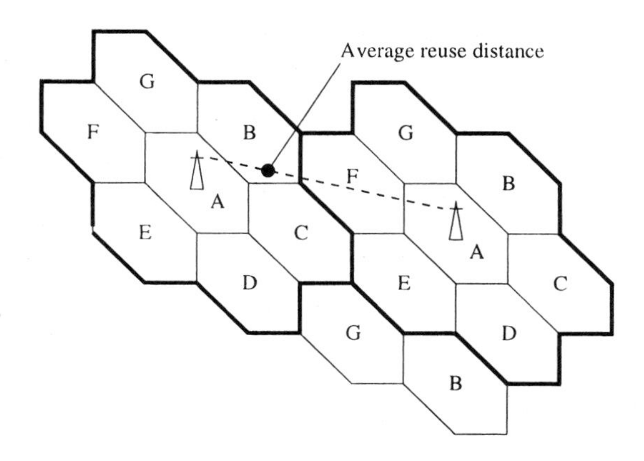

Figure 9.7. Concept of average reuse distance.

Dimension	D/R	Channels/cell	Traffic capacity	Trans. quality
1	1.73	360	Higher	Lower
3	3.00	120	↑	
4	3.46	90		
7	4.58	51		↓
12	6.00	30	Lower	Higher

Table 9.2. Traffic capacity and co-channel interference.

Sets of frequencies, with no frequency in common, must be allocated to neighboring cells, since a total radio coverage of a region requires superposition of cells. However, if a particular frequency is used in neighboring cells, the so called co-channel interference is likely to appear in the superposition areas.

Interference is the factor that pushes for a substantial increase in the separation distance between two cells using the same frequency. This distanced receives the denomination of *repetition distance* and the reutilization of the same frequency in different cells is called *frequency reuse.* Due to the use of the frequency reuse technique, it is concluded that the maximum number of voice channels in the AMPS system is a multiple of 395, in the so called extended AMPS.

9.4.1 Cell Division

Suppose that a new cellular system has been introduced, which covers a large city and its neighboring region. If it can be affirmed that in the near future the subscriber traffic density will be low, then there will be no need to expand the system quickly. This would avoid making available an excessive capacity at a high cost and without a quick return for the investment.

As a first step a base station could be installed, with a high antenna and an output power sufficient to cover a radius of 20 km, for example. Considering a cluster structure, the base station could allocate a group of frequencies like, for example, group A, and consequently the base station would be limited to approximately 45 voice channels and three control channels.

A possible way to increase capacity within a given coverage area is to place, for example, three more base stations superimposed. The three new cells could temporarily work as omnidirectional cells, but in future they can be replaced by sector cells. The frequency group must be defined according to then frequencies that the new base stations will use in future. For example, according to the designation shown in Section 9.4, groups C1, E1 and G1 can be chosen.

As time passes it may become necessary to increase the capacity in the central area. For that purpose, the coverage of the base station in group A is reduced, by means a reduction in its power level, which causes a reduction of the coverage area. In spite of cells in groups C1, E1 and G1 being designed to cover suburbs, they can in some case provide an auxiliary coverage to the central area.

When the demand for traffic grows within a given cell, there are two possible solutions to handle more traffic: addition of new cells or the sectoring of existing cells.

- Addition of new cells – The power of the transmitters in the existing cells is reduced enough to cover half of the original coverage area and new cells are added to meet the additional need for coverage. The addition of new cells implies in the installation of towers and antennae, with their respective costs;

- Cell sectoring – The set of omnidirectional antennas in a cell is replaced by directional antennas (with 60° or 120° aperture) and the cell is divided into sectors. Cell sectoring is a more economical solution because it uses exiting structures.

The cell size must be adequate to the telephone traffic density. The higher the traffic the smaller the corresponding cell size, since the number of voice channels available per cell is limited. This implies, for example, that in areas near the city center the cells are smaller than those in suburban areas.

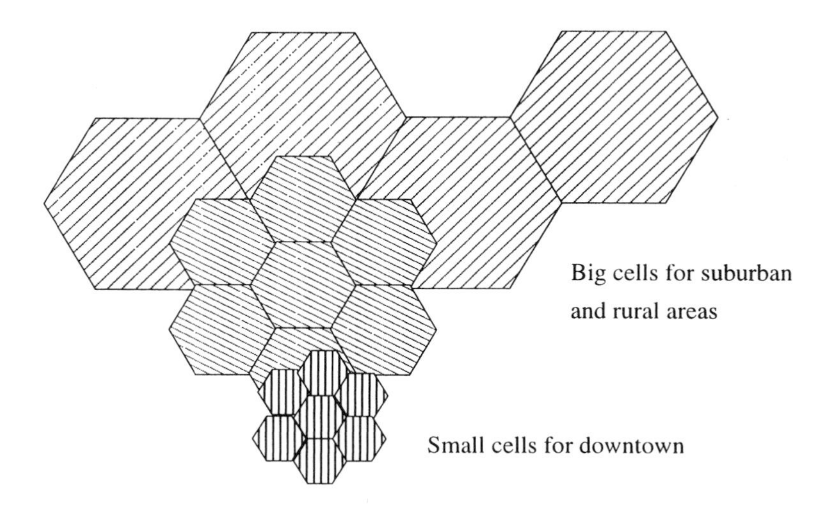

Figure 9.8. Cell division.

The addition of new cells is a more flexible manner to expand the system, but is also a most expensive one. It involves the installation of new base stations and antennas. The acquisition of land for the installation of base stations in city central areas became prohibitive for the operators, which prefer to rent a small area on the top of buildings or other commercial areas. Cell sectoring has a smaller cost for the operator, however it imposes restrictions to future expansions.

The subdivision of large cells into small cells implies that the frequency reuse distance becomes smaller and that the number of channels within the same geographic area is increased, thus increasing system capacity.

The technique of superimposing cells is employed in different situations to solve specific problems. The border region between large and small cells provides a typical example. It is possible to define a superimposed cell as a group of voice channels allocated in the same region of a common cell, which receives the name of sub-imposed cell. A superimposed cell differs from a sub-imposed cell because the latter not only has a smaller radius, but does not have a signal intensity receiver and a control channel.

9.5 Management of Channel Utilization

The spectrum allocation in the AMPS standard considers separate frequencies for the transmission from the base station and transmission from the mobile terminal (full duplex operation). The difference between the two frequencies is 45 MHz.

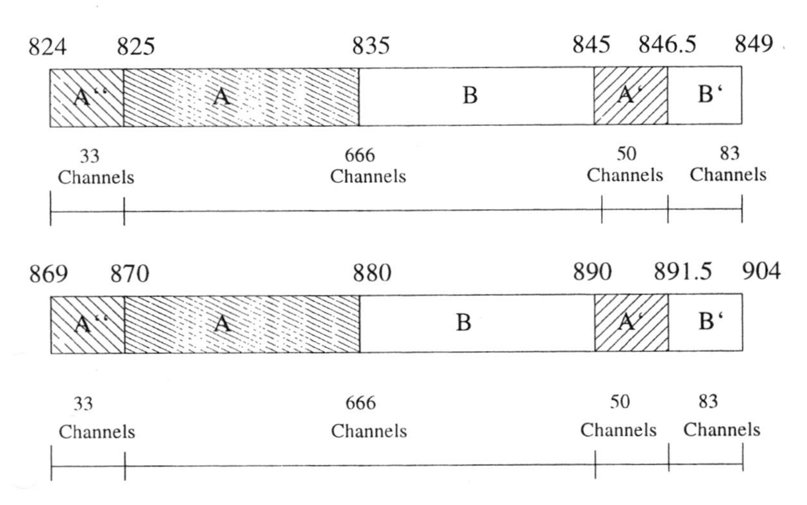

Figure 9.9. Frequency bands.

The AMPS system operates in the frequency band of 825 to 845 MHz (824 to 849 MHz in the extended system) for transmission from the mobile terminal to the base station, and the frequency band of 870 to 890 MHz (869 to 894 MHz in the extended system) for transmission in the reverse direction. Most of the channels (having 30 kHz bandwidth) is allocated for telephonic conversation. The remaining channels transmit signaling information in digital form. The channels used for signaling are called *setup* and are used for exchanging messages required to establish a call.

9.5.1 Channel Allocation in the AMPS System

In the AMPS system the channel frequency allocation is done as follows, where the figures in parenthesis refer to the extended system.

- 825 to 845 MHz: 666 channels with bandwidth 30 kHz each (824 to 849 MHz: 832 channels with bandwidth 30 kHz each) for communication from the mobile terminal to the base station;

- 870 to 890 MHz: 666 channels with bandwidth 30 kHz each (869 to 894 MHz: 832 channels with bandwidth 30 kHz each) for communication from the base station to the mobile terminal;

- The set of 666 (832) channels is divided into two independent systems, A and B, such that each can be exploited by a different operator;

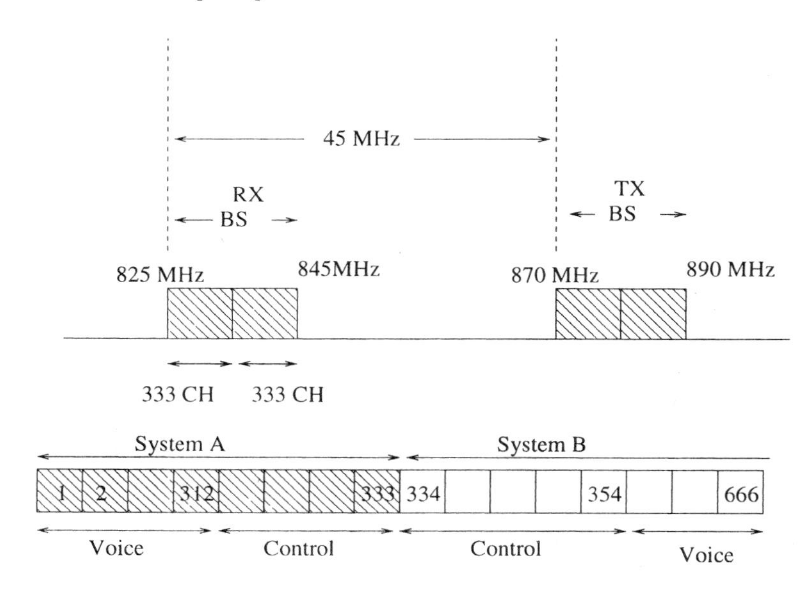

Figure 9.10. Channel allocation in the AMPS system.

- Each operator has available 333 (416) channels and can market 312 voice channels, because the remaining 21 channels are required for control purposes;
- In the original standard, channels are numbered from 1 to 333 in system A, and from 334 to 666 in system B, beginning from the lowest channel frequency available in each system.

A channel center frequency can be calculated from its corresponding channel number as follows.

- Direction from base station to mobile terminal: F = (channel number x 0.03) + 870;
- Direction from mobile terminal to base station :F = (channel number x 0.03) + 825.

9.6 Constitution of the Cellular System

A typical cellular system consists of three elements, including the connections between them. The basic components of a cellular system are as follows (CNTr, 1992).

- Mobile terminal (MT);

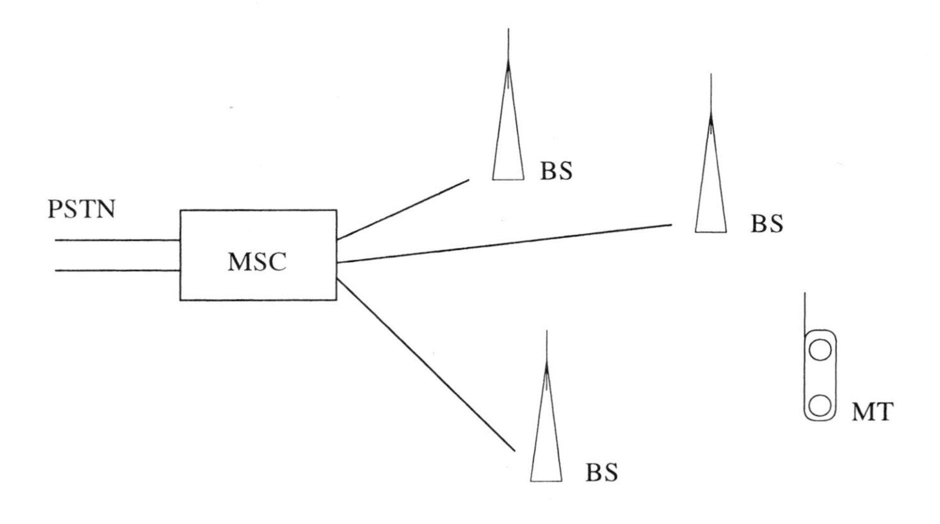

Figure 9.11. Components of a cellular system.

- Base station (BS);
- Mobile switching center (MSC);

The mobile terminal (MT) provides the air interface for the user of the cellular system, contains also a control unit, a transceiver and an antenna. The MT transmits and receives voice signals, thus making conversation possible, and transmits and receives control signals, thus allowing the establishment of a call. Mobile terminals are produced by a large number of manufacturers, which is the reason for a wide variety of designs and facilities for the subscribers. The MT's can be used in a variety of applications as, for example:

- Installed in a car (vehicular cellular telephone);
- Portable (portable cellular telephone);
- Installed in a rural area (rural cellular telephone).

The maximum output power of a unit installed in a car or in a rural area is 3 watts, while the power of a portable unit is 0.6 W for American standards, and 1 W for the European standard. When a mobile terminal assesses an MSC, it sends its station class in which is indicated

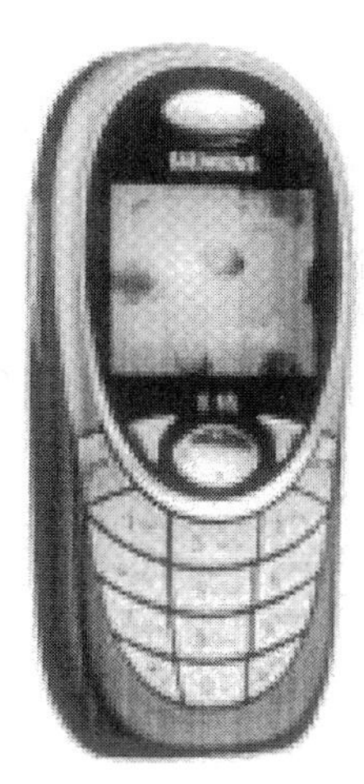

Figure 9.12. Example of a cellular telephone, Siemens C60 (Courtesy of Siemens).

its maximum output power. An MSC keeps a file with parameters of all registered cellular terminals, for call control purposes.

In order to be identified by the system, each mobile terminal has a *mobile identification number* (MIN) and, for security reasons, each mobile unit has an *electronic serial number* (ESN), which is defined by the manufacturer.

The signal quality monitoring which reaches the mobile terminals is performed by the audio monitoring tone, or supervision audio tone (SAT), and by the measurement of the intensity level of signals received by the mobile terminals, in the analog case, and by the value of the bit error rate (BER) in digital systems.

A base station encompasses the following functional units, as illustrated in Figure 9.14:

- Group of radio channels (GRC);
- Radio exchange interface (REI);
- Antenna combiner;
- Antennas and power sources.

The group of radio channels contains voice channels and control channels. The radio exchange interface (REI) operates with an adaptor for voice signals between the MSC and the base station, since the coding for the aerial interface is different from the coding used for communication

Figure 9.13. Example of a base station, BS 240 XS (Courtesy of Siemens).

between telephone exchanges. Thus, the equipment receives data and voice from the channel units and sends information to the MSC by means of a dedicated link from base station to MSC. In the opposite direction the equipment receives data and voice from the MSC by means of a link from MSC to base station, and sends them to the corresponding channel unit or control unit. Among other things a base station takes care of the following.

- Provides the interface between an exchange and control unit and mobile terminals;
- Contains a control unit, radio transceivers, antennas, power plants and data terminals;
- Transmits and receives control signals for establishing call monitoring;
- Transmits and receives voice signals from various mobile terminals within its coverage area.

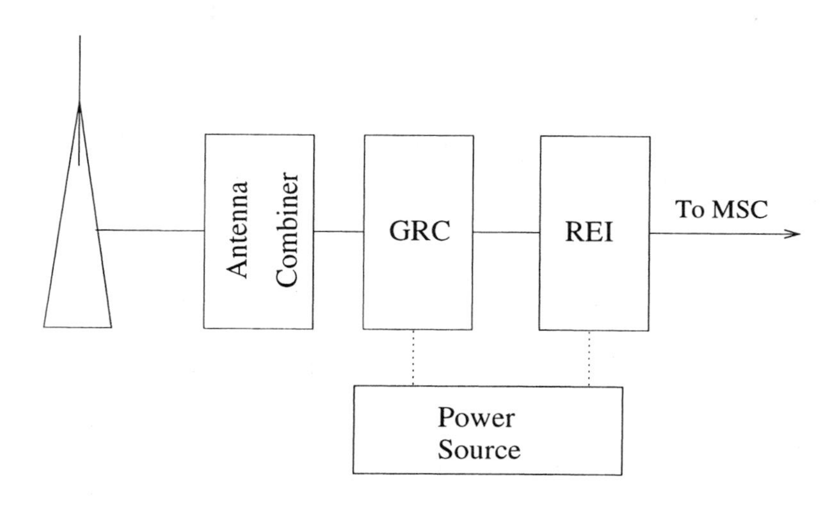

Figure 9.14. Composition of a base station.

In the connections between base stations and MSC's two types of facilities are employed.

- Trunks which provide a voice communication path. The number of trunks is decided based on the traffic and on the blocking probability desired (desired grade of service);

- Trunks which provide a signaling path to the establishment and monitoring of calls.

Since a base station transmits and receives control signals for the establishment and monitoring of calls, as well as voice signals from various mobile terminals within its coverage area, it can be said that the base station acts as a traffic concentrator for the MSC.

The switching and control center (MSC) is considered the main element of the mobile cellular system. Among its functions it is worth mentioning the following.

- Central coordination of the whole cellular network, managing all base stations within its control area, i.e., switching and controlling the cellular aggregate;

- It works as an interface with the public switched network;

- It switches calls originated or terminated in a mobile station;

- It allows the mobile station to have available the same services provided by the public network for fix subscribers;

  enlace

- The MSC connects to the base stations via four wire type connections, by PCM multiplex. Usually the connection is implemented with an optical cable.

  recebimento

9.7 Characteristic Functions of a Cellular Network

A complete cellular system may contain many MSC's, which can be seen as functional interfaces to the switched public telephone network, and the signaling employed for establishing calls will depend on that signaling defined by the switched public telephone network. Each mobile station is connected by software to a unique MSC, which is normally that in the subscriber home area. This MSC is called the home center (MSC-H).

The MSC operates on a trunk-to-trunk basis, having three types of interconnecting links.

- Type 1. Interconnects the MSC to a local public switching telephone exchange;

- Type 2A. Interconnects the MSC to a tandem public switching telephone exchange;

- Type 2B. Interconnects the MSC to a local exchange together with type 2A over a high usage alternative routing basis.

9.7.1 Handoff

Handoff is a function which allows the continuity of a conversation to be preserved when a user crosses from one cell to another. The handoff is centralized in the MSC and in the AMPS system it causes a communication interruption causa of less than 0.5 s.

9.7.2 Roaming

When a mobile station enters the control area of a MSC which is not his/her home MSC, this switching center is known as the visited center (MSC-V) and the subscriber is called the visitor. Calls for a subscriber in this condition are routed and switched by the MSC-V. The concept of a mobile station moving from one control area to another is called roaming.

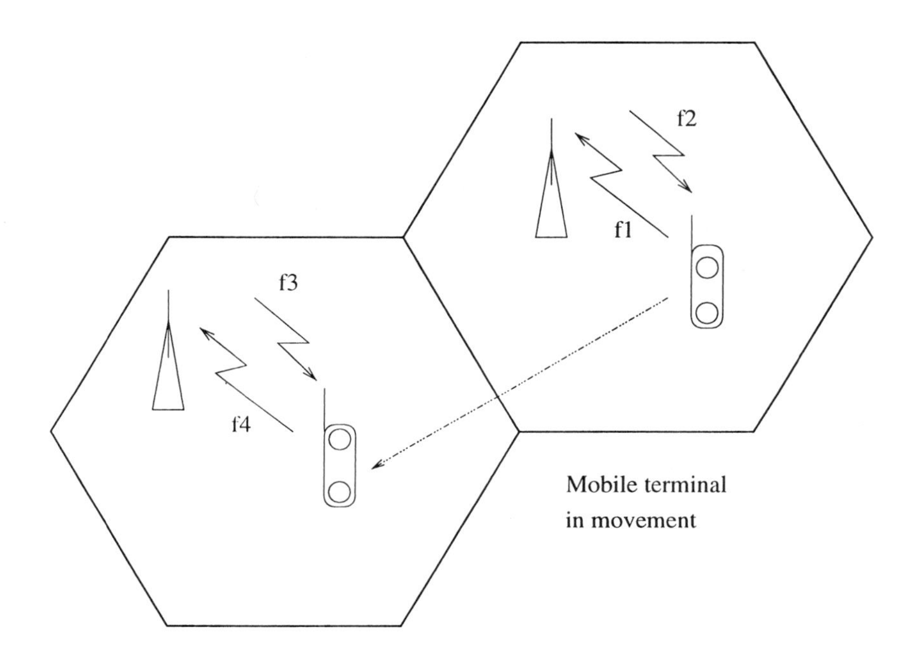

Figure 9.15. Example de of handoff.

Signaling between MSCs can be implemented, for example, according to a protocol called common channel signaling number 7 (CCS-7) of the ITU-T, by means of a direct link connecting two MSCs or by a switched public telephone network (SPTN) (Alencar, 2002).

The following procedures are employed when the roaming function is used.

1) When a mobile station enters a new control area it is automatically registered in the MSC controlling that area;

2) The visited MSC checks whether the mobile station had already been registered earlier. If that was not the case, the visited MSC will inform the home MSC of this new condition;

3) The home MSC registers the service area being visited by the subscriber. After this procedure is completed, the visiting subscriber is

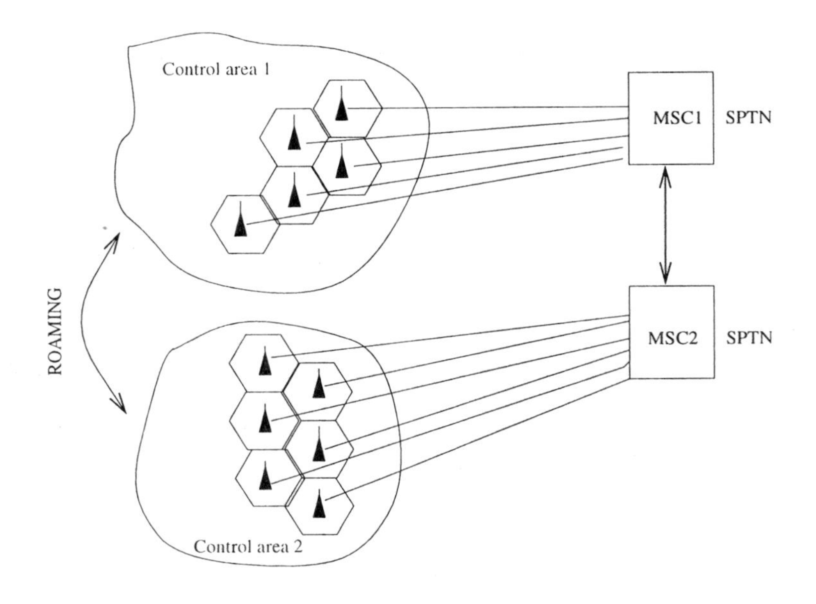

Figure 9.16. Example of roaming.

then able to originate and to receive calls, as if he was in his own home area.

The change from a base station connected to a MSC to another base station connected to another MSC, during a call in progress, is called *call handoff* between control centers. In some systems, the control areas can still be subdivided into location areas. When a mobile terminal passes from one location area to another, it must inform the MSC about its new location. This task is known as forced registration or as registration of the location area. In this case the search for the mobile terminal is done only in those cells that belong in the location area.

The forced registration implies in the updating of files in the MSC at predefined time intervals. Mobile terminals perform a routine of generation of a random time value (time instant), starting from a seed sent by the base station, for sending location information. The maximum period for sending data is defined by the control center operator. The periodic updating of registers might overload the central processor and degrade the grade of service of the MSC (Alencar, 2002).

The registration per location area is done every time a subscriber crosses the limit preestablished by the system operator. This limit can be, for example, the cell or the coverage area of the MSC. The larger the area to be covered, for locating a subscriber, the larger the amount of allocated resources of the mobile cellular system. This can block channels and also degrade the grade of service of the system. A combination of the two techniques is in general more efficient and several algorithms have been developed with this objective in mind.

9.7.3 Configuration of the Mobile Cellular System

A base station is connected to a MSC by means of point-to-point circuits. The base station has as its main function to work as a repeater for voice information and data, as well as to monitor the transmission link quality during a conversation.

The connections between a base station and a MSC are normally done by trunks of physical lines (PCM), but connections via radio or optical fibers is also possible. The MSC's are in general connected to the public switched telephone network via optical fibers.

9.7.4 Data Communication from the MSC to the Base Station

In a cellular system there is a constant exchange of information between the control central and the base station. The communication between the MSC and the base station occurs in the following cases.

- Whenever the MSC originates a message for a mobile station, to be sent through the control channel or through the voice channel;
- Whenever the MSC receives a message from a mobile station;
- Whenever the MSC receives a request from the base station as, for example, a handoff request;
- Whenever the MSC sends a message to the base station as, for example, requests for the values of measurements made in a process of handoff;
- Whenever the MSC receives an alarm message due to failure in the radio link as, for example, failure in the transmitter or a power failure;
- Whenever the MSC receives a message originated from a base station;
- Whenever a call of external alarm is detected, for example, due to the presence of an intruder in the base station or due to a fire;

- Whenever other maintenance routines must be executed as, for example, loading channel units from a memory bank in a MSC or a test of routines in these units.

9.8 Radio Channel Types

In the AMPS system there are two types of transmission channels between a base station and a mobile terminal as follows.

1) Voice channels – (VC);

2) Control channels, also known as setup channels – (CC).

9.8.1 Voice Channels

During the establishment of a call a MSC selects a voice channel, which will transport the conversation. When the conversation is finished, this voice channel is free and subsequently becomes part of a file managed by the MSC, always updated with various channels and their respective states. Whenever a voice channel is free the transmitter in the channel unit at the base station is switched off, being switched on again only when the that channel is again captured. These actions are all controlled by the MSC.

In order to control the establishment of a call the MSC needs to exchange information constantly with the mobile terminal, by means of the voice channel. Because of that other signals, besides voice signals, flow through this channel as follows.

- Supervisory audio tone (SAT);
- Signaling tone (ST);
- Wide band data.

SAT and ST signals and wide band data are transmitted outside the voice frequency band, while the voice channel is in use. From a transmission point of view, the voice channel starting from the base station is called the forward voice channel (FVC), while the channel starting at the mobile terminal is called the reverse voice channel (RVC).

Supervision Audio Tone

As was already mentioned, cellular systems reuse frequencies, thus originating large quantities of clusters. The probability of a mobile terminal suffering co-channel interference increases with the reuse of frequencies. During a conversation the mobile terminal receives signals from other base stations transmitting at the same frequency of the voice channel being used.

In order to avoid such situations, three different supervision audio tones (SAT) were implemented in the AMPS system. These tones are used to monitor transmission quality. The SAT is sent continuously at a frequency of approximately 6 kHz during voice transmission, in both directions, i.e., mobile terminal to base station and vice-versa.

A mobile terminal receives the tone, recognizes it and sends it back again to the base station. By means of feedback monitoring the base station can constantly monitor the communications conditions. The three possible values for the SAT have frequencies 5,970 Hz, 6,000 Hz and 6,030 Hz, respectively.

All voice channels in a given main cluster employ, for example, the tone SAT1 of 5,970 Hz, the neighboring cluster employs the tone SAT2 of 6,000 Hz and the voice channels in clusters neighboring the latter employ the tone SAT3 of 6,030 Hz. In this situation, voice channels with the same carrier frequency employ distinct SAT tones. That condition is easily identified by the mobile terminal.

Every time a mobile terminal has to use a given voice channel, it also informs which SAT will be transmitted in this channel. If during a conversation the SAT differs from the one expected, it is not returned to the base station. This action leads to an immediate handoff or to abandon the call. The co-channel interference can also be detected by the base station itself.

Signaling Tone

The signaling tone (ST), sent from the mobile terminal at the frequency of 10 kHz in the voice channel, serves the purpose of line signaling for establishing a call or handoff. This tone is transmitted during the time the mobile terminal emits the "ringing current", for 1.8 seconds after the conversation has finished. This tone is also transmitted during 15 to 50 ms when the conversation is interrupted for handoff.

Wide Band Data

Aiming at having complete control of the progress of call, in certain situations there is an exchange of messages between the base station and the terminal unit via the voice channel. Message transmission is done via a wide band data channel which can be characterized as a forward or reverse voice channel.

- Forward Voice Channel

The forward voice channel is established between a base station and a mobile terminal during a conversation, for sending messages to the mobile terminal. The messages sent by the forward voice channel can be of two types as follows.

- Orders (power level increase, call release);
- Designation of a voice channel during handoff.

- Reverse Voice Channel

  The reverse voice channel is activated between the mobile terminal and the base station during a conversation, with the objective of sending messages to the base station. The messages sent by the reverse voice channel can be of two types as follows.

  - Confirmation of orders;
  - Called address (to make a consultation, conference and activation/deactivation of a subscriber service).

9.8.2 Control Channel

The control channel is the means for exchanging signaling in the cellular system air interface. It is responsible for allowing a conversation to be established. Usually there is only one control channel in each cell. In general, the control channel is used for the following tasks.

a) To transmit general purpose messages with system parameters;

b) To transmit messages for search and general information;

c) To transmit messages for designating a voice channel;

d) To receive messages from the mobile terminal.

The mobile terminal, when moving from one cell to another, stops tuning in the control channel from the original cell and synchronizes the control channel of the new cell. This procedure characterizes the o handoff.

- Forward Control Channel

  The forward control channel is established between the base station and the mobile terminal for sending signaling messages. Each message is transmitted in the form of words. In the AMPS system each word contains 28 bits, which, for reliability reasons, are encoded using an error correcting code. This code is a Bose-Chaudhuri-Hocquenghem (BCH) code and, as discussed earlier, allows the correction of a single bit per word; patterns having two or three errors are simply detected. The error-correcting code adds 12 bits to each word, thus producing 40 bits long codewords.

The messages sent through the forward control channel are of five types as follows.

a) Paging for a mobile terminal;

b) Continuous transmission of messages to the mobile terminals;

c) Designation of a voice channel;

d) Directed trial, which indicates an adjacent cell to which the call must be directed;

e) Overhead, which allows the mobile terminals to adjust to various systems, including different manufacturers.

- Reverse Control Channel

  The connection between a mobile terminal and a base station (upstream uplink) is done eventually for sending information. Messages are transmitted when a mobile terminal tuned to the control channel has meaningful data to send, as follows.

  a) Access of a mobile terminal to originate a call;

  b) Confirmation of orders;

  c) Answer to the paging signal.

9.9 Digital Systems

Distinct philosophies guided the development of digital systems which constitute the basis for the second generation of mobile cellular systems. In USA the main concern was to maintain compatibility with the AMPS system and to allow roaming in the country. In Europe, since there many diverse systems, the aim was to develop a standard compatible with the fixed telephone network and with ITU standards. The main objectives of a digital mobile communication system are.

- Improvement of spectral efficiency;
- Interference rejection;
- Raise immunity to the propagation environment;
- Increase of receiver sensitivity;
- Use of error-correcting codes to protect voice signals.

By considering the previous objectives, the goals established for the transition between the analog standard and the digital cellular system, as defined by the Federal Communications Commission (FCC), in USA, were as follows.

- To allow the accommodation of more subscribers within the same frequency band, with the advantages of a time access (TDMA) or by code (CDMA);
- To ensure adequate capacity to expand the coverage for analog subscribers;
- To guarantee performance quality in the transmissions of either analog or digital subscribers;
- To minimize the impact hardware and other costs.

In Europe, as a consequence of the multiplicity of existing analog systems, from the beginning there was no compatibility requirement. The digital GSM technology made international roaming possible, which was not the case earlier due to the various diverse existing standards. The creation of new and more encompassing standards, like UTMS, can give the manufacturers the opportunity of, in a reasonable time, producing telephone sets which are smaller, more efficient and cheaper.

The USA and Japan opted for keeping a certain degree of compatibility with their first generation systems, thus allowing by means of dual telephone sets a soft transition into the digital technology. However, as can be seen in Table 9.3, such systems have some points in common, although with distinct ways of implementation (Falconer et al., 1995). For example, all systems use hybrid source encoders, benefitting from the combination of high quality waveform encoders and the high compression efficiency of parametric encoders (vocoders) (Steele, 1993).

In order to operate with high efficiency non-linear power amplifiers, variations of the four-level modulation by phase deviation Quaternary Phase Shift Keying (QPSK) are employed, which provide also a satisfactory compromise between the transmission channel bandwidth and tolerance to noise. The TDMA access technique is used in one of the American standards and in the European and Japanese standards. The other American standard adopted the CDMA access technique (Assis, 1994).

9.9.1 The European GSM Standard

The development of a cellular mobile telephone system for Europe, employing digital technology began in 1982 with the Conférence Eu-

System	**Europe**	**USA**	**USA**	**Japan**
Parameter	GSM	IS-54	IS-95	PDC
Band (MHz) Base → Mobile Mobile → Base	 935-960 890-915	 869-894 824-849	 869-894 824-849	 810-826 940-956
Multiple access	TDMA	TDMA	CDMA	TDMA
Channels/carrier	8	3	55/62	3
Voice encoder	RPE-LTP	VSELP	CELP	VSELP
Channel rate uncoded (kbit/s)	13.0	7.95	1.2/9.6	6.7
Channel rate coded (kbit/s)	22.8	13.0	19.2/28.8	11.2
Modulation	GMSK	$\pi/4$-DQPSK	QPSK/DPSK	$\pi/4$-DQPSK
Transmission rate (kbit/s)	270.8	48.6	1228.0	42.0
Min. cluster	4 (pessim.) 3 (optim.)	7 (pessim.) 4 (optim.)	1 1	7 (pessim.) 4 (optim.)
Adaptive equalizer	Mandatory	Mandatory	No	Optional
Bandpass (in kHz)	200	30	1250	25

Table 9.3. Comparison among various digital cellular systems.

ropéene des Administrations des Postes et Télécommunications (DEPT), in a work group called Groupe Spéciale Mobile (GSM).

The basic specification of the European digital standard, or GSM standard as it became known, including the TDMA access technique and the system architecture, was approved in June 1987. Later, GSM was adopted by the European Telecommunications Standards Institute (ERSI) created in March 1988. The field tests with GSM began in July 1991, and its introduction as a commercial service in Europe started in 1992. After the development of GSM was concluded, its meaning was changed to Global System for Mobile Communications.

Currently this standard is adopted in all of Europe, in Australia, in Brazil and in various countries in Asia and Africa. The GSM is the world standard *de facto* and already has about one billion users in the whole world. In Brazil GSM operates in the 1.8 GHz band.

It is noticed in Table 9.3 that the bandwidth per carrier is 200 kHz. In the TDMA access technique, each user has all of this bandwidth available for 577 μs, i.e. , for the period of a user time slot. In GSM a *frame* consists of a group of eight distinct users, the duration of which is 4.615 ms (8 x 0.577 ms). This structure simplifies the RF part of the base stations and of the mobile terminals, since an independent receiver for each user is not required, as occurs with analog systems employing the FDMA access technique.

The handoff process has also a characteristic distinct from that of the analog systems. It is noted that the mobile terminal uses only 1.18 ms of a total 4.615 ms of frame duration to transmit and to receive messages from the base station. The time duration (480 ms in case of GSM) and the six strongest levels are sent via a control channel to the base station which is connected to the mobile terminal. This procedure is known as *handoff* with the help of the mobile terminal (MAHO – *Mobile Assisted Handoff*).

The Gaussian Minimum Shift Keying (GMSK) modulation is derived from the minimum shift keying modulation. In GMSK the modulating signal is passed through a gaussian filter before entering the modulator. The advantage of this modulation technique comes from the fact that it has a constant envelope, which allow the use of power amplifiers operating in the non-linear region where their efficiency is higher.

The transmission rate at the output of the source encoder is 13 kbit/s (RPE-LTP – *Regular Pulse Excitation-Long Term Prediction*). By adding 9.8 kbit/s of the error correction circuitry (rate 1/2 convolutional code) and 11.05 kbit/s for the control channel and synchronism, a total rate of 33.85 kbit/s per user results and an overall rate of 270.8 kbit/s for a frame.

In order to compensate for the fading effects due to multipath propagation, the GSM standard employs channel equalizers. The convolutional code mentioned earlier, with bit interleaving and frequency hopping (meaning that each user time slot is transmitted with a distinct carrier frequency), are the techniques employed to improve GSM performance. Frequency hopping is an optional function in the base station but it is obligatory in the mobile terminal.

The GSM standard was designed so as to be compatible with the Integrated Services Digital Network (ISDN). Consequently GSM offers various call options as, for example, wait, call redirecting, calls with restriction plus a variety of data services up to 9,600 kbit/s, where specific modems are not required (Assis, 1994).

GPRS

The General Packet Radio Service (GPRS) was developed for data transmission in the GSM system. It uses the network only when there are packets ready for transmission, at a rate of up to 115 kbps.

UMTS

The Universal Mobile Telephone Standard (UMTS) proposes a data transmission rate of up tó 2 Mbps by means of a combination of TDMA and W-CDMA (Wide-Band CDMA), operating in the 2 GHz band. UMTS is the European member of the IMT-2000 family of third generation (3G) cellular systems. The data rates offered to users are as follows: 144 kbps for cars, 384 kbps for pedestrians, and 2 Mbps for stationary users (Dhir, 2004).

9.9.2 American Standards

The need to develop a system with higher capacity than that obtained with the AMPS standard became evident in USA in 1988, with the overload of the cellular mobile service in large cities like New York, Los Angeles and Chicago. A solution called narrow band AMPS (NAMPS) was proposed as an alternative for duplicating the number of users by means of a more efficient use of the frequency band occupied by a voice channel.

However, the effort spent in this solution was cancelled by the increase in interference and by the consequent loos of signal quality. On the other hand, adoption of a completely new standard could bring problems to millions of users of the AMPS system, besides requiring a high cost infra structure for its implementation. Thus, the Telecommunications Industry Association (TIA) opted for a system which allowed an increase in service capacity and was compatible with the AMPS system.

The TDMA (IS-54/136) Standard

Having as a requisite compatibility with the existing system, the TDMA standard using 30 kHz of bandwidth per carrier frequency was approved by the TIA. Originally this standard was designated as IS-54 (*Interim Standard*). Later it became known as D-AMPS, or digital AMPS. This standard employs a TDMA with three users per carrier. The frame has a duration of 20 ms, with each channel window corresponding to 6,67 ms.

The transmission rate at the voice encoder output (VSELP – *Vector Sum Excited Linear Predictive Coding*) is 7,95 kbit/s. The error correcting circuit (rate 1/2 convolutional code) contributes 5,05 kbit/s. There is still 3,2 kbit/s due to the control channel and the e synchronization bits. In this manner, the rate per user is 16,2 kbit/s, which leads to a total transmission rate of 48,6 kbit/s. At this rate the use of equalizers could be optional, being recommended only in situations where the delay is greater than 2 μs. However, the use of equalization is specified in the D-AMPS standard.

In the $\pi/4$-QPSK modulation the carrier phase transition is limited to $\pm\pi/4$ and $\pm3\pi/4$. Since there are no $\pm\pi$ transition as in a conventional QPSK modulator, the envelope fluctuation is substantially reduced, thus allowing the use of non-linear power amplifiers. Differential encoding is employed together with a filter with a 0.35 roll-off factor.

The handoff process is similar to that described for the GSM. The only difference is that in the D-AMPS the measurements are performed in at most 12 base stations, and the average of these measurements is taken over a period of 1 s. The result of all measurements is sent to the base station in use.

With respect to the capacity of a TDMA system, itn is possible to obtain a gain of three to six times the capacity of the analog AMPS, depending on the use of an optimistic plan (set of 4 cells) or a pessimistic one (7 cells) (Raith and Uddenfeldt, 1991). In practical terms, however, the accepted TDMA gain factor is between 3 and 4. The use of the channel dynamic allocation technique can raise this capacity by a factor between 2 and 3. In order to increase capacity, there is also a proposal called *Extended* TDMA (E-TDMA) which employs a 4 kHz voice channel (half the rate shown in Table 9.3) and incorporates a statistical multiplex to take advantage of the pauses in conversation. Theoretically it is possible to achieve a gain approximately twice that obtained with current technology, which however cannot exceed a factor of 8. Finally, as a comment, the GSM systems present a capacity gain with respect to the analog AMPS in the range of 2,5 to 3,5 (sets of 3 to 4 cells).

The IS-95 CDMA Standard

The spread spectrum technology was kept for a long time restricted just for military applications, taking advantage of its privacy characteristics (difficult interception) and resistance to interfering signals (intentional or not). Only in the 1980's the potential of this technology began to be exploited also for commercial applications.

In the case of mobile cellular systems, the use of spread spectrum as an access technique in a CDMA version is based on the high rejection of interfering signals, regarding both system inherent interference (co-channel and adjacent channel) as well as external interference. In this respect, the use of orthogonal codes and a careful control of the power levels transmitted by a base station are fundamental procedures. As a consequence of this property, the following characteristics are worth noting (Lee, 1991).

1) Reuse of all available frequencies – since all cells share the same RF channel, frequency coordination becomes unnecessary. According to

this characteristic it is worth noting a single radio equipment is used in each base station;

2) Use to advantage the multipath interference, avoiding the use of channel equalizers – in a multichannel receiver (called RAKE) used in the CDMA system, demodulation of each signal arriving by a different path (multipaths) is done individually, thus getting an efficient effect of diversity;

3) Soft handoff procedure – the various channels in the RAKE receiver gradually replace the code of the base station in operation by the corresponding code for the new base station. Thus, the handoff does not occur in an abrupt form as in systems that employ FDMA or TDMA techniques;

4) It naturally allows the use of the vocal activity factor to increase system capacity – considering that during a conversation on the average the channel is activated only 35% of the time, since the lack of vocal activity implies in no signal present in the system for transmission. This characteristic is exploited to increase system capacity;

5) Ease of sharing frequency with other systems – when under interference, as discussed earlier, the CDMA system offers high rejection to interfering signals. As a source of interference, within certain limits, the perturbation introduced by the CDMA system can be controlled, since the spectrum spreading produces a relatively low power density in the frequency band being used.

However, apart from using a more complex technology there are two more points usually criticized by opponents of the CDMA system .

1 Near-far interference – since all mobile terminals share the same channel, a strong signal from a user near the base station can mask a weak signal from a user far from the base station. There is need for an effective control of the power level in the mobile terminal. In practice this control covers a 100 dB range;

2 Need for synchronization – a synchronization failure can harm the system operation. In order to avoid such a situation occurring, the base stations are synchronized by means of signals coming form a GPS (*Global Positioning System*) satellite GPS whereas mobile terminals receive synchronization signals from the base stations.

In terms of capacity, theoretic studies show that the gain factor, relative to the analog systems, is of the order of 20 (Lee, 1991, Gilhousen et al., 1991). However, based on experiments performed in San Diego, California (Gilhousen et al., 1991), it can be said that currently this factor is situated between 8 and 10 (Cox, 1992).

The CDMA system has relevant qualities recognized by TIA in July 1992, which conferred to it the designation IS-95, becoming then the second digital American standard. The system commercial name is *cdmaOne*.

cdmaOne

The *cdmaOne* is considered the second generation (2G) of mobile wireless technology. The system is supported by the CDMA Development Group (CDG) and is compatible with the air interface IS-95 and with the network standard for exchange interconnection ANSI-41. The protocol, version IS-95A, operates in the frequency bands of 800 MHz and 1,9 GHz, and employs a data transmission of up to 14,4 kbps. Version IS-95B supports up to 115 kbps, by employing eight channels.

cdma2000

The *cdma2000* standards aims to offer users of the IS-95 standard a soft transition to the third generation (3G) cellular system. This st5andard is also by the designation ITU, IMT-CDMA Multi-Carrier (1X/3X). In the first phase (1X), the cdma2000 standard operates at 144 kbps. The second phase (2X) incorporates the 1X standard and adds support to all frequency bands, including 5 MHz and 10 MHz, transmission in circuit and packet mode up to 2 Mbps, advanced multimedia and new voice encoders.

9.9.3 The Japanese Standard

This standard (*Personal Digital Cellular – PDC*) is similar to the American TDMA system described earlier, but operates in the 800 MHz and 1.500 MHz frequency bands. A few differences such as frequency bands and transmission rate of the voice channel can be observed in Table 9.3. One distinct aspect in this standard is that the use of an equalizer is optional (Assis, 1994).

9.10 Problems

1) Define a mobile communication system and compare it with a wireless cellular communication system.

2) Describe the main types of multiple access systems and their characteristics. Which are the main advantages and disadvantages of each system?

3) Discuss the reason for the presence of a high level of interference in conventional mobile systems, which led to the development of the cell concept.

4) Which are the main existing mobile services? Cite a few examples.

5) Which is the first generation mobile cellular system adopted as a standard in USA, and which systems are derived from it? Comment on the main differences among such systems.

6) Describe the basic components of a mobile cellular system and its main characteristics.

7) The ratio D/R is used in assessing the performance of mobile systems. For a ratio $D/R = 6$, compute the number of cells per cluster, and give an example of frequency distribution $(f_1, \cdots, f_N)$ with two clusters. Frequency f_1 must be in a cell at the interface between clusters.

8) Explain why the traffic capacity decreases with an increase in the cluster size, while simultaneously transmission quality is improved.

9) Explain the difference between handoff and roaming in a cellular mobile communication system.

10) In the two-ray propagation model (Lee) consider the transmitted power is 10 W, the antenna gain at the base stations is $G_T = 1.6$ and at the mobile terminal is $G_R = 1.4$. The antenna at the base station is 30 m. high, while the mobile antenna is 1.8 m. high. For an average distance of 2 km, calculate the received power in dBm.

11) Explain the exchange of signals for handoff in the AMPS cellular mobile communication system. Which channels are employed?

12) Explain the exchange of signals for roaming in the AMPS cellular mobile communication system. Which subsystems are employed?

13) Using the hexagonal model, computer the intensity of the received power at the interface between two cells with radius R using the Lee model.

14) In the installation of an IS-136 system the first digital channel requires the removal of three analog channels in order to provide a 30 kHz separation between IS-136 and AMPS channels. Calculate how many digital channels can be installed in a system with seven base stations.

15) A certain city has its coverage area consisting of a seven cell cluster, with base stations having each a power of 10 W and 30 channels. The central cell is busy. Using Lee's formula, determine the new power for the central base station in order to alleviate this condition.

16) Analyze the convenience of making a forced registration, or registration by location region, considering the processing time at the MSC and the occupation of the hunting circuits in a mobile system having 6,000 users that makes forced registration every 8 minutes.

17) Which are the main approaches in the characterization of channels employed in mobile communication and which are the usual techniques to combat the eventual shortcomings and constraints found in these channels?

18) Using the two-ray model for a mobile channel it is possible to conclude that the received power is proportional to the square of the height of the antennas – and decreases with the fourth power of the distance between them. How much power is received by a mobile terminal with a 1 m. antenna, situated at a distance of 2 km from the base station, which couples 10 W to a 15m. transmitter antenna?

19) What types of antennas are used in mobile communication systems and what are their characteristics?

20) Why analog cellular mobile communication systems employ frequency modulation for the audio signal?

21) What is meant by a *cell* in mobile communications and why a structure with hexagonal cells is used to split a given coverage area? What are the basic types of cells?

22) Why does the traffic capacity decreases and transmission quality improves as between the reuse distance and the cell radius increases?

23) Describe the usual solutions to compensate for the increase of traffic in a given cell.

24) Which are the main components of a typical cellular system? Describe the role of each of these components.

25) What is handoff? What characterizes automatic roaming?

26) Describe the role of the following:

- Supervision audio tone;
- Signaling tone;

- Wide band data transmission.

27) Explain the role of the control channels in a cellular mobile communication system.

28) Which digital systems are adopted in Europe and in USA, and what are the basic characteristics of each one of them?

29) What are pseudo-random sequences? How can they be generated? How are they used in practice? What are their properties?

30) Define outage for both time varying channels and time invariant channels.

31) An FM signal is demodulated by a system which employs a local carrier with a frequency error $\Delta\omega$ with respect to the transmitted carrier. Show the effect of this error on the demodulated signal. How is it possible to compensate for this effect?

32) A signal with binary FSK modulation, used in mobile communication and transmitted in narrowband, has a modulating signal power of $P_{\Theta} = \frac{V^2}{2}$. Plot the power spectral density of the modulated signal. What is the approximated value of V that allows the signal to be transmitted without a carrier?

33) The error probability for the FSK modulation scheme used for signaling in the AMPS system is given by $P_e = \frac{1}{2}e^{-\frac{1}{2}\eta}$, where η represents the signal to noise ratio. Knowing that the system is affected by Rayleigh fading, compute its effect on the error probability.

34) The AMPS cellular mobile communication system uses frequency modulation in its air interface. Explain the reason for that. Knowing that the voice signal in telephony occupies the frequency range of 300 Hz to 3.400 Hz and that the band reserved for each channel is 30 kHz, what is the AMPS modulation index due exclusively to the voice signal? For a power of 0 dBm for the modulating signal, what is the equipment frequency deviation index?

Appendix A
Fourier Series and Fourier Transform

This Appendix presents some of the properties of Fourier series and Fourier transform used in the preceeding chapters, along with Parseval's theorem (Hsu, 1973) (Lathi, 1989).

Trigonometric series: $f(t) = a_0 + \sum_{n=1}^{\infty}(a_n \cos n\omega_o t + b_n \sin n\omega_0 t)$

Cosine series: $f(t) = C_0 + \sum_{n=1}^{\infty} C_n \cos(n\omega_0 t + \theta_n)$

Complex exponential series: $f(t) = \sum_{n=-\infty}^{\infty} F_n e^{jn\omega_0 t}$

For all previous series, consider

$$f(t+T) = f(t), \qquad \omega_0 = \frac{2\pi}{T}$$

Conversion formulas:

$$F_0 = a_0, \qquad F_n = \frac{1}{2}(a_n - jb_n), \qquad F_{-n} = \frac{1}{2}(a_n + jb_n);$$

$$F_n = |F_n|e^{j\phi_n}, \qquad |F_n| = \frac{1}{2}\sqrt{a_n^2 + b_n^2};$$

$$a_0 = F_0, \qquad a_n = F_n + F_{-n}, \qquad b_n = j(F_n - F_{-n}),$$

$$C_0 = a_0, \qquad C_n = 2|F_n| = \sqrt{a_n^2 + b_n^2}, \qquad \theta_n = -\arctan\left(\frac{b_n}{a_n}\right).$$

Definition of the Fourier transform and some of its properties:

- Definition: $F(\omega) = \int_{-\infty}^{\infty} f(t)e^{-j\omega t}dt$
- Inverse: $f(t) = \frac{1}{2\pi}\int_{-\infty}^{\infty} F(\omega)e^{j\omega t}d\omega$
- Magnitude and phase: $F(\omega) = |F(\omega)|e^{j\theta(\omega)}$

- Even function $f(t)$: $F(\omega) = 2\int_0^{\infty} f(t)\cos\omega t dt$
- Odd function $f(t)$: $F(\omega) = -2j\int_0^{\infty} f(t)\sin\omega t dt$
- Area under function in time: $F(0) = \int_{-\infty}^{\infty} f(t)dt$
- Area under transform: $f(0) = \frac{1}{2\pi}\int_{-\infty}^{\infty} F(\omega)d\omega$
- Linearity: $\alpha f(t) + \beta g(t) \leftrightarrow \alpha F(\omega) + \beta G(\omega)$

Parseval's theorem:

$$\int_{-\infty}^{\infty} f(t)g(t)dt = \frac{1}{2\pi}\int_{-\infty}^{\infty} F(\omega)G^*(\omega)d\omega,$$

$$\int_{-\infty}^{\infty} |f(t)|^2 dt = \frac{1}{2\pi}\int_{-\infty}^{\infty} |F(\omega)|^2 d\omega,$$

$$\int_{-\infty}^{\infty} f(\omega)G(\omega)d\omega = \int_{-\infty}^{\infty} F(\omega)g(\omega)d\omega.$$

A.1 Useful Fourier Transforms

This section contains a set of Fourier transforms, and properties, that are useful to solve problems (Hsu, 1973) (Spiegel, 1976) (Lathi, 1989) (Gradshteyn and Ryzhik, 1990) (Oberhettinger, 1990).

$f(t)$	$F(\omega)$
$f(at)$	$\frac{1}{\lvert a \rvert}F(\frac{\omega}{a})$
$f(-t)$	$F(-\omega)$
$f^*(t)$	$F^*(-\omega)$
$f(t-\tau)$	$F(\omega)e^{-j\omega\tau}$
$f(t)e^{j\omega_0 t}$	$F(\omega-\omega_0)$
$f(t)\cos\omega_0 t$	$\frac{1}{2}F(\omega-\omega_0)+\frac{1}{2}F(\omega+\omega_0)$
$f(t)\sin\omega_0 t$	$\frac{1}{2j}F(\omega-\omega_o)-\frac{1}{2j}F(\omega+\omega_0)$
$F(t)$	$2\pi f(-\omega)$
$f'(t)$	$j\omega F(\omega)$
$f^{(n)}(t)$	$(j\omega)^n F(\omega)$
$\int_{-\infty}^{t} f(x)dx$	$\frac{1}{j\omega}F(\omega)+\pi F(0)\delta(\omega)$
$-jtf(t)$	$F'(\omega)$
$(-jt)^n f(t)$	$F^{(n)}(\omega)$
$f(t)*g(t)=\int_{-\infty}^{\infty} f(\tau)g(t-\tau)dx$	$F(\omega)G(\omega)$
$\delta(t)$	1
$\delta(t-\tau)$	$e^{-j\omega\tau}$
$\delta'(t)$	$j\omega$
$\delta^{(n)}(t)$	$(j\omega)^n$

$f(t)$	$F(\omega)$
$f(t)g(t)$	$\frac{1}{2\pi}F(\omega)*G(\omega)=\frac{1}{2\pi}\int_{-\infty}^{\infty}F(\phi)G(\omega-\phi)d\phi$
$e^{-at}u(t)$	$\frac{1}{a+j\omega}$
$e^{-a\|t\|}$	$\frac{2a}{a^2+\omega^2}$
e^{-at^2}	$\sqrt{\frac{\pi}{a}}e^{-\omega^2/(4a)}$
te^{-at^2}	$j\sqrt{\frac{\pi}{4a^3}}\omega e^{-\omega^2/(4a)}$
$p_T(t)=\begin{cases}0 & \text{for } \|t\|>T/2\\ A & \text{for } \|t\|\leq T/2\end{cases}$	$AT\frac{\sin(\frac{\omega T}{2})}{(\frac{\omega T}{2})}$
$\frac{\sin at}{\pi t}$	$p_{2a}(\omega)$
$te^{-at}u(t)$	$\frac{1}{(a+j\omega)^2}$
$\frac{t^{n-1}}{(n-1)!}e^{-at}u(t)$	$\frac{1}{(a+j\omega)^n}$
$e^{-at}\sin bt\, u(t)$	$\frac{b}{(a+j\omega)^2+b^2}$
$e^{-at}\cos bt\, u(t)$	$\frac{a+j\omega}{(a+j\omega)^2+b^2}$
$\frac{1}{a^2+t^2}$	$\frac{\pi}{a}e^{-a\|\omega\|}$
$\frac{t}{a^2+t^2}$	$j\pi e^{-a\|\omega\|}[u(-\omega)-u(\omega)]$
$\frac{\cos bt}{a^2+t^2}$	$\frac{\pi}{2a}[e^{-a\|\omega-b\|}+e^{-a\|\omega+b\|}]$
$\frac{\sin bt}{a^2+t^2}$	$\frac{\pi}{2aj}[e^{-a\|\omega-b\|}-e^{-a\|\omega+b\|}]$
$\sin bt^2$	$\frac{\pi}{2b}\left[\cos\frac{\omega^2}{4b}-\sin\frac{\omega^2}{4b}\right]$
$\cos bt^2$	$\frac{\pi}{2b}\left[\cos\frac{\omega^2}{4b}+\sin\frac{\omega^2}{4b}\right]$
$\operatorname{sech} bt$	$\frac{\pi}{b}\operatorname{sech}\frac{\pi\omega}{2b}$
$\ln\left[\frac{x^2+a^2}{x^2+b^2}\right]$	$\frac{2e^{-b\omega}-2e^{-a\omega}}{\pi\omega}$

$f(t)$	$F(\omega)$
$f_P(t) = \frac{1}{2}[f(t) + f(-t)]$	$\mathrm{Re}\,(\omega)$
$f_I(t) = \frac{1}{2}[f(t) - f(-t)]$	$j\mathrm{Im}\,(\omega)$
$f(t) = f_P(t) + f_I(t)$	$F(\omega) = \mathrm{Re}\,(\omega) + j\mathrm{Im}\omega)$
$e^{j\omega_0 t}$	$2\pi\delta(\omega - \omega_0)$
$\cos\omega_0 t$	$\pi[\delta(\omega - \omega_0) + \delta(\omega + \omega_0)]$
$\sin\omega_0 t$	$-j\pi[\delta(\omega - \omega_0) - \delta(\omega + \omega_0)]$
$\sin\omega_0 t u(t)$	$\frac{\omega_0}{\omega_0^2 - \omega^2} + \frac{\pi}{2j}[\delta(\omega - \omega_0) - \delta(\omega + \omega_0)]$
$\cos\omega_0 t u(t)$	$\frac{j\omega}{\omega_0^2 - \omega^2} + \frac{\pi}{2}[\delta(\omega - \omega_0) + \delta(\omega + \omega_0)]$
$u(t)$	$\pi\delta(\omega) + \frac{1}{j\omega}$
$u(t - \tau)$	$\pi\delta(\omega) + \frac{1}{j\omega}e^{-j\omega\tau}$
$tu(t)$	$j\pi\delta'(\omega) - \frac{1}{\omega^2}$
1	$2\pi\delta(\omega)$
t	$2\pi j\delta'(\omega)$
t^n	$2\pi j^n\delta^{(n)}(\omega)$
$\|t\|$	$\frac{-2}{\omega^2}$
$\frac{1}{t}$	$\pi j - 2\pi j u(\omega)$
$\frac{1}{t^n}$	$\frac{(-j\omega)^{n-1}}{(n-1)!}[\pi j - 2\pi j u(\omega)]$
$u(t) - u(-t)$	$\frac{2}{j\omega}$
$\frac{1}{e^{2t} - 1}$	$\frac{-j\pi}{2}\coth\frac{\pi\omega}{2} + \frac{j}{\omega}$
$\delta_T(t) = \sum_{n=-\infty}^{\infty}\delta(t - nT)$	$\omega_0\delta_{\omega_0}(\omega) = \omega_0\sum_{n=-\infty}^{\infty}\delta(\omega - n\omega_0)$

$f(t)$	$F(\omega)$
$\cos\left(\frac{t^2}{4a} - \frac{1}{4}\pi\right)$	$2\sqrt{\pi a}\cos(a\omega^2)$
$\sin\left(\frac{t^2}{4a} + \frac{1}{4}\pi\right)$	$2\sqrt{\pi a}\sin(a\omega^2)$
$\frac{\Gamma(1-s)\sin\left(\frac{1}{2}s\pi\right)}{\lvert t\rvert^{1-s}}$	$\pi\lvert\omega\rvert^{-s}$ $0 < \mathrm{Re}\,(s) < 1$
$\frac{1}{\lvert t\rvert}$	$\frac{\sqrt{2\pi}}{\lvert\omega\rvert}$
$\frac{\sqrt{\sqrt{\alpha^2+t^2}+\alpha}}{\sqrt{\alpha^2+t^2}}$	$\sqrt{\frac{2\pi}{\lvert\omega\rvert}}e^{-\alpha\lvert\omega\rvert}$
$\frac{\cos(\frac{1}{2}\alpha)\cosh(\frac{1}{2}t)}{\cosh(t)+\cos(\alpha)}$	$\frac{\pi\cosh(\alpha\omega)}{\cosh(\pi\omega)}$ $-\pi < \alpha < \pi$
$\frac{\sin(\alpha)}{\cosh(t)+\cos(\alpha)}$	$\frac{\pi\sin,(\alpha\omega)}{\sin,(\pi\omega)}$ $-\pi < \alpha < \pi$
$J_0(\alpha t)$	$\frac{2}{\sqrt{\alpha^2-\omega^2}}$ $\lvert\omega\rvert < \alpha$
	0 $\lvert\omega\rvert > \alpha$
$J_0(\alpha\sqrt{b^2-t^2})\lvert t\rvert < b$	$2\frac{\sin[b(\alpha^2+\omega^2)^{\frac{1}{2}}]}{\sqrt{\alpha^2+\omega^2}}$
0 $\lvert t\rvert > b$	
$j^n J_{n+\frac{1}{2}}(t)$	$\sqrt{2}P_n(\omega)\lvert\omega\rvert < 1$
	0 $\lvert\omega\rvert > 1$
$J_0(\alpha\sqrt{t^2+b^2})$	$\frac{2\cos(b\sqrt{\alpha^2-\omega^2})}{\sqrt{\alpha^2-\omega^2}}$ $\lvert\omega\rvert < \alpha$
	0 $\lvert\omega\rvert > \alpha$
$J_0(\alpha\sqrt{t^2-b^2})$	$\frac{2\cosh(b\sqrt{\alpha^2-\omega^2})}{\sqrt{\alpha^2-\omega^2}}$ $\lvert\omega\rvert < \alpha$
	0 $\lvert\omega\rvert > a$

$f(t)$	$F(\omega)$
$\frac{1}{(\alpha-jt)^{\nu}}$	$2\pi\omega^{\nu-1}e^{-\alpha\omega}/\Gamma(\nu), \quad \omega>0$
$\mathrm{Re}\,\alpha>0, \quad \mathrm{Re}\,\nu>0$	$0, \quad \omega<0$
$\frac{1}{(\alpha+jt)^{\nu}}$	$0, \quad \omega>0$
$\mathrm{Re}\,\nu>0, \quad \mathrm{Re}\,\alpha>0$	$-2\pi(-\omega)^{\nu-1}e^{\alpha\omega}/\Gamma(\nu), \quad \omega<0$
$\frac{1}{(t^2+\alpha^2)(jt)^{\nu}}$	
$\lvert\nu\rvert<1, \quad \mathrm{Re}\,\alpha>0$	$\pi\alpha^{-\nu-1}e^{-\lvert\omega\rvert\alpha}$
$\arg(jt)=1/2\pi \quad ,(t>0)$	
$\arg(jt)=-1/2\pi \quad ,(t<0)$	
$\frac{1}{(t^2+\alpha^2)(\beta+jt)^{\nu}} \quad \mathrm{Re}\,\nu>-1$	$\frac{\pi e^{-\alpha\omega}}{\alpha(\alpha+\beta)^{\nu}}, \quad \omega>0$
$\mathrm{Re}\,\alpha>0, \quad \mathrm{Re}\,\beta>0$	
$\frac{1}{(x^2+\alpha^2)(\beta-jt)^{\nu}}$	$\frac{\pi(\beta-\alpha)^{\nu}e^{\alpha\omega}}{\alpha}, \quad \omega>0$
$\mathrm{Re}\,\nu>-1, \quad \mathrm{Re}\,\alpha>0$	
$\mathrm{Re}\,\beta>0, \quad \alpha\neq\beta$	
$\frac{1}{(\alpha-e^{-t})e^{\lambda t}}$	$\pi\alpha^{\lambda-1+j\omega}\cot(\pi\lambda+j\pi\omega)$
$0<\mathrm{Re}\,\lambda<1, \quad \alpha>0$	
$\frac{1}{(\alpha+e^{-t})e^{\lambda t}}$	$\pi\alpha^{\lambda-1+j\omega}\mathrm{cossec}(\pi\lambda+j\pi\omega)$
$0<\mathrm{Re}\,\lambda<1, \quad -\pi<\arg\alpha<\pi$	
$\frac{t}{(\alpha+e^{-t})e^{\lambda t}}$	$\pi\alpha^{\lambda-1+j\omega}\mathrm{cossec}(\pi\lambda+j\pi\omega)$
$0<\mathrm{Re}\,\lambda<1, \quad -\pi<\arg\alpha<\pi$	$\times[\log\alpha-\pi\cot(\pi\lambda+j\pi\omega)]$

$f(t)$	$F(\omega)$
$\frac{t^2}{(1+e^{-t})e^{\lambda t}}$	$\pi^3 \mathrm{cossec}^3(\pi\lambda + j\omega\pi)[2 - \sin(\pi\lambda + j\omega\pi)]$
$0 < \mathrm{Re}\,\lambda < 1$	
$\frac{1}{(\alpha+e^{-t})(\beta+e^{-t})e^{\lambda t}}$	$\pi(\beta - \alpha)^{-1}(\alpha^{\lambda-1+j\omega} - \beta^{\lambda-1+j\omega})$
$0 < \mathrm{Re}\,\lambda < 2, \quad \beta \neq \alpha$	$\times \mathrm{cossec}\,(\pi\lambda + j\omega\pi)$
$\lvert \arg\alpha \rvert < \pi, \quad \lvert \arg\beta \rvert < \pi$	
$\frac{t}{(\alpha+e^{-t})(\beta+e^{-t})e^{\lambda t}}$	$\frac{\pi(\alpha^{\lambda-1+j\omega}\log\alpha - \beta^{\lambda-1+j\omega}\log\beta)}{(\alpha-\beta)\sin(\lambda\pi+j\omega\pi)}$
$0 < \mathrm{Re}\,\lambda < 2, \quad \alpha \neq \beta$	$+\frac{\pi^2(\alpha^{\lambda-1+j\omega} - \beta^{\lambda-1+j\omega})\cos(\lambda\pi+j\omega\pi)}{(\beta-\alpha)\sin^2(\lambda\pi+j\omega\pi)}$
$\lvert \arg\,\alpha \rvert < \pi, \quad \lvert \arg\,\beta \rvert < \pi$	
$\frac{1}{(1+e^{-t})^n e^{\lambda t}}$	$\pi \mathrm{cossec}\,(\pi\lambda + j\omega\pi)\prod_{j=1}^{n-1}(j - \lambda - j\omega)/(n-1)!$
$n = 1, 2, 3, \cdots, \quad 0 < \mathrm{Re}\,\alpha < n$	
$e^{-\lambda t}\log\lvert 1 - e^{-t}\rvert$ $-1 < \mathrm{Re}\,\lambda < 0$	$\pi(\lambda + j\omega)^{-1}\cot(\pi\lambda + j\omega\pi)$
$e^{-\lambda t}\log(1 + e^{-t})$	$\pi(\lambda + j\omega)^{-1}\mathrm{cossec}\,(\pi\lambda + j\omega\pi)$
$-1 < \mathrm{Re}\,\lambda < 0$	
$e^{-\lambda t}\log\left(\frac{\lvert 1+e^{-t}\rvert}{\lvert 1-e^{-t}\rvert}\right)$	$\pi(\lambda + j\omega)^{-1}\tan(\frac{1}{2}\pi\lambda + \frac{1}{2}j\omega\pi)$
$\lvert \mathrm{Re}\,\lambda \rvert < 1$	
$\frac{1}{(\sin t+\sin\alpha)} \quad \alpha > 0$	$-\pi j e^{j\alpha\omega}\mathrm{sech}\,\alpha\,\mathrm{cossech}\,(\pi\omega) \times [\cosh(\pi\omega) - e^{-2j\alpha\omega}]$
$\frac{1}{[\Gamma(\nu-t)\Gamma(\mu+t)]}$	$[2\cos(\frac{1}{2}\omega)]^{\mu+\nu-2}e^{\frac{1}{2}j\omega(\mu-\nu)} \times [\Gamma(\mu+\nu-1)]^{-1}, \quad \lvert\omega\rvert < \pi$
	$0, \quad \lvert\omega\rvert > \pi$

A.2 Some Hilbert Transforms

This section presets some Hilbert Transforms. Assume, for the third and fourth properties that the signal is bandpass (Haykin, 1987) (Baskakov, 1986).

$m(t)$	$\hat{m}(t)$
$m(t)$	$\hat{m}(t) = \frac{1}{\pi t} * m(t)$
$\alpha m(t) + \beta n(t)$	$\alpha \hat{m}(t) + \beta \hat{n}(t)$
$m(t) \cos \omega_c t$	$m(t) \sin \omega_c t$
$m(t) \sin \omega_c t$	$-m(t) \cos \omega_c t$
1	0
$\cos \omega_c t$	$\sin \omega_c t$
$\sin \omega_c t$	$-\cos \omega_c t$
$\frac{\sin t}{t}$	$\frac{1-\cos t}{t}$
$\delta(t)$	$\frac{1}{\pi t}$
$\frac{1}{t}$	$-\pi\delta(t)$
$\frac{1}{1+t^2}$	$\frac{1}{1+t^2}$
$u(t+1/2) - u(t-1/2)$	$-\frac{1}{\pi} \ln \left\| \frac{t-1/2}{t+1/2} \right\|$

Appendix B
Formulas and Important Inequalities

Schwartz inequality

$$\left|\text{Re}\int_{-\infty}^{\infty} x(t)y^*(t)dt\right| \leq \left[\int_{-\infty}^{\infty} |x^2(t)|dt \int_{-\infty}^{\infty} |y^2(t)|dt\right]^{1/2}$$

equality verifies for $x(t) = ky(t)$, where k is a constant.

Holder inequality

$$\left|\sum_{i=0}^{\infty} a_i b_i\right| \leq \left[\sum_{i=0}^{\infty} |a_i^2| \sum_{i=0}^{\infty} |b_i^2|\right]^{1/2}$$

equality holds for $a_i = kb_i$.

Other inequalities

1 Consider $w_1, w_2, \ldots, w_N$ positive arbitrary numbers and let $q_1, q_2, \ldots, q_N$ be positive numbers, such that $\sum_{i=1}^{N} p_i = 1$. Then,

$$\sum_{i=1}^{k} q_i w_i \log \sum_{i=1}^{k} q_i w_i \leq \sum_{i=1}^{k} q_i w_i \log q_i,$$

this equality holds, because of the continuity and convexity of the function $x \log x$ in the closed interval $(0, 1)$, if $q_i = q$, for all $i = 1, \ldots, N$ (Nedoma, 1957).

2 Let $p_1, p_2, \ldots, p_N$ be positive arbitrary numbers and let $q_1, q_2, \ldots, q_N$ be positive numbers, such that $\sum_{i=1}^{N} p_i = 1$. Then,

$$q_1^{p_1} \cdots q_k^{p_k} \leq \sum_{i=1}^{k} p_i q_i, \ \sum_{i=1}^{N} q_i = 1, \ q_i \geq 0,$$

equality occurs for $k = N$ and $q_i = q$ (Ash, 1990).

3 If $p_1 \geq p_2 \geq \cdots \geq p_k$ and $q_1 \geq q_2 \geq \cdots \geq q_k$, then (Chebyshev's inequality)

$$\left(\frac{p_1 + p_2 \cdots p_k}{k}\right)\left(\frac{q_1 + q_2 \cdots q_k}{k}\right) \leq \frac{1}{k}\sum_{i=1}^{k} p_i q_i,$$

or

$$\sum_{i=1}^{k} q_i \sum_{j=1}^{k} p_j \leq k \sum_{i=1}^{k} p_i q_i,$$

equality holds if and only if $p_i = p$ or $q_i = q$, for $1 \leq i \leq k$ (Gradshteyn and Ryzhik, 1990).

4 Consider $w_1, w_2, \ldots, w_N$ positive arbitrary numbers and let $p_1, p_2, \ldots, p_N$ be positive numbers, such that $\sum_{i=1}^{N} p_i = 1$. Then,

$$w_1^{p_1} \cdots w_k^{p_k} \leq \sum_{i=1}^{k} p_i w_i, \quad \sum_{i=1}^{N} p_i = 1, \; p_i \geq 0,$$

equality verifies for $k = N$ and $w_i = w$. The inequality holds if any p_i is zero, as long as $\sum_{i=1}^{N} p_i = 1$ (Ash, 1990).

Final Value theorem

$$\lim_{t\to\infty} x(t) = \lim_{\omega\to 0}[j\omega X(\omega)]$$

MacLaurin series

$$f(x) = f(0) + f^{'}(0)x + \frac{1}{2!}f^{''}(0)x^2 + \frac{1}{3!}f^{'''}(0)x^3 + \cdots$$

Bessel's identities

$$J_n(x) = \left(\frac{x}{2}\right)^n \sum_{k=0}^{\infty} \frac{(-x^2/4)^k}{k!(n+k)!}$$

$$= \frac{j^{-n}}{\pi}\int_0^{\pi} e^{jx\cos\theta}\cos(n\theta)d\theta$$

$$J_n(xe^{jm\pi}) = e^{jnm\pi}J_n(x)$$

$$I_n(x) = j^n J_n(x/j)$$

$$= \frac{1}{\pi}\int_0^{\pi} e^{x\cos\theta}\cos(n\theta)d\theta$$

Complex Identities

$$z = x + jy = \sqrt{x^2 + y^2} e^{j \tan^{-1}(y/x)}$$

$$z^* = x - jy = \sqrt{x^2 + y^2} e^{-j \tan^{-1}(y/x)}$$

$$|z|^2 = zz^* = x^2 + y^2$$

$$\mathrm{Re}\{z\} = \frac{1}{2}[z + z^*]$$

$$\mathrm{Im}\{z\} = \frac{1}{2j}[z - z^*]$$

$$\mathrm{Re}\{z_1 z_2\} = \mathrm{Re}\{z_1\}\mathrm{Re}\{z_2\} - \mathrm{Im}\{z_1\}\mathrm{Im}\{z_2\}$$

$$\mathrm{Im}\{z_1 z_2\} = \mathrm{Re}\{z_1\}\mathrm{Im}\{z_2\} + \mathrm{Im}\{z_1\}\mathrm{Re}\{z_2\}$$

Trigonometric Identities

$$\sin\theta = \frac{1}{2j}(e^{j\theta} - e^{-j\theta})$$

$$\cos\theta = \frac{1}{2}(e^{j\theta} + e^{-j\theta})$$

$$e^{\pm j\theta} = \cos\theta \pm j\sin\theta$$

$$\sin(\alpha \pm \beta) = \sin\alpha\cos\beta \pm \cos\alpha\sin\beta$$

$$\cos(\alpha \pm \beta) = \cos\alpha\cos\beta \mp \sin\alpha\sin\beta$$

$$\tan(\alpha \pm \beta) = \frac{\tan\alpha \pm \tan\beta}{1 \mp \tan\alpha\tan\beta}$$

$$\sin\alpha\sin\beta = \frac{1}{2}\cos(\alpha - \beta) - \frac{1}{2}\cos(\alpha + \beta)$$

$$\cos\alpha\cos\beta = \frac{1}{2}\cos(\alpha - \beta) + \frac{1}{2}\cos(\alpha + \beta)$$

$$\sin\alpha\cos\beta = \frac{1}{2}\sin(\alpha - \beta) + \frac{1}{2}\sin(\alpha + \beta)$$

$$e^{\pm j\theta} = \cos\theta \pm j\sin\theta$$

$$\cos\theta = \frac{1}{2}(e^{j\theta} + e^{-j\theta})$$

$$\sin\theta = (e^{j\theta} - e^{-j\theta})/2j$$

$$\sin^2\theta + \cos^2\theta = 1$$

$$\cos^2\theta - \sin^2\theta = \cos 2\theta$$

$$\cos^2\theta = \frac{1}{2}(1 + \cos 2\theta)$$

$$\cos^3\theta = \frac{1}{4}(3\cos\theta + \cos 3\theta)$$

$$\sin^2\theta = \frac{1}{2}(1 - \cos 2\theta)$$

$$\sin^3\theta = \frac{1}{4}(3\sin\theta - \sin 3\theta)$$

Expansions

$$(1+x)^n = 1 + nx + \frac{n(n-1)}{2!}x^2 + \cdots |nx| < 1$$

$$e^x = 1 + x + \frac{1}{2!}x^2 + \cdots$$

$$a^x = 1 + x\ln a + \frac{1}{2!}(x\ln a)^2 + \cdots$$

$$\ln(1+x) = x - \frac{1}{2}x^2 + \frac{1}{3}x^3 + \cdots$$

$$\sin x = x - \frac{1}{3}!x^3 + \frac{1}{5}!x^5 - \cdots$$

$$\cos x = 1 - \frac{1}{2!}x^2 + \frac{1}{4!}x^4 - \cdots$$

$$\tan x = x + \frac{1}{3}x^3 + \frac{1}{5}x^5 + \cdots$$

$$\sum_{m=0}^{M} x^m = \frac{(x^M - 1)}{(x-1)}$$

$$e^{a\cos b} = \sum_{i=0}^{\infty} \epsilon_i I_i(a)\cos(ib), \quad \epsilon_0 = 1,\ \epsilon_i = 2,\ i \geq 1$$

$$\cos(x\sin\theta) = J_0(x) + 2\sum_{k=1}^{\infty} J_{2k}(x)\cos(2k\theta)$$

$$\sin(x\sin\theta) = 2\sum_{k=0}^{\infty} J_{2k+1}(x)\sin[(2k+1)\theta]$$

$$\cos(x\cos\theta) = J_0(x) + 2\sum_{k=0}^{\infty}(-1)^k J_{2k}(x)\cos(2k\theta)$$

$$\sin(x\cos\theta) = 2\sum_{k=0}^{\infty}(-1)^k J_{2k+1}(x)\cos[(2k+1)\theta]$$

$$\int_0^{\infty} \frac{dx}{1+x^n} = \frac{(\pi/n)}{\sin(\pi/n)}, \quad n > 1 \qquad \int_0^{\infty} \frac{dx}{(a^2+x^2)^2} = \frac{\pi}{4a^3}$$

$$\int_0^{\infty} \frac{x^u\,dx}{1+x^n} = \left(\frac{\pi}{n}\right)\operatorname{cossec}\left[\frac{(u+1)\pi}{n}\right] \qquad \int_0^{\infty} \frac{dx}{1+x^2} = \tan^{-1}(b)$$

Undefined integrals

$$\int \sin ax\,dx = -\frac{1}{a}\cos ax$$

$$\int \cos ax\,dx = \frac{1}{a}\sin ax$$

$$\int \sin^2 ax\,dx = \frac{x}{2} - \frac{\sin 2ax}{4a}$$

$$\int \cos^2 ax\,dx = \frac{x}{2} + \frac{\sin 2ax}{4a}$$

$$\int \sin ax\cos ax\,dx = \frac{1}{2a}\sin^2(ax)$$

$$\int x\sin ax\,dx = \frac{1}{a^2}(\sin ax - ax\cos ax)$$

$$\int x\cos ax\,dx = \frac{1}{a^2}(\cos ax + ax\sin ax)$$

$$\int x^2 \sin ax \, dx = \frac{1}{a^3}(2ax \sin ax + 2 \cos ax - a^2x^2 \cos ax)$$

$$\int x^2 \cos ax \, dx = \frac{1}{a^3}(2ax \cos ax - 2 \sin ax + a^2x^2 \sin ax)$$

$$\int \sin ax \sin bx \, dx = \frac{\sin(a-b)x}{2(a-b)} - \frac{\sin(a+b)x}{2(a+b)} \qquad a^2 \neq b^2$$

$$\int \cos ax \cos bx \, dx = \frac{\sin(a-b)x}{2(a-b)} + \frac{\sin(a+b)x}{2(a+b)}x \qquad a^2 \neq b^2$$

$$\int \sin ax \cos bx \, dx = \frac{\cos(a-b)x}{2(a-b)} - \frac{\cos(a+b)x}{2(a+b)} \qquad a^2 \neq b^2$$

$$\int e^{ax} dx = \frac{1}{a}e^{ax}$$

$$\int xe^{ax} dx = \frac{1}{a^2}e^{ax}(ax-1)$$

$$\int x^2e^{ax} dx = \frac{1}{a^3}e^{ax}(a^2x^2 - 2ax + 2)$$

$$\int e^{ax} \sin bx \, dx = \frac{1}{a^2+b^2}e^{ax}(a \sin bx - b \cos bx)$$

$$\int e^{ax} \cos bx \, dx = \frac{1}{a^2+b^2}e^{ax}(a \cos bx + b \sin bx)$$

$$\int \left[\frac{\sin ax}{x}\right]^2 dx = a \int \frac{\sin 2ax}{x} dx - \frac{\sin^2 ax}{x}$$

$$\int \frac{dx}{a^2+b^2x^2} = \frac{1}{ab} \tan^{-1}\left(\frac{bx}{a}\right)$$

$$\int \frac{x^2 dx}{a^2+b^2x^2} = \frac{x}{b^2} - \frac{a}{b^3} \tan^{-1}\left(\frac{bx}{a}\right)$$

$$\int \frac{dx}{(a^2+b^2x^2)^2} = \frac{x}{2a^2(a^2+b^2x^2)} + \frac{1}{2ab^3} \tan^{-1}\left(\frac{bx}{a}\right)$$

$$\int \frac{x^2 dx}{(a^2+b^2x^2)^2} = \frac{-x}{2b^2(a^2+b^2x^2)} + \frac{1}{2ab^3} \tan^{-1}\left(\frac{bx}{a}\right)$$

$$\int \frac{dx}{(a^2+b^2x^2)^3} = \frac{x}{4a^2(a^2+b^2x^2)^2} + \frac{3x}{8a^4(a^2+b^2x^2)} + \frac{3}{8a^5b}\tan^{-1}\left(\frac{bx}{a}\right)$$

Defined integrals

$$\int_0^\infty \frac{\sin ax}{x}dx = \begin{cases} \pi/2 & a > 0 \\ 0 & a = 0 \\ -\pi/2 & a < 0 \end{cases}$$

$$\int_0^x \frac{\sin u}{u}du \triangleq \mathrm{Si}(x) \quad \text{the integral appears in tables as a function of } x$$

$$\int_0^\infty \frac{\sin^2 ax}{x^2}dx = |a|\pi/2$$

$$\int_0^\infty e^{-ax^2}dx = \frac{1}{2}\sqrt{\pi/a}$$

$$\int_0^\infty xe^{-ax^2}dx = \frac{1}{2a}$$

$$\int_0^\infty x^2e^{-ax^2}dx = \frac{1}{4a}\sqrt{\pi/a}$$

$$\int_0^\infty \frac{dx}{(x^2+a^2)(x^2+b^2)} = \frac{\pi}{2ab(a+b)} \qquad a > 0,\ b > 0$$

$$\int_0^\infty \frac{dx}{ax^4+b} = \frac{\pi}{2\sqrt{2b}}\left(\frac{b}{a}\right)^{1/4} \qquad ab > 0$$

$$\int_0^\infty \frac{dx}{ax^6+b} = \frac{\pi}{3b}\left(\frac{b}{a}\right)^{1/6} \qquad ab > 0$$

Appendix C
Use of the Radiofrequency Spectrum

The standardization organizations and regulatory agencies are responsible for establishing regulations and enforcing them. Table C.1 presents some regulatory agencies, institutes and consortia.

ETSI	*European Telecommunication Standardization Institute*
RACE	*Research and Development into Advance Communications Technologies for Europe*
CEPT	*Council of European PTTs*
ITU	*International Telecommunication Union*
TIA	*Telecommunications Industry Association*
ANSI	*American National Standard Institute*
JTC	*Joint Technical Committee*
RCR	*Research Center for Radio*
EIA	*Electronics Industry Association*
IEEE	*Institute of Electrical and Electronics Engineers*
WARC	*World Allocation Radio Conference*, approved the allocation the frequency range between 800 and 900 MHz for cellular telephony
CDG	*(CDMA Development Group)*, is composed of several companies that exploit the cellular service, including BELL Atlantic, Ameritech, NYNEX and GTE, and vendors, including Motorola, Qualcomm, Northern Telecom, Nokia, Oki and Sony

Table C.1. Standardization organizations.

This appendix presents the allocation of services in part of the radiofrequency spectrum, including some of the power levels standardized by the Federal Communications Commission (FCC) in the United States (Stremler, 1982).

VLF (*very low frequencies*)	3Hz up to 30kHz
LF (*low frequencies*)	30kHz up to 300kHz
MF (*medium frequencies*)	300kHz up to 3MHz
HF (*high frequencies*)	3MHz up to 30MHz
VHF (*very high frequencies*)	30MHz up to 300MHz
UHF (*ultrahigh frequencies*)	300MHZ up to 3GHz
SHF (*superhigh frequencies*)	3GHz up to 30GHz
EHF (*extra-high frequencies*)	30GHz up to 300GHz

Table C.2. Frequency bands.

Banda	Frequency bands	wavelength, cm
HF	3 - 30 MHz	10,000 - 1,000
VHF	30 - 300 MHz	1,000 - 100
UHF	300 - 1000 MHz	100 - 30
L	1.0 - 2.0 GHz	30 - 15
S	2.0 - 4.0 GHz	15 - 7.50
C	4.0 - 8.0 GHz	7.50 - 3.75
X	8.0 - 12.0 GHz	3.75 - 2.50
K_u	12.0 - 18.0 GHz	2.50 - 1.67
K	18.0 27.0 GHz	1.67 - 1.11
K_a	27.0 - 40.0 GHz	1.11 - 0.75
Millimeter	40 - 300 GHz	0.75 - 0.10

Table C.3. Detail of the frequency bands for the radar range.

Banda AM:	Tranmission class	Average power, kW
	Local	0.1 - 1.0
	Regional	0.5 - 5.0
	Line of sight	0.25 - 50

FM band: 0.25, 1, 3, 5, 10, 25, 50, 100kW, it depends on service class (size of comminuty) and coverage area.

Television (video):	Channels	Efective iradiated power (average), kW
	VHF (2 - 6)	100
	VHF (7 - 13)	316
	UHF	5000

Television (audio): 10% (-10dB) - 20% (-7dB) of the video carrier power.

Table C.4. Allowed permitted power levels.

Channel number	Frequency band (MHz)	Video carrier (MHz)
1		Not used
2	54 - 60	55.25
3	60 - 66	61.25
4	66 - 72	67.25
5	76 - 82	77.25
6	82 - 88	83.25
		(Commercial FM, band 88 - 108)
7	174 - 180	175.25
8	180 - 186	181.25
9	186 - 192	187.25
10	192 - 198	193.25
11	198 - 204	199.25
12	204 - 210	205.25
13	210 - 216	211.25

Table C.5. Frequency allocation for VHF television.

Channel number	Frequency band (MHz)	Video carrier (MHz)	Channel number	Frequency de band (MHz)	Video carrier (MHz)
14	470 - 476	471.25	42	638 - 644	639.25
15	476 - 482	477.25	43	644 - 650	645.25
16	482 - 488	483.25	44	650 - 656	651.25
17	488 - 494	489.25	45	656 - 662	657.25
18	494 - 500	495.25	46	662 - 668	663.25
19	500 - 506	501.25	47	668 - 674	669.25
20	506 - 512	507.25	48	674 - 680	675.25
21	512 - 518	513.25	49	680 - 686	681.25
22	518 - 524	519.25	50	686 - 692	687.25
23	524 - 530	525.25	51	692 - 698	693.25
24	530 - 536	531.25	52	698 - 704	699.25
25	536 - 542	537.25	53	704 - 710	705.25
26	542 - 548	543.25	54	710 - 716	711.25
27	548 - 554	549.25	55	716 - 722	717.25
28	554 - 560	555.25	56	722 - 728	723.25
29	560 - 566	561.25	57	728 - 734	729.25
30	566 - 672	567.25	58	734 - 740	735.25
31	572 - 578	573.25	59	740 - 746	741.25
32	578 - 584	579.25	60	746 - 752	747.25
33	584 - 590	585.25	61	752 - 758	753.25
34	590 - 596	591.25	62	758 - 764	759.25
35	596 - 602	597.25	63	764 - 770	765.25
36	602 - 608	603.25	64	770 - 776	771.25
37	608 - 614	609.25	65	776 - 782	777.25
38	614 - 620	615.25	66	782 - 788	783.25
39	620 - 626	621.25	67	788 - 794	789.25
40	626 - 632	627.25	68	794 - 800	795.25
41	632 - 638	633.25	69	800 - 806	801.25

Table C.6. Frequency allocation for UHF television.

Band designation	Frequency allocation (MHz)
160 meters	1.800 - 2.000
80 meters	3.500 - 4.000
40 meters	7.000 - 7.300
20 meters	14.000 - 14.350
15 meters	21.000 - 21.450
10 meters	28.000 - 29.700
6 meters	50.0 - 54.0
2 meters	144 - 148
	220 - 225
	420 - 450
	1,215 - 1,300
	2,300 - 2,450
	3,300 - 3,500
	5,650 - 5,925
	10,000 - 10,500
	24,000 - 24,500
	48,000 - 50,000
	71,000 - 76,000
	165,000 - 170,000
	240,000 - 250,000
	above 300,000

Table C.7. Frequency bands for radio amateur service.

Station class	Service	Band (MHz)	Maximum power at transmitter input	Output mean power	Frequency tolerance, %
A	General use	460 - 470	60 W	48 W	0.0005
B	General use	460 - 470	5 W	4 W	0.5
C	Romote control	26.96 - 27.23	5 W	4 W	0.005
	Hobby control	72 - 76	5 W	4 W	0.005
D	Radio-telephone	26.96 - 27.41	5 W	4 W	0.005

Table C.8. Radioamateur bands.

Appendix D
The CDMA Standard

D.1 Introduction

In September 1988, as a consequence of the growing demand for cellular mobile communication services, the Cellular Telecommunications Industry Association (CTIA), in USA, published the document called *User's Performance Requirements* (UPR) specifying basic requirements of the digital technology, for the operators of cellular systems. The requirements indicated directions for the new proposed systems.

- a ten times increase with respect to the capacity of analog systems;
- long life and an adequate technology growth for the second generation;
- capacity to allow the introduction of new facilities;
- quality improvement;
- privacy;
- ease of transition and compatibility with the existing analog system;
- availability, at low cost, of radios and cells which operate on both systems;
- Cellular Open Network Architecture – CONA.

In 1989, the subcommittee for digital cellular systems of the Telecommunications Industry Association (TIA), in USA, formulated the specification EIA/TIA/IS-54 (*Dual-Mode Subscriber Equipment – Network Equipment Compatibility Specification*), which adopted the TDMA as the multiple access technique.

In December 1991, during the presentation of results of field tests for the next generation systems, the Qualcomm company, together with various equipment manufacturers and operators, presented the results of their field tests with a CDMA system.

The American manufacturers of infrastructure equipment which took part in the development of the CDMA system were: AT&T, Motorola and Northern Telecom. The manufacturers of subscriber equipment included Motorola, OKI Telecom, Clarion, Sony, Alps Electric, Nokia and Matsushita-Panasonic. The operators that took part in the efforts to validate the cellular CDMA system were PacTel Cellular, Ameritech Mobile, Nynex Mobile, GTE Mobile Communications, Bell Atlantic Mobile Systems, US West New Vector Group and Bell Cellular of Canada (Qualcomm, 1992).

On 6th January 1992, CTIA adopted,unanimously, a resolution to accept contributions in terms of wide band cellular systems. In February of the same year, TIA also adopted a motion to recommend the standardization of wide band systems with spread spectrum technologies. The efforts of Qualcomm and associates resulted in the adoption of the IS-95 standard for cellular CDMA. A few advantages of CDMA are described next

1 increases the capacity of telephone traffic;

2 improves voice quality;

3 eliminates audible effects of multipath;

4 reduces the number of call drops due to failures in handoff;

5 provides secure data transportation;

6 reduces operation costs because only a few cells are required;

7 reduces the average value of the transmitted power;

8 reduces interference in other electronic devices;

9 presents higher privacy and security.

D.2 Spread Spectrum

The first developments using concepts related with current spread spectrum techniques occurred in the 40's, involving applications to military problems (Scholtz, 1982). From the beginning until the present date, there has been a process of evolution due mainly to improvements in electronic components.

Spread spectrum techniques became very popular in the communications community, over the last 20 years, since technical articles on this subject began to appear in the specialized literature, after almost two decades of secrecy due to military interests. Diverse commercial spread spectrum applications are today in a regulatory phase in various countries (Newman Jr., 1986).

The main characteristic of spread spectrum systems is concerned with its bandwidth, which is far greater than that of baseband signals, or modulated signals produced by conventional techniques. The manner by which the spectrum is spread is done with the help of a code, independent of the transmitted data. The receiver needs an identical copy of the code, synchronized in time, in order to recover the data.

The process of spreading introduces a large amount of redundancy in the transmitted signal. This allows the receiver to suppress a considerable portion of the interference introduced by the channel in the transmitted signal. This feature constitutes one of the major advantages of the process. However, it is possible to name other advantages of spread spectrum systems (Pickholtz et al., 1982, Scholtz, 1982) as follows.

- Low intercept probability (*jamming*) of the signal, due to the random-like nature of the spreading code;
- Allows the addressing of a select group of users in the same channel, i.e., code division multiple access (CDMA);
- Use of the signal to measure propagation delay, which is important for radar systems to determine the position and direction of targets;
- It is possible to hide the signal in background noise, so as to make it difficult to detect by those for which the spreading code is unknown.

- Useful as a means of achieving diversity in channels with multipath.

One of the current strategies in use to spread the spectrum is the technique of direct sequence – DS modulation, where a fast pseudo-noise – PN causes amplitude switching in the data signal. Figure D.1 illustrates this technique.

Another used technique is called frequency hopping – FH, where the signal is subjected to switching in frequency, according to a pre-determined pseudo-random sequence. This technique is illustrated in Figure D.2.

It is also possible to produce a spread spectrum by means of transmission of signal bursts at random time instants. The technique is known as time hopping – TH and is illustrated in Figure D.3.

Other systems may be conceived by combining the techniques presented earlier. In this manner, the spread spectrum system is the antithesis of the usual communication system. This follows because, contrasting with the usual assumptions for designing communication systems, spread spectrum systems require a very large bandwidth and a very low power spectral density for the transmitted signal.

However, what happened then was that spread spectrum was recommended sometimes to applications were apparently it was not very effective, as pointed out by Viterbi (Viterbi, 1985):

> "The mystique of spread spectrum communications is such that commercial enterprises, as well as academics, are often attracted by the novelty and cleverness of the technique. Also, in small artificially aided markets, there may be temporary economic advantages. In the long run, though, it's the author's opinion that we must stand back and question the wisdom of squandering a precious resource such as bandwidth for reasons of expediency".

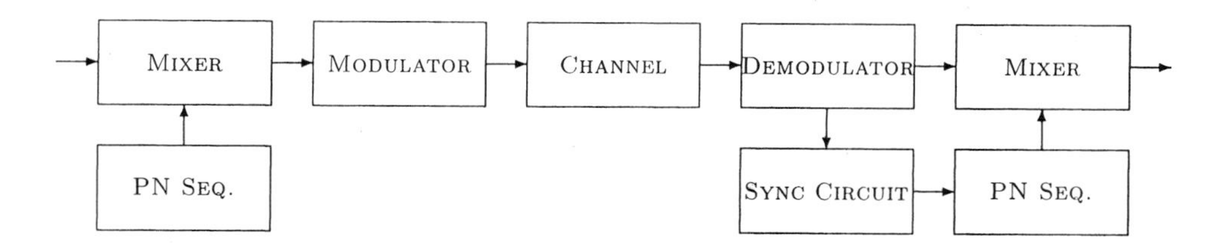

Figure D.1. Direct sequence spread spectrum system.

Transmission of the spread signal is done by means of digital modulation of a carrier. The digital modulator is referred to as a mixer in the above figures. Many modulation techniques can be employed, e.g., BPSK and QPSK.

D.2.1 Generation of Pseudo-Random Sequences

In principle pseudo-random sequences could be stored in a memory and a search algorithm could be used for selecting a particular sequence. However, usually sequential techniques are fast enough and offer lower implementation costs for producing such sequences.

Using Galois field theory, a branch of Abstract Algebra, it is possible to generate pseudo-random sequences in a sequential manner.

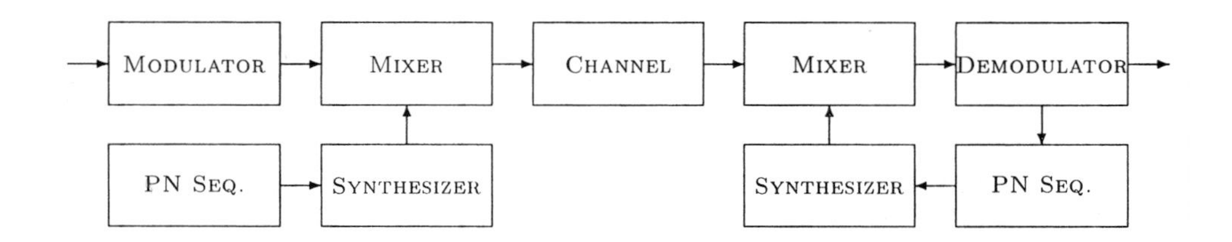

Figure D.2. Frequency hop spread spectrum system.

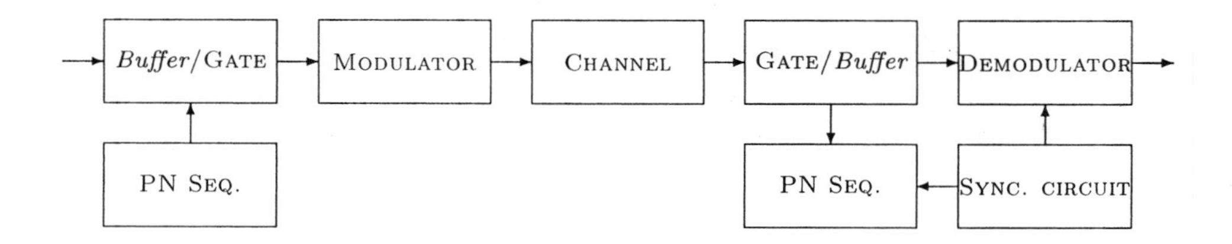

Figure D.3. Spread spectrum using random time windows.

In the sequel only unipolar sequences will be considered, i.e., sequences the elements of which are 0's and 1's. Figure D.4 shows a general purpose pseudo-random sequence generator.

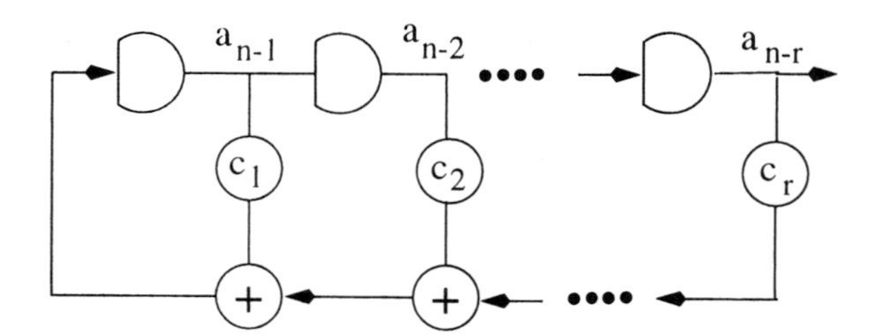

Figure D.4. Pseudo-Random Sequence Generator.

The shift-register is started with some set of bits as the initial state. At each clock period the contents a_j of the shift-register are shifted to the right. There are r stages in the shift-register, the outputs of which may be connected to a modulo-2 adder. The modulo-2 adders are part of the shift-register feedback circuit. The feedback connections depend on the value of c_j; if $c_j = 1$ there is a connection present, if $c_j = 0$ there is no connection to the feedback circuit. At each clock pulse a bit of the pseudo-random sequence is produced at the output of the generator by means of a_{n-r}.

Note that all used processes are linear and each bit entering the first stage of the shift-register influences the generation of the next r bits, characterizing the memory property of the shift-register.

For each generator there is a characteristic polynomial the degree of which is equal to the number of shift-register stages, and the polynomial coefficients are indicate the connections between stages and the modulo-2 adders. The characteristic polynomial corresponding to the generator in Figure D.4 is given by

$$f(D) = 1 + c_1 D + c_2 D^2 + \ldots + c_r D^r, \tag{D.1}$$

where D^2 is associated with the second shift-register stage and c_2 indicates a connection of the second stage with a modulo-2 adder. An example of a generator with 42 stages is given in the simulation presented in Figure D.8.

Considering an r-stage shift-register, there is a maximum of 2^r possible binary r-tuples. However, because of the linear feedback, the all-zero (r 0 bits) initial state would lead to an all-zero sequence. Therefore, for the sequences of interest there are at most $2^r - 1$ distinct r-tuples which can be used as initial state in the shift-register.

Suppose then that a primitive characteristic polynomial is available, i.e., a feedback connection polynomial for which all non-zero $2^r - 1$ r-tuples appear in the r stages of the shift-register. In this case a pseudo-random sequence of period $2^r - 1$ is generated. It is remarked that the only truly random data in this sequence is represented by the initial state of the shift-register. The initial state together with the primitive polynomial are sufficient to generate the corresponding complete pseudo-random sequence.

Primitive polynomials and their properties are studied in abstract algebra. A necessary but not sufficient condition for a given polynomial to be primitive is that it must be irreducible in the field containing its coefficients.

An important property of irreducible polynomials is that any irreducible polynomial $p(x)$ of degree r, with binary ($\{0,1\}$) coefficients divides the polynomial

$$f(x) = x^{2^r - 1} - 1. \tag{D.2}$$

Let $p(x)$ be a binary irreducible polynomial of degree r. The polynomial $p(x)$ is called a primitive polynomial if and only if the least degree polynomial $x^n + 1$ divisible by $p(x)$ has degree $n = 2^r - 1$. In general, it is not an easy matter to tell whether a polynomial is primitive, let alone to find one. Usually one resorts to tables of primitive polynomials in order to select a primitive polynomial of a desired degree.

The following equation gives the number of binary primitive polynomials $N_p(r)$ of a given degree r:

$$N_p(r) = \frac{2^r - 1}{r} \prod_{i=1}^{J} \frac{P_i - 1}{P_i}, \tag{D.3}$$

where $P_i, P_i > 1, i = 1, 2, \ldots, J$ are the natural numbers that divide $2^r - 1$.

D.3 Convolutional Coding

A convolutional encoder is a device that inserts redundancy to the binary data, before the spread spectrum process. As a consequence of the insertion of redundant digits usually the transmission rate is increased. The insertion of redundancy is a way to achieve error-correction. Besides convolutional coding, there is at least one more coding technique for error-correction. That technique is called block coding, in which k-bit blocks of data are encoded into n-bit long codewords by means of a predefined encoding algorithm, where $n > k$.

Convolutional codes were first studied by Elias in 1955. As a result of the effort of many researchers convolutional coding experienced great development, culminating in 1967, when Viterbi developed the decoding algorithm for convolutional codes now known as the Viterbi decoding algorithm. Later Forney showed that the Viterbi decoder is a maximum likelihood decoder.

D.3.1 The Structure of Convolutional Codes

The implementation of a convolutional encoder is simple for it employs procedures similar to those adopted in the generation of pseudo-random sequences, in Section D.2.1. Decoding, however, is a more elaborate issue and will be discussed later.

In the generation of pseudo-random sequences, see Figure D.4, the output of the furthest right stage is the only output, and there are no external inputs in this circuit. On the other hand, in a convolutional encoder, see Figure D.5 for example, the encoder has at least one external input and more than one output.

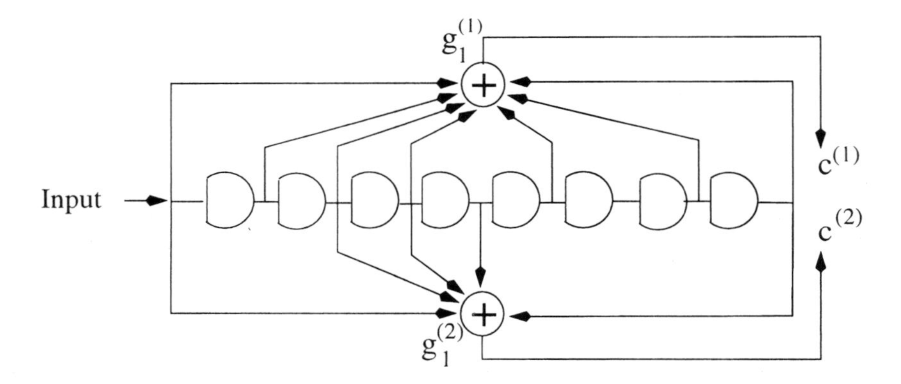

Figure D.5. Convolutional encoder diagram.

Since the job of the encoder is to introduce redundancy, it follows that the number of external inputs to the encoder is less than the number of its outputs.

The encoder shift-register stages are assumed to contain zeros initially. If an information bit equal to 1 is present at the encoder input, then at the next clock pulse the shift-register contents are shifted one stage to the right and the information bit enters the first stage of the encoder shift-register. Simultaneously, the encoder outputs are generated with the aid of modulo-2 adders; see Figure D.5 as an example.

For every k input information bits the encoder outputs n bits (a sub-block) and retains r bits in its memory. The code rate is defined as the ratio between the number of input bits and the number of output bits. Therefore if transmission ends after N sub-blocks, the code rate is $kN/(nN + r)$, which asymptotically approaches k/n for large values of N. The code constraint length K is an important parameter defined as the number of encoder shift-register stages plus 1, i.e., $K = r + 1$.

In order to fully characterize the identification of the encoder it is necessary to specify the connections between the shift-register and the modulo-2 adders, i.e., it is necessary to specify each encoder output as a linear function of the input and current state; see Figure D.5.

The encoder connections are specified by a set of sequences given by

$$g_i^{(j)} = (g_{i,0}^{(j)}, g_{i,1}^{(j)}, \dots, g_{i,r}^{(j)})$$

where $i = 1, \dots, n$ refers to the inputs, and $j = 0, \dots, r$ refers to the outputs. If $g_{i,m}^{(j)} = 1$, then there is a connection from the output of the mth shift-register stage to a modulo-2 adder, to produce an output bit at output $c^{(j)}$. If $g_{i,m}^{(j)} = 0$ then there is no connection present at the output of the mth shift-register stage. For $m = 0$ there is just a connection from the input of the shift-register.

¿From the example illustrated in Figure D.5 it follows that.

$$g_1^{(1)} = (1, 1, 1, 1, 0, 1, 0, 1, 1)$$

$$g_1^{(2)} = (1, 0, 1, 1, 1, 0, 0, 0, 1)$$

thus characterizing outputs $c^{(1)}$ and $c^{(2)}$.

The structure of the convolutional encoder used in the reverse link, in the IS-95 standard, is shown in Figure D.13.

D.4 CDMA and the IS-95 Standard

The technology of spread spectrum appeared initially in connection with military applications and only in the 1980's it began to be considered for commercial applications.

Spread spectrum techniques were used in military applications mainly due to its privacy properties, since the spread signal looks like noise and its detection is dependent on the sequence employed to perform the spreading. Resistance to interference is another advantage, because even if a signal is transmitted with the same carrier frequency as the spread signal, depending on the signal to noise ratio it may be still possible to recover the original signal by employing the spreading sequence.

In wireless communication, each spread spectrum system has a frequency band to perform its transmissions. Since, in a given frequency band, there are many users trying to send their signals, a technique is required to discriminate users so as to allow transmissions to operate with a prescribed quality. Operation of many users in this manner characterizes what is known as multiple access.

Various multiple access digital systems were proposed and commercially used, as the GSM system in Europe, the TDMA in USA, and PDC in Japan. CDMA was first commercially used in USA and became a standard under the name IS-95, later changed by Qualcomm to CDMA2000.

D.4.1 Processing Gain and CDMA Capacity

From an analysis of the code division multiple access system, it is possible to derive an equation for the number of calls per cell as a function of the data rate, final bandwidth, signal to noise ratio, voice activity factor, frequency reuse and sectorization.

An important parameter when determining capacity is the ratio between the power of the received signal, $C = R \cdot E_b$, and the interference, $I = W \cdot N_0$, i.e.,

$$\frac{C}{I} = \frac{RE_b}{WN_0}, \tag{D.4}$$

where R denotes the data rate in kbit/s, W denotes the bandwidth spread in frequency and E_B/N_0 is the ratio between the bit energy and the noise power spectral density.

For a system using CDMA, the interfering power is given by $I = C \cdot (N-1)$, where N is the number of users transmitting in the bandwidth W, and thus $C/I = 1/(N-1)$. It is assumed that all signals are transmitted with a controlled power level, which is the case for the Qualcomm systems.

Using the expressions given earlier it follows that the capacity of a CDMA system, in a non-cell environment is given by

$$N = 1 + \frac{W}{R}\frac{1}{E_b/N_0}, \tag{D.5}$$

where $\frac{W}{R}$ is called the capacity gain of the CDMA scheme, denoted by G_P.

The voice activity factor, V, reflects the fact that in a full-duplex conversation, the fraction of time actually used by the voice signal is usually less than 0,35. With CDMA it is possible to reduce the transmission rate when there is no signal to transmit and, in this manner, to minimize the interference caused to other users. The consideration of the voice activity factor, thus practically doubles capacity, since the interfering power is given by NV. The modified equation for capacity then becomes

$$N = 1 + \frac{W}{R}\frac{1}{E_b/N_0}\frac{1}{V}. \tag{D.6}$$

In a cellular system, as a consequence of interference by adjacent cells, the signal to interference ratio becomes

$$\frac{C}{I} = \frac{1}{N + 6Nk_1 + 12Nk_2 + 18Nk_3 + \cdots} \tag{D.7}$$

where N is the number of mobile stations per cell and $k_1,\ k_2,\ k_3, \ldots$ represent the contributions of individual cells in the layers around the cell under consideration. By defining

$$F = \frac{1}{1 + 6k_1 + 12k_2 + 18k_3 + \cdots} \tag{D.8}$$

as the efficiency of frequency reuse, Equation D.7 can be rewritten as

$$\frac{C}{I} = \frac{F}{N}. \tag{D.9}$$

Finally, capacity can still be increased as a function of the sectorization gain, G, thus yielding the full capacity equation

$$N = 1 + \frac{W}{R}\frac{1}{E_b/N_0}\frac{1}{V}FG, \tag{D.10}$$

where N denotes the number of calls per cell (assuming Rayleigh fading), W is the bandwidth spread in frequency (1,25 MHz), R denotes the data rate in kbp/s (assuming 9,6 kbps), E_B/N_0 is the ratio between bit energy and noise power spectral density (assuming 7,0 dB), V is the voice activity factor (assuming 40 %), F is the efficiency of frequency reuse (assuming 60 %), G_S denotes the sectorization gain (assuming 3 sectors: 2.55).

In order to increase capacity it is a usual practice to reduce both the transmitting power level and the date rate whenever there is no voice activity. The capacity is then proportionally incremented with transmission rate reduction. This proportionality factor is called (as mentioned earlier) the voice activity gain V.

As mentioned earlier, another way to increase capacity is to sectorize an area of coverage, thus reducing interference by a factor factor due to the antenna. Other factors like interference from other cells and noise in the channel itself can be incorporated to the equation for channel capacity, in order to investigate by simulation the effect of these factors in the system.

D.4.2 Channel Layout for IS-95

In each cell there are two data links, one for direct channels (transmission from the base station to the mobile) and the other for the reverse channels (transmission from the mobile to the base station).

The direct link consists of a pilot channel, a synchronization channel, seven paging channels and at most 64 channels of direct traffic. The reverse link consists of access channels and channels for traffic.

There are then six distinct types of channels in the IS-95 standard. The task of each type of channel will be described in the following sections. In (Rhee, 1998) one finds a quantitative description of these channels with the associated transmission procedures. Figures D.6 and D.7 illustrate the structure of these channels.

Pilot Channel

The pilot channel transmits continuously a non-modulated spread spectrum signal from the base station in a given cell. The structure of this channel is illustrated in Figure D.6. The mobile station monitors the pilot channel practically all the time, as a phase reference for demodulation.

Synchronization Channel

Communication between the base station and the mobile is synchronous, and is performed by the pilot channel and by the synchronization channel. The mobile station uses the synchronization channel only when it is entering the system, at the beginning of its connection to the base station. After that the mobile station will hardly use the synchronization channel again.

After the synchronization channel is received, the mobile station adjusts its timing and initiates monitoring the paging channels (described next). Figure D.7 shows the structure of the data link for reverse channels, which will be discussed later.

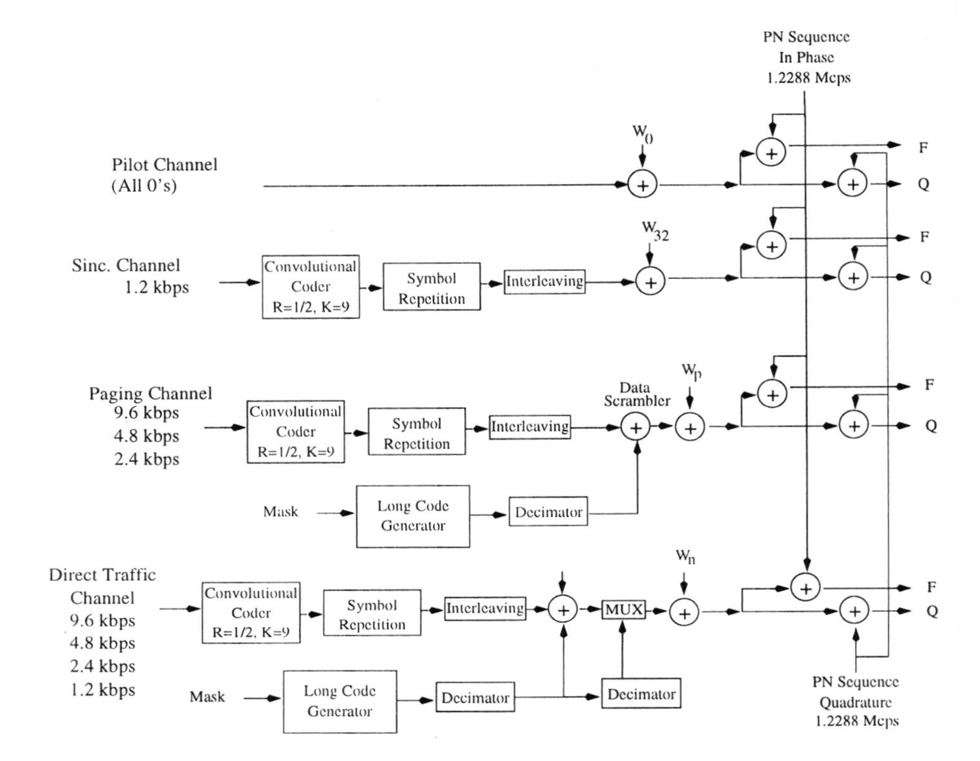

Figure D.6. CDMA direct channel diagram.

Paging Channel

The base station uses the paging channel to transmit system information and specific information to a given mobile station. Information like, for example, beginning and end of a call and register of a mobile station flow in this channel.

Each mobile station monitors only one paging channel. Figure D.6 also shows the structure of the paging channel.

Channels for the direct and for the reverse traffic

Both channels, for direct and for reverse traffic, use control structures consisting of frames of 20 ms. Transmission can be done at rates of 9,600, 4,800, 2,400 and 1,200 bps. In these channels flow conversations and the following five types of messages.

- Call control messages;
- Handoff control messages;
- Power control messages;
- Security and authentication messages;
- Messages to obtain/provide special information from/to a mobile station.

The structure of channels of direct traffic and channels of reverse traffic are illustrated in Figures D.6 and D.7, respectively.

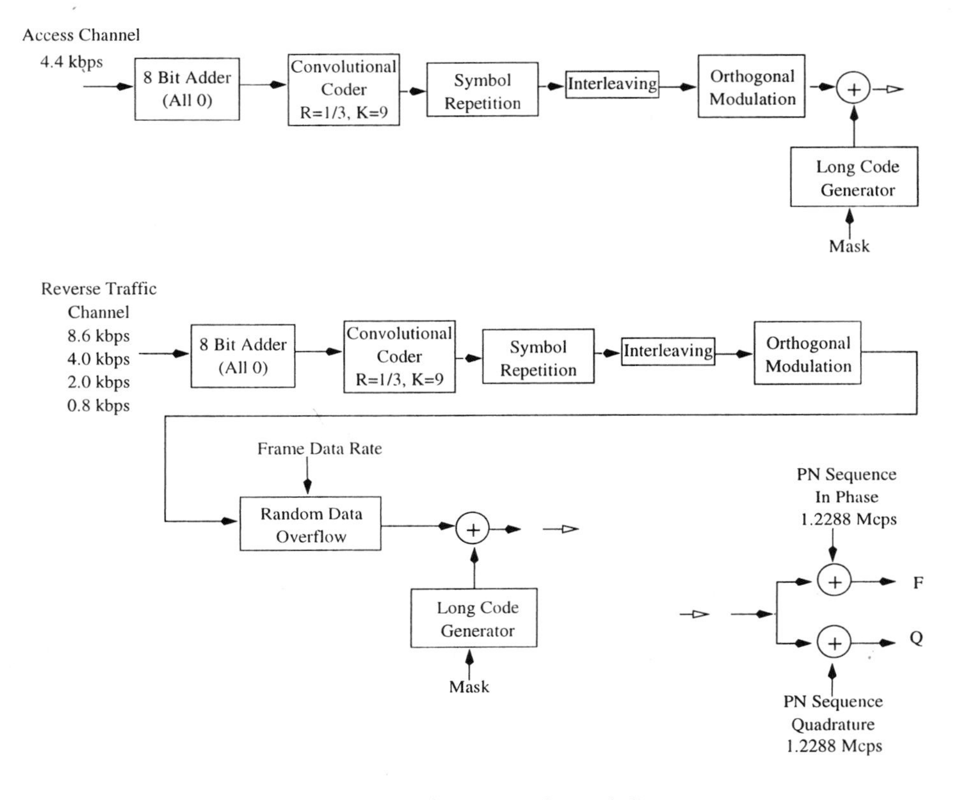

Figure D.7. CDMA reverse channel diagram.

Access Channel

The access channel provides communication between a mobile station and the base station, when the mobile station is not using a traffic channel. The mobile station randomly chooses an access channel, which is distinguished from other channels through its associated pseudo-random sequence.

This channel is used to send short signaling messages from the mobile station to the base station, as for example messages for originating a call, channel response, state of the mobile station and register.

Each frame in an access channel has 96 bits and 20 ms duration, which corresponds to a transmission rate of

Out of the 96 bits only 88 bits are data bits. The remaining 8 bits are all-zeros and are used for synchronization with the convolutional encoder. Figure D.7 shows the structure of the access channel.

Transmission on the Access Channel

In the sequel a description is presented of the process of generating digits for transmission on the access channel. According to the structure of the access channel, illustrated in Figure D.7, there are 88 information bits at the input and 24,576 chips

MSI	Walsh sequences
0	0000000000000000000000000000000000000000000000000000000000000000
1	0101010101010101010101010101010101010101010101010101010101010101
2	0011001100110011001100110011001100110011001100110011001100110011
3	0110011001100110011001100110011001100110011001100110011001100110
4	0000111100001111000011110000111100001111000011110000111100001111
5	0101101001011010010110100101101001011010010110100101101001011010
6	0011110000111100001111000011110000111100001111000011110000111100
7	0110100101101001011010010110100101101001011010010110100101101001
8	0000000011111111000000001111111100000000111111110000000011111111
9	0101010110101010010101011010101001010101101010100101010110101010
10	0011001111001100001100111100110000110011110011000011001111001100
11	0110011010011001011001101001100101100110100110010110011010011001
12	0000111111110000000011111111000000001111111100000000111111110000
13	0101101010100101010110101010010101011010101001010101101010100101
14	0011110011000011001111001100001100111100110000110011110011000011
15	0110100110010110011010011001011001101001100101100110100110010110
16	0000000000000000111111111111111100000000000000001111111111111111
17	0101010101010101101010101010101001010101010101011010101010101010
18	0011001100110011110011001100110000110011001100111100110011001100
19	0110011001100110100110011001100101100110011001101001100110011001
20	0000111100001111111100001111000000001111000011111111000011110000
21	0101101001011010101001011010010101011010010110101010010110100101
22	0011110000111100110000111100001100111100001111001100001111000011
23	0110100101101001100101101001011001101001011010011001011010010110
24	0000000011111111111111110000000000000000111111111111111100000000
25	0101010110101010101010100101010101010101101010101010101001010101
26	0011001111001100110011000011001100110011110011001100110000110011
27	0110011010011001100110010110011001100110100110011001100101100110
28	0000111111110000111100000000111100001111111100001111000000001111
29	0101101010100101101001010101101001011010101001011010010101011010
30	0011110011000011110000110011110000111100110000111100001100111100
31	0110100110010110100101100110100101101001100101101001011001101001
32	0000000000000000000000000000000011111111111111111111111111111111

Table D.1. The 64 Walsh sequences (First part).

at the output, of which 12,288 phase modulate the carrier and 12,288 modulate the carrier in quadrature. The following sections describe in more detail the main blocks illustrated in Figure D.7.

D.4.3 Orthogonal Modulation

A set of 64 orthogonal sequences is employed, each sequence being 64 bits long, called Walsh sequences, which guarantee orthogonality among the transmitted data. Each sequence has a number called modulation symbol index – MSI.

MSI	Walsh sequences
33	0101010101010101010101010101010110101010101010101010101010101010
34	0011001100110011001100110011001111001100110011001100110011001100
35	0110011001100110011001100110011010011001100110011001100110011001
36	0000111100001111000011110000111111110000111100001111000011110000
37	0101101001011010010110100101101010100101101001011010010110100101
38	0011110000111100001111000011110011000011110000111100001111000011
39	0110100101101001011010010110100110010110100101101001011010010110
40	0000000011111111000000001111111111111111000000001111111100000000
41	0101010110101010010101011010101010101010010101011010101001010101
42	0011001111001100001100111100110011001100001100111100110000110011
43	0110011010011001011001101001100110011001011001101001100101100110
44	0000111111110000000011111111000011110000000011111111000000001111
45	0101101010100101010110101010010110100101010110101010010101011010
46	0011110011000011001111001100001111000011001111001100001100111100
47	0110100110010110011010011001011010010110011010011001011001101001
48	0000000000000000111111111111111111111111111111110000000000000000
49	0101010101010101101010101010101010101010101010100101010101010101
50	0011001100110011110011001100110011001100110011000011001100110011
51	0110011001100110100110011001100110011001100110010110011001100110
52	0000111100001111111100001111000011110000111100000000111100001111
53	0101101001011010101001011010010110100101101001010101101001011010
54	0011110000111100110000111100001111000011110000110011110000111100
55	0110100101101001100101101001011010010110100101100110100101101001
56	0000000011111111111111110000000011111111000000000000000011111111
57	0101010110101010101010100101010110101010010101010101010110101010
58	0011001111001100110011000011001111001100001100110011001111001100
59	0110011010011001100110010110011010011001011001100110011010011001
60	0000111111110000111100000000111111110000000011110000111111110000
61	0101101010100101101001010101101010100101010110100101101010100101
62	0011110011000011110000110011110011000011001111000011110011000011
63	0110100110010110100101100110100110010110011010010110100110010110

Table D.2. The 64 Walsh sequences (Continuation).

The Walsh sequences can be recursively generated Hadamard matrices

$$H_{2n} = \begin{bmatrix} H_n & H_n \\ H_n & \overline{H_n} \end{bmatrix}, \tag{D.11}$$

with $n = 1, 2, 4, 8, 16, 32$ e $H_1 = 0$. Matrix H_{64} is shown in Tables D.1 and D.2.

In orthogonal modulation 6 input bits are used to calculate a MSI value. Once calculated, the output has the 64 bits of the Walsh sequence corresponding to the MSI value. This value is determined by the equation

$$MSI = C_0 + 2C_1 + 4C_2 + 8C_3 + 16C_4 + 32C_5. \tag{D.12}$$

D.4.4 Direct Sequence Spreading

As shown in Figure D.7, scattering results from the modulo-2 addition of the output sequence of the modulation block with the sequence at the output of the PN generator; Figure D.8.

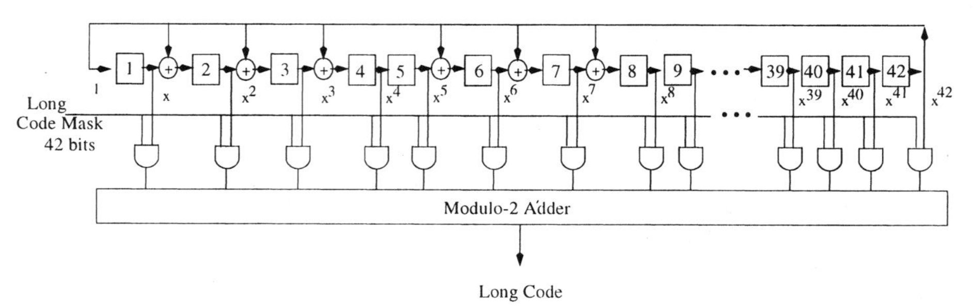

Figure D.8. Generator of long code sequence.

The PN sequence has a period of $2^{42} - 1$. The generator is characterized by the polynomial

$$p(x) = 1 + x + x^2 + x^3 + x^5 + x^6 + x^7 + x^{10} + x^{16} + x^{17} + x^{18} + x^{19}$$

$$+x^{21} + x^{22} + x^{25} + x^{26} + x^{27} + x^{31} + x^{33} + x^{35} + x^{42}.$$

At the initial state of the process the register contents are 1000000000000000000000 00000000000000000000. At each clock pulse there is a displacement of the register contents to the right, and those register cells which receive as input the output of a modulo-2 adder receive the sum of the contents of the previous cell and that of the furthest cell to the right.

At each time instant there are 42 bits in the register. Each one of the bits is fed to one input of a two-input AND gate, and the other input is fed with bits from another 42 bit long sequence called mask. The mask is formed from data belonging to the particular mobile station. The description of the mask is illustrated in Figure D.9.

41 … 33	32 … 28	27 … 25	24 … 9	8 … 0
110001111	ACN	PCN	BASE-ID	PILOT-PN

Highest Significant Bit — Least Significant Bit

ACN: Access Channel Number
PCN: Paging Channel Number
BASE-ID: Base Station Identification
PILOT-PN: Direct Channel PN Offset

Figure D.9. Mask code for the access channel.

The output of each AND gate is connected to a modulo-2 adder the output of which is a bit of the long code.

In the access channel, the long code sequence is four times faster than the orthogonally modulated sequence. Each input bit is added (modulo-2) successively with four bits of the long code, thus resulting in four bits at the output of the direct sequence spreading circuit.

D.4.5 Quadrature Scattering

Quadrature scattering, Figure D.10, is similar to direct sequence scattering. The bits received from the scattering block are divided into two sequences, one for phase scattering (even bits) and the other in quadrature (odd bits).

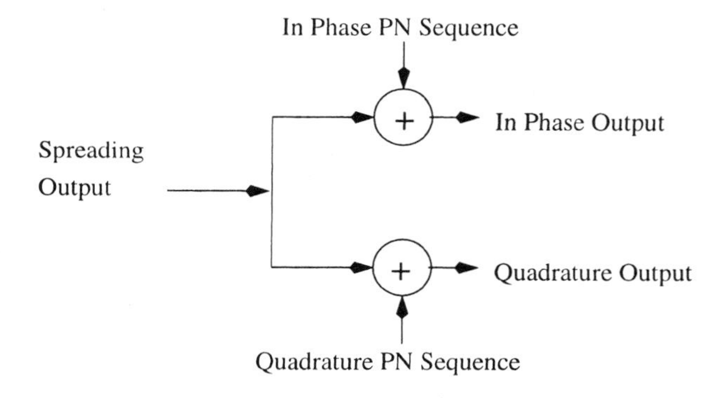

Figure D.10. Quadrature scattering.

The phase sequence is added bit-by-bit with the PN phase spreading sequence. The scattering sequence is obtained from the generator shown in Figure D.11. The associated characteristic polynomial is $P_I = 1 + x^5 + x^7 + x^8 + x^9 + x^{13} + x^{15}$.

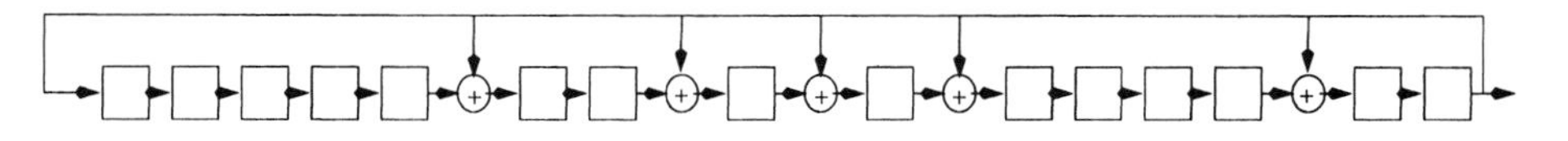

Figure D.11. Generator for PN phase spreading sequences.

The quadrature sequence is added bit-by-bit with the quadrature PN spreading sequence. The spreading sequence is obtained from the generator shown in Figure D.12. The associated characteristic polynomial is $P_Q = 1 + x^3 + x^4 + x^5 + x^6 + x^{10} + x^{11} + x^{12} + x^{15}$.

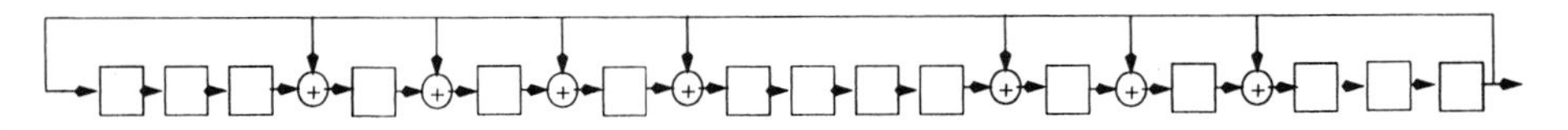

Figure D.12. Generator for quadrature PN spreading sequences.

D.4.6 Convolutional Encoder

This encoder has 96 bits input bits at a transmission rate of 4,8 kbps, producing at the output 288 bits at 14,4 kbps. Its asymptotic coding rate is $R = 1/3$ and the sub-block length is 9. Figure D.13 shows the encoder structure (Rhee, 1998).

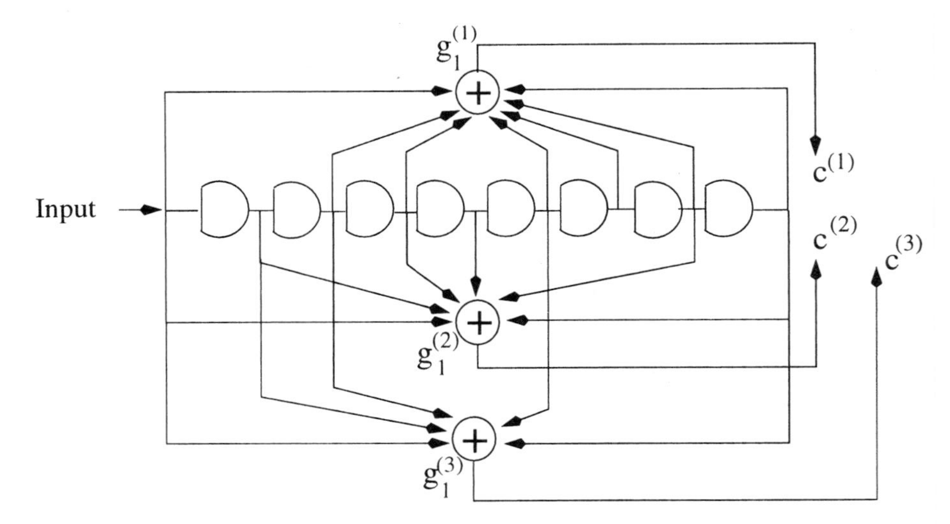

Figure D.13. Diagram of convolutional encoder for the access channel.

For each bit $b_k, k = 0, 1, \ldots, 95$ at the encoder input, three bits are generated at the output, denoted as C_0^k, C_1^k, C_2^k. Initially the shift-registers $D_i, i = 1, 2, \ldots, 8$ are all set to zero.

At each clock pulse the register contents of a cell is moved to its neighbor cell on the right. Register cell D_1 is fed with the input bit, according to the order from bit b_0 to b_{95}.

The outputs are calculated in the following manner.

$$
\begin{aligned}
C_0^k &= d_8 + d_6 + d_5 + d_3 + d_2 + d_1 + b_k, \\
C_1^k &= d_8 + d_7 + d_5 + d_4 + d_1 + b_k, \\
C_2^k &= d_8 + d_7 + d_6 + d_3 + +b_k,
\end{aligned}
$$

where $k = 0, \ldots, 95$ and d_i is the output of register cell D_i. This procedure is repeated until b_{95} is encoded. The output is read out in series, from the first to the last bit, as follows.

$$
C_0^0, C_1^0, C_2^0, C_0^1, C_1^1, C_2^1, \ldots, C_0^{95}, C_1^{95}, C_2^{95}.
$$

Symbol Repetition Block

At the input to the symbol repetition block there are 288 bits and at the output there are 576 bits. There is no further processing of the input bits, apart from their repetition. The bits are read out in the following order.

$$C_0^0, C_0^0, C_1^0, C_1^0, C_2^0, C_2^0, \ldots$$

Interleaving Block

The purpose of the interleaving block is to help correct burst errors, during data transmission through a channel with multipath fading. Another reason is to reduce the excess in redundancy, thus improving system efficiency.

The interleaving block receives 576 bits and changes their order. Suppose that the bits are ordered in matrix form 32x18, column by column. The output is then obtained by reading the matrix in a row-by-row manner. Reading is performed alternately, i.e., first line 1 is read out then line 17 is read out, and so on successively in the following order.

1 17 9 25 5 21 13 29 3 19 11 27 7 23 15 31
2 18 10 26 6 22 14 30 4 20 12 28 8 24 16 32

D.5 CDMA 1x EV-DO

The 1x EV-DO is the data optimized third generation (3G) cellular telecommunications system for CDMA2000. CDMA2000 1x EV-DO cell phone system is a standard that has evolved from the CDMA2000 mobile phone system. The letters EV-DO stand for Evolution Data Only or Data Optimized. It is a data only mobile telecommunications standard that can be run on CDMA2000 networks.

The EV-DO cell phone system is capable of providing the full 3G data rates up to 3.1 Mbps. The first commercial CDMA2000 1x EV-DO network was deployed by SK Telecom (Korea) in January 2002. Now operators in several countries, including Brazil, Ecuador, Indonesia, Jamaica, Puerto Rico, Taiwan and the USA launched networks.

The CDMA2000 1xEV-DO cell phone system is defined under IS-856 rather than IS-2000 that defines the other CDMA2000 standards, and as the name indicates it only carries data. In Release 0 of the standard the maximum data rate was 2400 Mbps in the forward (downlink) with 153 kbps in the reverse (uplink) direction, the same as CDMA2000 1X. However, in the later release of the standard, Release A, the forward data rate rises to 3.1 Mbps, and 1.2 Mbps in the reverse direction.

The forward channel forms a dedicated variable-rate, packet data channel with signalling and control time multiplexed into it. The channel is itself time-divided and allocated to each user on a demand and opportunity driven basis. A data only format was adopted to enable the standard to be optimized for data applications. If voice is required then a dual mode phone using separate 1X channel for the voice call is needed. The phones used for data only applications are referred to as Access Terminals or ATs.

The EV-DO RF transmission is very similar to that of a CDMA2000 1X transmission. It has the same final spread rate of 1.228 Mcps and it has the same modulation bandwidth because the same digital filter is used. Although 1xEV-DO has many similarities with 1X transmissions, it cannot occupy the same channels simultaneously, and therefore requires dedicated paired channels for its operation.

The forward link possesses many features that are specific to EV-DO, having been optimized for data transmission, particularly in the downlink direction. Average continuous rates of 600 kbps per sector are possible. This is a six fold increase over CDMA2000 1X and is provided largely by the ability of 1xEV-DO to negotiate increased data rates for individual ATs because only one user is served at a time.

The forward link is always transmitted at full power and uses a data rate control scheme rather than the power control scheme used with 1X, and the data is time division multiplexed so that only one AT is served at a time.

In order to be able to receive data, each EV-DO AT measures the signal to noise ratio (S/N) on the forward link pilot every slot (1.667 ms). Based on the information this provides the AT sends a data rate request to the base station.

Accurate time synchronization is required between the EV-DO Access Nodes. This time information is taken from the Global Positioning System (GPS).

A number of channels are transmitted in the forward direction to enable signalling, data and other capabilities to be handled. These channels include the Traffic channel, MAC channel, Control channel and Pilot. These are time division multiplexed.

The Traffic Channel uses Quadrature Phase Shift Keying (QPSK) modulation for data rates up to 1.2288 Mbps. For higher data rates, higher order modulation techniques are used in the form of 8 PSK with 3 bits per symbol or 16 QAM with 4 bits per symbol. The levels of the I and Q symbols are chosen so that the average power is normalized.

The Incoming data to be used as the modulation comes from the from the turbo coder and is scrambled by mixing it with a Pseudo Random Number (PN) sequence. The initial state of the PN is derived from known parameters, and is unique for each user. Every packet starts at the same initial value of the PN sequence.

The Control Channel carries the signalling and overhead messages. What makes the difference between the cell and the sector is the PN offset of the pilot channel and the pilot signal is only gated on for 192 chips per slot.

The Medium Access Control (MAC) Channel carries a number of controls including the Reverse Power Control (RPC), the Data Rate Control (DRC) Lock, and the reverse activity (RA) channels.

The reverse link for 1xEV-DO has a structure similar to that for CDMA2000. In EV-DO all signalling is performed on the data channel and this means that there is no Dedicated Control Channel. The data channel can support five data rates which are separated in powers of two from 9.6 to 153.6 kbps. These rates are achieved by varying the repeat factor. The highest rate uses a Turbo coder with lower gain. The following channels are transmitted in addition to those used with 1X:

- Reverse Rate Indicator (RRI) Channel – This indicates the data rate of the Reverse Data Channel.
- Acknowledgement (Ack) Channel – This channel is transmitted after the AT detects a frame with the preamble detailing it to be the recipient of the data.
- Data Rate Control (DRC) Channel – This channel contains a four bit word in each slot to allow the choice of 12 different transmission rates.

D.6 Problems

1 Which types of diversity are present in the CDMA system, IS-95 standard?

2 What characteristic can be singled out in the CDMA standard (IS-95), that distinguishes it from other adopted systems? Which are the negative aspects of the CDMA standard?

3 Which are the basic requirements, established by CTIA (*Cellular Telecommunications Industry Association*), for second generation mobile communication systems?

4 Which technique is used by Qualcomm to combat the near-far problem? Explain.

5 What is meant by processing gain? What is the processing gain of the CDMA system adopted in USA?

6 Why is a RAKE receiver used in the cdmaOne system?

7 Calculate the capacity, in terms of calls per cell, of the CDMA system, using the following data: spread bandwidth (1,25 MHz); data rate in kbp/s (assume 9,6 kbps); ratio between the bit energy and the noise power density (assume 7,0 dB); voice activity factor (assume 40 %); efficiency of frequency reuse (assume 60 %); sectorization gain (assume 2.55 for three sectors).

Appendix E
The GSM Standard

E.1 Introduction

In the early 1980's administrations meeting at the *Conference of European Posts and Telecommunications (CEPT)* created the *Groupe Spéciale Mobile* (GSM) with the aim of developing a pan-European system for replacing various systems until then in use in Europe (Yacoub, 2002). The GSM was thought initially as a European standard. A few years after its introduction it was noticed that GSM could offer global coverage, and for that reason the services initially offered by the system had to be modified along its deployment.

When GSM was released in 1991 it was clear that the standardization process could not be finished before its services were deployed, mainly due to economic reasons. Consequently, a sequence of *phases* were adopted for the introduction of new services from time to time. The first set of services offered by GSM was called *GSM Phase* 1. The additional features that could not be made available initially became known as *GSM Phase 2*. At present there are new services there were devised after the GSM system became operational. These new services became known as *Phase* 2+.

The GSM system has a rather ambitious set of objectives, including for example (Yacoub, 2002):

- international *roaming* ;
- open architecture;
- high degree of flexibility;
- easy installation;
- integrated operation with ISDN (Integrated services digital network), CSPDN (Circuit switching public digital network), PSPDN (Public switched packet digital network) and PSTN (Public switched telephone network);
- availability of high quality signals and link integrity guarantee;
- high spectral efficiency;
- low-cost infra-structure;
- small low-cost terminals;
- security characteristics.

E.2 System Architecture

The GSM architecture consists of three interconnected subsystems which interact between themselves and with the users through network interfaces. The three subsystems are (Yacoub, 2002): the base station subsystem (BSS), the network and switching subsystem (NSS) and the operation support subsystem (OSS). The mobile station is also a subsystem but is usually considered part of the base station subsystem, for the architecture purpose. In GSM, equipment and services are designed to support one or more subsystems. Figure E.1 illustrates the GSM system architecture (Barbosa, 2002).

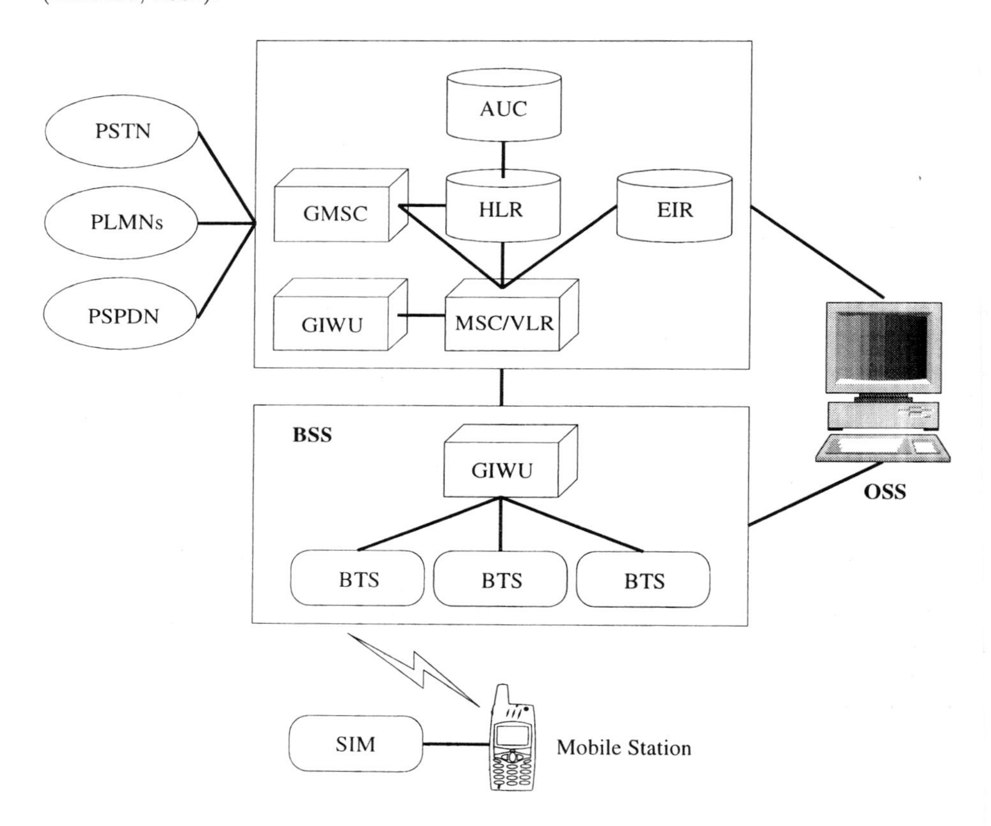

Figure E.1. GSM system architecture.

E.2.1 The Mobile Station

The most widely known of a cellular network is certainly the mobile station (MS), consisting of two elements:

- The mobile equipment or terminal;
- The subscriber identification module (SIM).

Different types of terminals are distinguished by their power and application. Mobile stations permanently installed in cars can have a maximum output power of 20 W. Portable GSM units which can also be installed in vehicles can radiate up to 8 W. The pocket terminals have their success guaranteed due to their reduced weight and volume and can radiate up to 1 W of output power. Evidently, the average transmitted power is usually well below this upper limit.

E.2.1.1 The Subscriber Identification Module

The Subscriber Identification Module (SIM) supplies an identification to the mobile equipment. Without the SIM, the mobile does not work except for emergency calls. The SIM is an intelligent card which contains a processor and a memory chip permanently installed in a plastic card of the size of a credit card. The SIM must be inserted in a reader of a mobile station before being used for its routine purposes. For pocket terminals, considerably smaller, a SIM of the size of a credit card would be too big. For that reason there is a smaller version of the SIM, called *plug in SIM* for this type of terminal.

Certain parameters of a subscriber are stored in a SIM card together with some personal data as for example personal telephone numbers. The SIM card identifies the subscriber to the network. Since the SIM alone identifies a terminal, it is possible to travel carrying only the SIM card, and to rent a mobile terminal at the destination and then use this mobile terminal (with the SIM card inserted) as if it was the original mobile terminal.

Short messages received from the network can also be stored in the SIM card. The introduction of larger memories and faster microprocessors will make the SIM card more flexible and powerful in future.

In order to protect the SIM card from improper use a security feature was installed. Before using a mobile, a user needs to supply a four digit *personal identification number* (PIN). The PIN is stored in the SIM card. If a wrong PIN is entered for three consecutive times, the SIM card is blocked and it can be unblocked with an eight digit *personal unblocking key* (PUK), also stored in the SIM card.

These are the main characteristics of the SIM card, however since GSM is an evolving standard, in future new SIM card related services may appear, for example, priority access for special clients or even restrict access to certain areas.

E.2.2 The Base Station Subsystem

The base station subsystem (BSS) provides and supervises transmissions between mobile stations and the mobile service switching center (MSC). The BSS supervises also the radio transmission interface between mobile stations and all GSM subsystems. Each BSS consists of a set of *base station controllers* (BSCs) which connect the mobile terminal to the NSS, via MSCs. The BSS is subdivided into two elements:

- The base transceiver station (BTS);
- The base station controller (BSC).

E.2.2.1 The Base Transceiver Station

The counterpart of a mobile station within a cellular network is the *base transceiver station* (BTS) or base station, which is the interface between the mobile station and the network. The BTS corresponds to the transceivers and antennae used in each cell of the network. The BTS is usually placed in the center of a cell. The BTS transmission power determines the absolute size of the cell. A base station has between one and sixteen transceivers, each one representing a separate RF channel. Some of the intelligent features that were incorporated in analog base stations, as for example measurements in the radio transmission channels as a criterion for handoff, were recently adopted for mobile stations. As a result the structure of GSM became less expensive than that of some analog systems. One consequence was that in some not so wealthy countries digital cellular systems were installed, instead of analog systems (AMPS, NMT or TACS, for example).

E.2.2.2 The Base Station Controller

The *base station controller* (BSC) monitors and controls various base stations. The number of base stations controlled by a BSC depends on the equipment manufacturer and can range from tens to hundreds of stations. The main tasks of a BSC are: *handover*, frequency hop, change and control of power levels of the BTSs. The hardware of a BSC can be installed in the same place as the BTSs, in a place of its own or in the same place as the MSC.

E.2.3 The Network Switching Subsystem

The *network switching subsystem* (NSS) plays a central role in the whole mobile network. While the BSS provides access to a mobile station, the various network elements in the NSS take responsibility for the complete set of control functions and data banks required for establishing call connections using one or more of the following: cryptography, authentication and roaming. The NSS consists of the following:

- Mobile service switching center (MSC);
- Home location register (HLR);
- Authentication Center (AuC);
- Visitor location register (VLR);
- Equipment identity register (EIR).

E.2.3.1 The Mobile Service Switching Center

The *mobile service switching center* (MSC) is the central component of a NSS, also known as the communications and control center (CCC). The MSC performs all switching functions of the network. From a technical perspective the MSC plays the role of a telephone exchange in an ISDN with some modifications required specifically to manage the mobile application. These modifications for mobile services affect in particular the allocation of user channels at the BSS, for which the MSC is responsible, as well as the functionality for executing and controlling the handover. These are two main functions of the MSC. Besides those there is the inter-networking function (IWF), which is necessary for connections of voice or data with external networks. There is a specific unit for the IWF in the same place as the MSC, it is called the GSM cooperation (GIWU).

E.2.3.2 The Gateway for Mobile Switching Center

An MSC with an interface for other networks is called a *gateway for mobile switching center* (GMSC) or MSC gateway. The GMSC has some additional tasks during the establishment of a call to a mobile, originated from an external network. The call must enter the network via a GMSC, which examines the HLR and then forwards the call to the MSC where the called mobile is situated. The GMSC is in general implemented in conjunction with the MSC.

E.2.3.3 The Local Register

The local register (HLR) is a data bank which stores the identity and the user data of all subscribers belonging to the area of its GMSC.

The HLR is considered the most important data bank since it permanently stores the subscribers' data, including a profile of user services, location information and the situation of a subscriber's activities. When someone gains access to a mobile network of a cellular telephony operator, he or she are registered in the operator's HLR.

In order to reduce the number of tasks performed by the HLR, the visitor's register (VLR) was created to help various requests related to the subscribers.

E.2.3.4 The Authentication Center

The authentication center (AuC) is always implemented as an integral part of the HLR. The authentication center is used for security reasons. Its task is to calculate and provide three parameters required for authentication, namely the signaled response (SRES), a random number (RAND) and the parameter K_c. For each subscriber up to five triplets of parameters can be simultaneously calculated and sent to the HLR. The HLR in turn sends such triplets to the VLR which uses them as input parameters for authentication and cryptography.

E.2.3.5 Visitor's Register

The visitor's register (VLR) is a data bank which contains temporary information about subscribers. The information stored in the VLR is necessary for the MSC to serve visiting subscribers. The VLR is always integrated with the MSC. When a mobile station is visiting (*roaming*) a new area (with a new MSC), the VLR connected to the new MSC requests to the HLR data about the mobile station. Later, if the mobile station makes a call, the VLR has the necessary information for the call setup without need to interrogate the HLR again.

E.2.3.6 The Equipment Identification Register

The separation between the subscriber's identity from that which identifies the mobile station is a potential source of problems for GSM users. It turns out that it is possible to operate any GSM mobile station with any valid SIM. There exists in this case an opportunity for a black market of stolen equipment. To combat this illicit practice the equipment identification register (EIR) locates and bars the equipment from being used in the network.

Each GSM terminal GSM has a unique identifier which represents its identity of international mobile equipment (IMEI), which can not be altered without destroying the terminal. The IMEI contains a serial number and a type identifier.

As well as the HLR and the VLR, the EIR consists basically of a data bank which stores three lists: (1) the "white list" which contains all the types of approved mobile stations; (2) the "black list" which contains the IMEIs of stolen terminals or that must be barred for technical reasons; and (3) the "grey list" which allows locating certain mobile stations.

E.2.4 The Operation and Support Subsystem

The operation and support subsystem (OSS) supports the operation and maintenance of the GSM system, allowing engineers to monitor, diagnose and solve problems of faults of any kind in the system. This subsystem interacts with other subsystems.

The OSS supports one or many Operations Maintenance Centers (OMCs), which are used to monitor and maintain the performance of each mobile station (MS),base station (BTS), base station controller (BSC) and mobile switching center (MSC) in the GSM system. The OSS has three main functions as follows.

1) Maintain the operation of all hardware and of all telecommunications network in a given area;
2) Manage the whole tariff process;
3) Manage all mobile terminals in the system. Inside each GSM system, an OMC is dedicated to the functions previously mentioned and can still adjust all parameters of a base station and the procedures for charging calls, as well as to offer the system operators the ability to determine the performance and the integrity of each part of the user equipment in the whole system.

E.3 Procedure for Registering the Mobile Station

After the mobile station is switched on, it carefully examines the whole GSM frequency band with an algorithm the objective of which is to detect the presence of a network in this time interval. When the network is detected the mobile station reads the system information in the base channel. Once having this information the mobile station is capable of determining its current position in the network. If the current location is not the same as it was the last time the mobile station was switched off, a procedure of registration is initiated.

First, the mobile station requests a network channel, which is designated by the base station. Before the channel is allocated the BSC has to activate the channel in the BTS, which in return has to validate the activation. After being connected to the network infrastructure, the mobile station informs the system that it wants to perform a location update. This request is passed also to the (G)MSC by the BSC, however since the (G)MSC is usually very busy it requires authentication of the mobile station before taking any action. Once the correct authentication parameters are received, the (G)MSC accepts the mobile station in its new location. A temporary mobile station identity (TMSI) is then attributed to the mobile station, the validation of which must be returned by the mobile station. When this procedure is ended, the channel is released by the BSC, by means of the BTS.

The registration of the mobile station could be performed by the network. This will be the case whenever the system wants to know exactly which mobile stations are currently available in the system. The registration procedure is a means of limiting the flux of messages within the network and besides to provide the network with a virtual control capability. The network always knows the contents of the HLR. All that is known to the HLR is also known to the GMSC, and whether the mobile station is on or off is a common knowledge within the network. If somebody dials a mobile station that is switched off, the GMSC immediately sends a message to the person originating the call, indicating that the specific mobile station is not available.

E.4 Establishing a Call

Before a call can be established, the mobile station must be switched on and registered in the system. There are two distinct procedures for the parts involved in establishing a call. The first procedure concerns the mobile which originated the call (MOC) and the second procedure concerns the mobile which receives the call (MTC). These two procedures are very similar, and for this reason only the first of the two will be described.

In a manner similar to that of location updating the mobile first makes a request for a channel, which is answered by the system with a channel allocation. The mobile station informs the system the reason for the channel request, which in this case is for establishing a call. Before this procedure is continued the mobile has to be authenticated again. In order to protect any further signaling message against an eavesdropper the network can ask the mobile station to start using cryptography to encipher its data in the next message.

To use cryptography means that messages will be transmitted in an enciphered form which can be read only by the mobile station and the base station. In the *setup* message the mobile transmits the number of the telephone it wants to reach. While the call is being processed the BSC (via BTS) allocates a traffic channel through which user data exchange is performed. If the mobile station at the destination is not busy, the terminal signals this situation and a connection is established when the telephone is lifted "off-hook".

E.5 Handoff

The procedure called handoff or handover is a form of continuing a call even when a mobile station crosses the boundary between two adjacent cells. Before this technique was implemented the call would simply be abandoned whenever the mobile station crossed the boundary between two adjacent cells.

In a cellular network each cell has a number of neighbor cells. The system needs to know which of the neighbor cells must be allocated in a handoff procedure. The method used to determine the next cell in analog systems is different from that used by the digital systems. The difference is also present in the names employed in each situation. The term handoff appeared in the analog world while the term handover was presented by GSM.

In analog systems the base station monitors the quality of the link between itself and the mobile station. When the base station detects that the quality is degraded (for example, the distance to the mobile station is too far) the base station asks neighboring cells to inform the power level that they are receiving from the mobile station. Possibly the highest power level comes from the cell that is the closest to the mobile station. The network then decides which channel the base station must use in the new cell and in which corresponding frequency the mobile station must be tuned. Generally the mobile station receives the command to change channel.

The mobile station plays a passive role in the handoff process. All measurements and subsequent tasks are performed in the base stations and in the network. The cell sites where the base stations are located are equipped with power meters that measure the various power levels associated with different mobile stations operating at various channel frequencies.

In the GSM system this situation is different. The mobile station must continuously monitor the power levels received from neighbor cells. For that purpose, the base station provides the mobile station with a list of base stations (and respective channels) the power of which must be measured. This list is transmitted through the base channel, which is the first channel to be tuned in by the mobile station, the moment it is switched on. The mobile station continuously performs measurements of the power level in the signal received from the cell where it is located, as well as power levels of neighboring cells. The measurement results are placed in a measurement report, which is periodically sent to the base station. The base station itself can also perform measurements of power levels of its link with the mobile station. If these measurements indicate the need for handover, then handover can be done immediately without the need to wait for the measurements performed by the mobile station, since the most appropriate base station to receive the mobile station is already known.

The GSM system distinguishes between two types of handover, depending on the type of cell boundary being crossed by the mobile station. If a handover needs to be done inside the area of a BSC, then this BSC can manage the operation without resorting to the MSC which, in any case, has at least to be notified. This type of handover is called simple handover between BTSs. On the other hand, if a mobile station is crossing the boundary of a BSC, then the MSC must control the procedure so as to ensure a soft transition for the conversation taking place. The conversation can continue uninterrupted even with the change of MSCs. The only difference is that, in spite of the mobile station being now controlled by the new MSC, the call management continues being done by the old MSC. Figure E.2 illustrates the two types of handover.

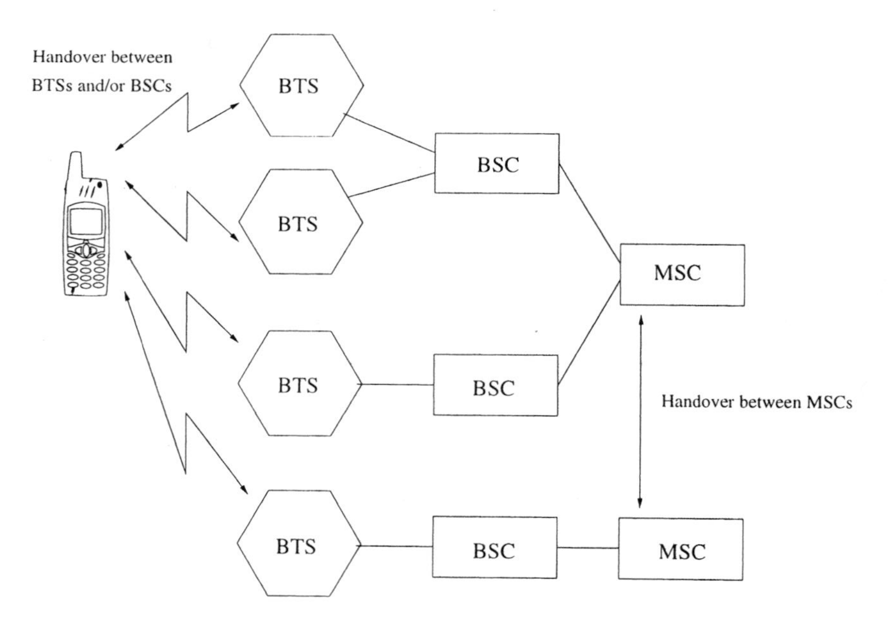

Figure E.2. Handoff hierarchy.

E.6 Security Parameters

The GSM system provides means to authenticate subscribers, in order to avoid mobile set aliases or the undue use of the system by unauthorized persons.

E.6.1 Authentication

The authentication procedure checks the validity of the subscriber's SIM card and whether they are allowed to use a particular network. The authentication is based on the authentication algorithm $A3$, which is stored in the SIM card and in the authentication center.

The $A3$ algorithm uses two input parameters. One is the authentication key, K_i, which is stored only in the SIM card and in the network. The second value, a randomly generated number (RAND), is transmitted to the mobile station. The mobile station passes the RAND to the SIM card, where RAND is used as an input value to the $A3$ algorithm. The result, denoted by SRES, is returned from the mobile station to the network where its value is compared with that calculated by the authentication center. A set of authentication parameters (RAND and SRES) is stored in HLR and in VLR for use by the authentication center. In general these parameters are stored because a distinct set is used for each call setup or register and is discarded after each use. If the HLR or the VLR have only a few sets of parameters then some new sets of parameters must be asked to the authentication center.

E.6.2 Temporary Identity of the Mobile Subscriber

In order to avoid that a possible invader identifies itself as a legitimate GSM user by means of the international mobile subscriber identity (IMSI), which is a number permanently assigned, a temporary mobile subscriber identity (TMSI) is assigned to each user as long as they are using the network. The TMSI is stored together with the true identity (IMSI) in the network. The temporary identity is assigned during the procedure of location updating and is used during the time that the subscribers stay active in the network. The mobile station uses the temporary identity when it communicates with the network or when it originates a call. Similarly, the network uses the temporary identity when paging the mobile station. The assignment, management and updating of the TMSI is done by the VLR. When the mobile station is switched off it stores the TMSI in the SIM card to make sure that this number will be available next time, when it is switched on again.

E.7 The GSM Radiotransmission Interface

The radiotransmission interface is the interface between the mobile stations and the fix infrastructure. This is one of the most important interfaces of the GSM system.

One of the features of GSM is roaming. Therefore, in order to obtain complete compatibility between the mobile stations and networks produced by different manufacturers and operators, the radiotransmission interface must be completely defined.

The spectral efficiency depends on the radiotransmission interface and on the transmission, particularly on aspects like system capacity and the techniques used to reduce interference and improve the frequency reuse scheme. The specification of the radiotransmission interface then has a significant influence on spectral efficiency.

E.7.1 The Primary GSM

The primary GSM system refers to the first generation of GSM systems, which were installed in Europe. Two 25 MHz frequency bands were allocated around 900 MHz. The mobile station transmits in the frequency band between 890 MHz and 915 MHz, while the base station transmits in the frequency band between 935 MHz and 960 MHz. The end points in the physical layer are the mobile station and the BTS. The transmission direction *mobile station to base station* is referred as the uplink while the direction *base station to mobile station* is referred as the downlink. Similar to analog systems, the frequency spacing used for duplex operation is 45 MHz. The base station transmits always in the higher frequency band of the duplex frequency pair.

E.7.2 The Multiple Access Scheme

The multiple access scheme defines how simultaneous communication, between different mobile stations situated in distinct cells, share the GSM frequency spectrum. A mixture of Frequency Division Multiple Access (FDMA) and Time Division Multiple Access (TDMA), combined with frequency hopping, were adopted as the multiple access schemes for GSM.

E.7.2.1 FDMA and TDMA

In the FDMA scheme, a frequency is assigned to each user. The available frequency band is split into sub-bands separated by guard bands. An important characteristic of FDMA is that, once a channel is assigned to a user, the frequency band of this channel is used exclusively by that user until this connection is no longer necessary. The TDMA scheme allows many users to share the same channel in time. Each user, sharing the common TDMA channel, is assigned a time slot within a group of non-overlapping time slots called a frame. In general the TDMA is used with an FDMA structure.

Considering that channel spacing in GSM is 200 kHz, it would constitute a loss for the system not to subdivide this resource even further. GSM achieves that by resorting to TDMA techniques by which each channel is subdivided into eight distinct time windows numbered from 0 to 7. Each one of these eight time windows is assigned to a unique user. The set of eight time windows is called a TDMA frame and all users of a common frequency share a common time frame. If to a mobile station, for example, it is assigned time window number 1, it will transmit only in this time window and will be idle during the time corresponding to the other seven time windows, with its transmitter switched off. This periodic and regular change of the transmitter status between on and off is called bursting. The time duration of a time window, which is equivalent to the duration of a pulse of a mobile station, is 577 μs, and the duration of a TDMA frame is 4.615 ms.

In typical cellular analog systems, allocated channels occupy the spectrum between 12.5 MHz and 30 kHz. The use of TDMA techniques in GSM in effect made the band allocated to traffic channels to be 200 kHz/8 = 25 kHz, which is equivalent to that of typical cellular analog systems. This fact demonstrates that the specifications of GSM were selected by their specific characteristics and additional qualities and not just by the possibility of increasing capacity.

E.7.2.2 Time Division Duplexing

When the TDMA technique is used it is not necessary to transmit and receive signals all the time in a full-duplex mode. Time division duplexing (TDD) is employed in GSM. The TDD presents some advantages to the mobile station as follows.

- It requires half the spectrum, since only one frequency is used;
- There is no need for a dedicated duplex stage (duplexer);
- The battery lifetime is increased, or battery weight is reduced;
- More robust and lower cost terminals .

E.7.3 Frame Structure

The synchronization of transmission bursts in GSM is of extreme importance. The information carried consists of different types of data, and need to be delivered at the destination according to the network plan for traffic and signaling, otherwise the system collapses. In order to ensure that transmission will occur on prescribed time instants and that the data packets are sent to the correct destinations, the data bursts are organized in frames. The frame structure helps the adequate flow of traffic and control signals.

Information, concerning both traffic and signaling, are organized in frames before they are mapped into time windows. The frames are carefully organized into structures consisting of combination of channels, one after the other, such that the receivers are able to recognize the data types that must be present at any time, with minimum delay or errors.

E.7.3.1 Channel Combination

Channels are always combined as physical channels. According to GSM specifications, there are seven types of channel combination, as shown in Table E.1 (the CCCH entry represents the three types of usual control channels: RACH, PCH and AGCH).

Each channel combination requires a unique physical channel. The TDMA technique employed creates eight physical channels for a carrier. It is then possible to place different channel combinations into one carrier. One carrier distinct carrier for each designated time window.

In a manner similar to the frame structure of TDMA, which allows the time windows to be ordered in one carrier, there are also some multi-frame structures consisting of a fixed number of TDMA frames, which allow logical channels to be ordered in time windows. There is a great difference between the logical channels that carry voice data and those that transport signaling data. A 26 multi-frame structure is used for the traffic channel combinations and a 51 multi-frame structure is used for signaling channel combinations (Jeszensky, 2004).

E.7.3.2 26 Multi-frame structure

E.7.3.3 *TCH/FS (Combination I)*

The first 12 frames (Figure E.3) are used to transmit traffic data. They can also be used for data transmission at 9.6, 4.8 or 2.4 kbps. Next comes the frame associated

I	TCH/FS + FACCH/FS + SACCH/FS;
II	TCH/HS(0,1) + FACCH/HS(0,1) + SACCH/HS(0,1);
III	TCH/HS(0) + FACCH/HS(0) + SACCH/HS(0) + TCH/HS(1) + FACCH/HS(1) + SACCH/HS(1);
IV	FCCH + SCH + CCCH + BCCH;
V	FCCH + SCH + CCCH + BCCH + SDCCH/4 + SACCH/4;
VI	CCCH + BCCH;
VII	SDCCH/8 + SACCH/8.

Table E.1. Channel combinations in GSM.

to the SACCH channel, and then 12 more frames for more traffic data. The last frame is kept idle (O), and nothing is transmitted in it. The idle frame gives the mobile station the time to perform other tasks like the measurement of signal power in neighboring cells or in the cell where it is situated. The total length of the 26 multi-frame structure is 26 × 4,615 ms = 120 ms.

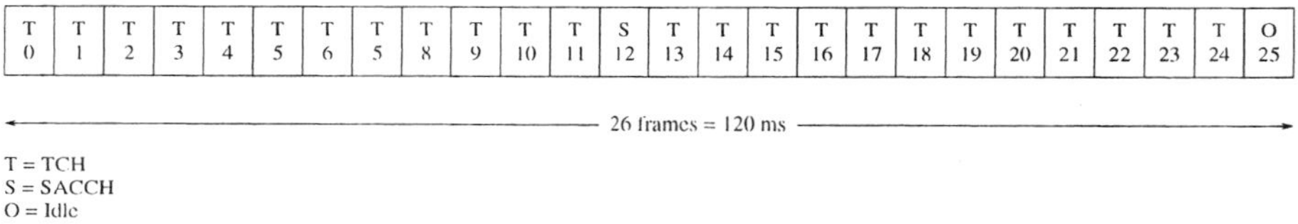

T	T	T	T	T	T	T	T	T	T	T	T	S	T	T	T	T	T	T	T	T	T	T	T	T	O
0	1	2	3	4	5	6	5	8	9	10	11	12	13	14	15	16	17	18	19	20	21	22	23	24	25

26 frames = 120 ms

T = TCH
S = SACCH
O = Idle

Figure E.3. Frame structure for TCH/FS.

E.7.3.4 *TCH/HS (Combinations II and III)*

For the transmission of half rate voice channels it is possible to join two channels in the 26 multi-frame structure (Figure E.4). Within single frames, these two channels are transmitted alternately. Frame number 25 in this case is reserved for the SACCH channel associated with the second half rate channel. Each half rate channel has its own SACCH channel. This structure is used whenever two half rate channels are simultaneously assigned according to combination III. If only one half rate channel is required then combination II is applicable. In combination II, the second frame is always idle, and during this time the mobile station is free to perform other tasks.

The FACCH channel is a signaling channel which has higher priority then the routine traffic channels. If necessary, the network replaces part of a traffic channel by an FACCH channel. This means that the FACCH channel takes the place of a traffic channel as long as it is required to transmit the signaling data. The more signaling data there is, the longer the time duration of the FACCH. The receiver decoder detects the presence of an FACCH channel simply by observing the stealing flags.

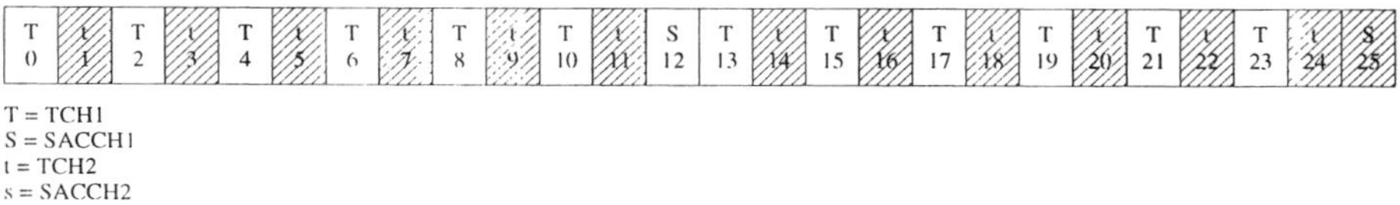

T	t	T	t	T	t	T	t	T	t	T	t	S	T	t	T	t	T	t	T	t	T	t	T	t	s
0	1	2	3	4	5	6	7	8	9	10	11	12	13	14	15	16	17	18	19	20	21	22	23	24	25

T = TCH1
S = SACCH1
t = TCH2
s = SACCH2

Figure E.4. Frame structure for TCH/HS.

E.7.3.5 The 51 multi-frame structure

The signaling do not carry user data. The 51 multi-frame structure is a little more complex than the 26 multi-frame structure because it incorporates four distinct channel combinations, each combination requiring a different structure.

E.7.3.6 *FCCH + SCH + CCCH + BCCH (Combination IV)*

All channel types used in this combination occur either in the direction of BTS to mobile station (downlink) or from mobile station to BTS (uplink). There are different structures for the two cases, one for the downlink and another for the uplink. For the downlink direction, combination IV offers much room for the CCCH channel, which can take the form of a PCH (to call a mobile station), or an AGCH (to assign a channel to the mobile station). The actual location of each channel in the 51 multi-frame structure is not important. The channel location will depend only on the signaling needs of the cell where it is located. However, the FCCH and SCH channels are always in consecutive frames. This simple construction eases frequency synchronization between a mobile station and the base channel, before time synchronization for the data takes place. The uplink path is used by mobile stations only for the transmission of the RACH channel in random access bursts. This requires other signaling channel from the BTS.

This channel combination is normally used by cells having many carriers and a large expected traffic load on the CCCHs (paging, channel requests and channel assignments). The combination allocation can only be made to a single cell, each time, since the FCCH and the SCH channels are particular to the base channel. The combination can be transmitted in any available frequency in the cell, using time window number 0. The frequency, in which the combination is transmitted, is used as a reference in the neighboring cells to mark it as an adjacent cell, i.e., mobile stations in neighboring cells periodically execute their measurements in this frequency, during time window number 0.

E.7.3.7 *FCCH + SCH + CCCH + BCCH + SDCCH/4 + SACCH/4 (Combination V)*

Combination V is the minimum combination for small cells having only one or two transceivers. The use of this combination follows the same rules as those of combination IV (i.e., only once per cell and always in time window number 0). Combinations IV and V are mutually exclusive. The expression SDCCH/4 + SACCH/4 is used to mean that up to four DCCHs channels can be assigned with their respective required

F	S	BCCH	CCCH	F	S	CCCH	F	S	CCCH	F	S	CCCH	F	S	CCCH	O
0	1	2 – 5	6 – 9	10	11	12 – 19	20	21	22 – 19	30	31	32 – 39	40	41	42 – 49	50

Downlink: F = FCCH
S = SCH
B = BCCH
C = CCCH (PCH, AGCH)
O = Idle

R	R		R	R		R	R		R	R		R	R		R
0	1		10	11		20	21		30	31		40	41		50

Uplink: R = RACH

← 51 frames = 235.38 ms →

Figure E.5. Frame structure for channel combination IV.

channels. These are called subchannels; that is, SDCCH(2) is called "subchannel 2 of SDCCH". Figure E.6 illustrates this combination.

There are two important aspects in the above diagram. One is that the location of corresponding subchannels in the SDCCH channel keep a certain distance from each other: 15 frames for the uplink and 36 frames for the downlink. The objective of doing this is to reduce the cycle of command response for a multi-frame. If, for example, the base station commands the mobile station to authenticate itself, the response can only be sent back 15 frames later. The same applies to the other transmission direction, with the only difference that the network has more time to give an answer. This is a need that occurs because the total distance traveled by downlink signals is much larger than the distance traveled by bone uplink signal. The same comment applies for combination VII, with eight SDCCHs channels.

Another important factor is that there exist two multi-frame structures together in Figure E.6. The reason for that is because one SACCH channel is transmitted only at every new multi-frame. One might think that there is a possibility of transmitting only two frames of the SACCH at each multi-frame, but the information is spread over four frames, and they belong to each other. The SACCH time is half the time of the SDCCH channel.

E.7.3.8 *CCCH + BCCH (Combination VI)*

If a base station manages a large number of transceivers, it is likely that the number of CCCH channels provided by combination IV will not be sufficient to do the job. In order to serve a large number of base stations, it is possible to assign additional control channels in combination VI. This assignment to combination VI makes sense only when combination IV is present, since it adds additional control capacity to the resources in combination IV. While combination IV occupies always time window 0, combination VI is assigned to time window 2, 4 or 6. The multi-frame structure in combination VI is similar to that of combination IV. The only difference is that there are no FCCH and SCH channels in combination VI.

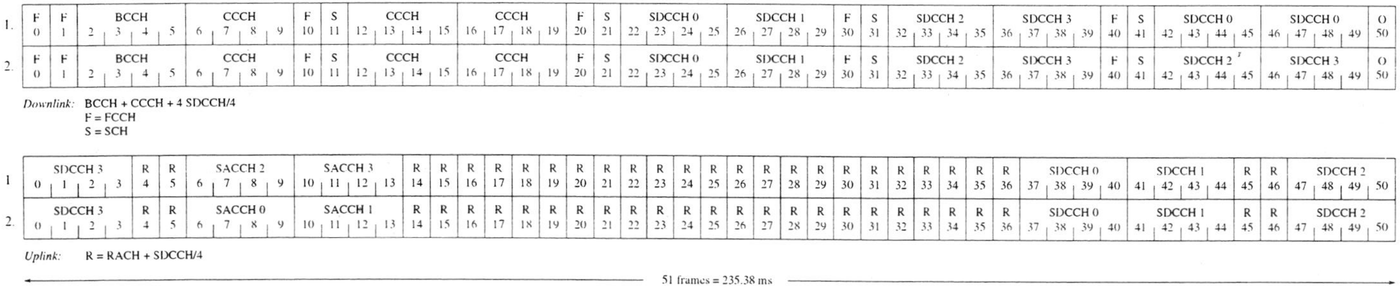

Frames	1.	2.
0	F	F
1	F	F
2, 3, 4, 5	BCCH	BCCH
6, 7, 8, 9	CCCH	CCCH
10	F	F
11	S	S
12, 13, 14, 15	CCCH	CCCH
16, 17, 18, 19	CCCH	CCCH
20	F	F
21	S	S
22, 23, 24, 25	SDCCH 0	SDCCH 0
26, 27, 28, 29	SDCCH 1	SDCCH 1
30	F	F
31	S	S
32, 33, 34, 35	SDCCH 2	SDCCH 2
36, 37, 38, 39	SDCCH 3	SDCCH 3
40	F	F
41	S	S
42, 43, 44, 45	SDCCH 0	SDCCH 2'
46, 47, 48, 49	SDCCH 0	SDCCH 3
50	O	O

Downlink: BCCH + CCCH + 4 SDCCH/4
F = FCCH
S = SCH

Frames	1	2.
0, 1, 2, 3	SDCCH 3	SDCCH 3
4	R	R
5	R	R
6, 7, 8, 9	SACCH 2	SACCH 0
10, 11, 12, 13	SACCH 3	SACCH 1
14	R	R
15	R	R
16	R	R
17	R	R
18	R	R
19	R	R
20	R	R
21	R	R
22	R	R
23	R	R
24	R	R
25	R	R
26	R	R
27	R	R
28	R	R
29	R	R
30	R	R
31	R	R
32	R	R
33	R	R
34	R	R
35	R	R
36	R	R
37, 38, 39, 40	SDCCH 0	SDCCH 0
41, 42, 43, 44	SDCCH 1	SDCCH 1
45	R	R
46	R	R
47, 48, 49, 50	SDCCH 2	SDCCH 2

Uplink: R = RACH + SDCCH/4

51 frames = 235.38 ms

Figure E.6. Frame structure for channel combination V.

E.7.3.9 *SDCCH/8 + SACCH/8 (Combination VII)*

If a cell uses the signaling from combination IV (together with combination VI), it still lacks channel signaling by which the mobile stations can perform basic tasks like a call setup or registration. Combination VII provides the capacity for routine signaling to the cell. The expression SDCCH/8 + SACCH/8 is used to mean that eight distinct DCCH channels can be used with eight SACCH channels in this combination and, therefore, can serve eight signaling links in parallel in a single physical channel. The frame structure for combination VII is illustrated in Figure E.7.

E.7.3.10 Combination of the 26 and 51 multi-frame structures

26 TDMA frames are required to transmit all 26 segments of a 26 multi-frame, a time window at a time. Similarly, 51 TDMA frames are required to send all 51 segments of a 51 multi-frame, a time window at a time. Until now these two structures were treated independently, however it is necessary to explain how they can be combined in the TDMA frames. For example, how the combination is done for a 51 multi-frame assigned to time window number 0 and a 26 multi-frame assigned to time window 5. This is apparently a difficult situation because 26 does not divide 51 and it would be inefficient to have empty slots (empty time windows). A new frame format was introduced for this combination, called a super frame. The super frame has length $51 \times 26 = 1,326$ frames, which represents the least common multiple of 26 and 51, representing also the least number of TDMA frames (having eight time windows each) that can handle the contents of all 26 and 51 multi-frames, without empty time windows. A super frame can accommodate both 26, 51 multi-frames as well as 51, 26 multi-frames. Figure E.8 illustrate how multi-frames are organized.

There is still another type of frame called hyper frame. The hyper frame does not contain anything special. It consists of 2,048 super frames. The hyper frame represents the system largest structure and lasts for approximately 3.5 hours before it is repeated.

When the signaling frame is described, it is important to know precisely which frame is currently being transmitted, otherwise there would be no way to know whether SDCCH(0) or SDCCH(2) is being received. In order to avoid ambiguity, the frames are numbered in a special manner, using three counters called $T1$, $T2$ and $T3$. $T1$ counts super frames. Whenever a super frame is completed, $T1$ is incremented by 1. $T1$ has values between 0 and 2,047. $T2$ counts voice frames, which occur only in 26 multi-frame structures. The value of $T2$, however, varies from 0 to 25. $T3$ counts signaling frames, which occur in 51 multi-frame structures. Similarly to the traffic counter, the contents of $T3$ can vary from 0 to 50. At some initial point the contents of all counters are set to all zeros, and then the transmission of frames begins.

Whenever then transmission of a multi-frame structure for voice or signaling is completed, the respective counters ($T2$ and $T3$) are set to zero and start counting again. After transmission of 1,326 TDMA frames, which is the duration of a super frame, $T2$ and $T3$ are restarted together and begin counting from the all zero state. When the transmission of the first super frame is completed, $T1$ is incremented by 1. $T1$ only restarts counting after reaching the count 2,047, which occurs more than 3 hours after it started counting. Knowing the contents of counters $T1$, $T2$ and $T3$ and which types of multi-frames were assigned in each one of the eight time windows

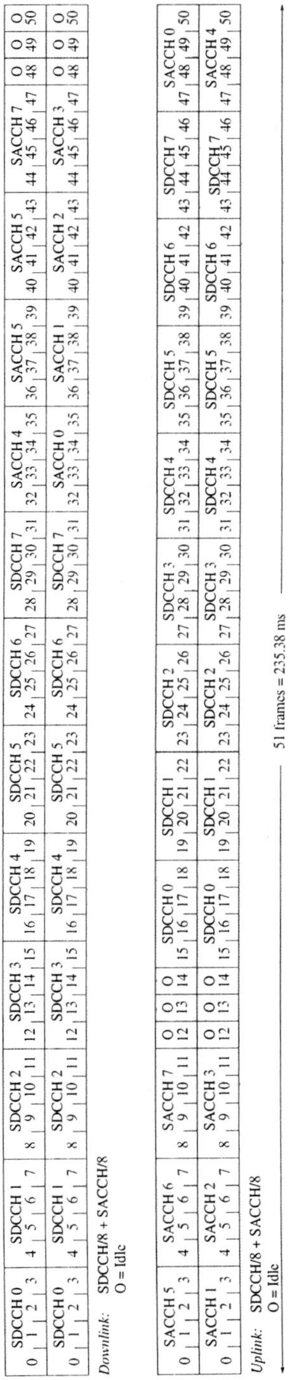

1	SDCCH 0 0 1 2 3	SDCCH 1 4 5 6 7	SDCCH 2 8 9 10 11	SDCCH 3 12 13 14 15	SDCCH 4 16 17 18 19	SDCCH 5 20 21 22 23	SDCCH 6 24 25 26 27	SDCCH 7 28 29 30 31	SACCH 4 32 33 34 35	SACCH 5 36 37 38 39	SACCH 5 40 41 42 43	SACCH 7 44 45 46 47	O 48	O 49	O 50
2.	SDCCH 0 0 1 2 3	SDCCH 1 4 5 6 7	SDCCH 2 8 9 10 11	SDCCH 3 12 13 14 15	SDCCH 4 16 17 18 19	SDCCH 5 20 21 22 23	SDCCH 6 24 25 26 27	SDCCH 7 28 29 30 31	SACCH 0 32 33 34 35	SACCH 1 36 37 38 39	SACCH 2 40 41 42 43	SACCH 3 44 45 46 47	O 48	O 49	O 50

Downlink: SDCCH/8 + SACCH/8
O = Idle

1.	SACCH 5 0 1 2 3	SACCH 6 4 5 6 7	SACCH 7 8 9 10 11	O 12	O 13	O 14	SDCCH 0 15 16 17 18	SDCCH 1 19 20 21 22	SDCCH 2 23 24 25 26	SDCCH 3 27 28 29 30	SDCCH 4 31 32 33 34	SDCCH 5 35 36 37 38	SDCCH 6 39 40 41 42	SDCCH 7 43 44 45 46	SACCH 0 47 48 49 50
2.	SACCH 1 0 1 2 3	SACCH 2 4 5 6 7	SACCH 3 8 9 10 11	O 12	O 13	O 14	SDCCH 0 15 16 17 18	SDCCH 1 19 20 21 22	SDCCH 2 23 24 25 26	SDCCH 3 27 28 29 30	SDCCH 4 31 32 33 34	SDCCH 5 35 36 37 38	SDCCH 6 39 40 41 42	SDCCH 7 43 44 45 46	SACCH 4 47 48 49 50

Uplink: SDCCH/8 + SACCH/8
O = Idle

51 frames = 235.38 ms

Figure E.7. Frame structure for channel combination VII.

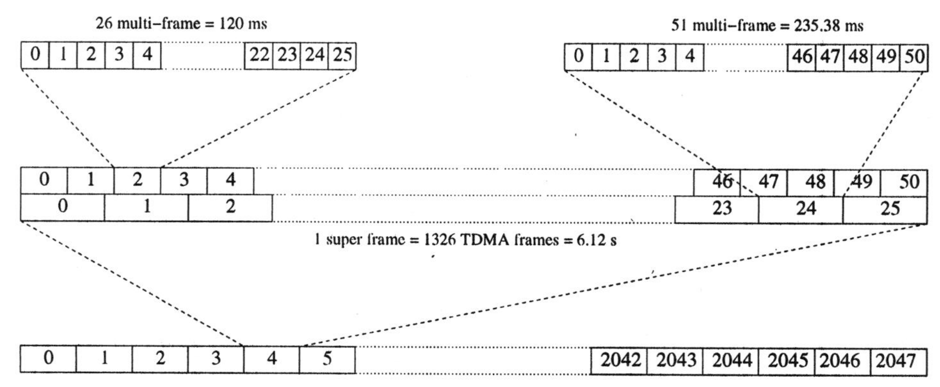

Figure E.8. GSM frame structures.

available in the TDMA frame, it is possible to know exactly what is being displayed in each time window at the instant considered.

The frame identification number consists of the current values registered $T2$ and $T3$, being transmitted in the SCH channel. This number gives the mobile station an initial indication of the prevailing frame structure at that moment. Having the two numbers, $T2$ and $T3$, it is an easy task for the mobile station to look for the BCCH channel and for information about the system. The mobile station knows that information is transmitted when counter $T3$ is, for example, between 2 and 5 (see Figure E.5). It is very important for a mobile station to have accurate information about the frame structures. Having this knowledge, the mobile station knows how long it has to wait for some data and when to make appropriate transmissions.

In general it is an intricate matter to understand how the transition from a signaling channel to a traffic channel is processed, when these channels are situated in distinct frame structures. It is important to notice that the 26 multi-frame structures or 51 multi-frame structures refer to a single time window within a TDMA frame. If a mobile station is ordered to halt receiving a signaling structure in a time window x and, instead, return to the traffic channel in time window y, it is necessary to take into consideration the different frame structures in the corresponding time windows. The counters keep always running. In this manner, it is easy to move from one structure to another. Figure E.9 shows an example of the mapping of different structures. In this example, three time windows were assigned to the mobile station: 0 for the BCCH channel (combination IV), 2 for the traffic channel (combination I) and 5 for the SDCCH/8 channel (combination VII). Time windows 0 and 5 are in the 51 multi-frame structure while time window 2 is in the 26 multi-frame structure. These two structures are mapped together in a super frame structure, in which different time window still keep their original identities.

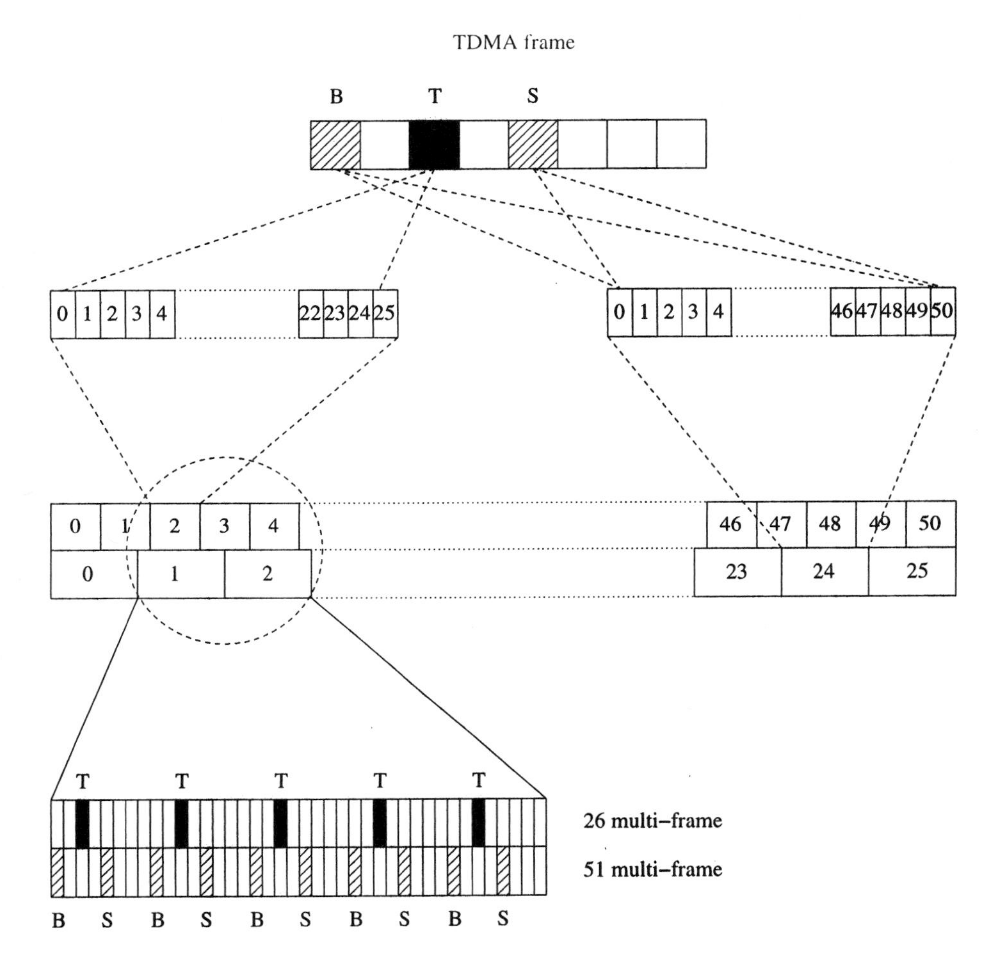

Figure E.9. Time window allocation with mapping in distinct frame structures.

E.7.4 Voice Coding in GSM

The so far most important service offered to the user of a mobile cellular network is the transmission of voice signals. It is this service that justifies great efforts and investments which are required to install and maintain the network. The general technical requirement is simple, that is, to transmit voice signals with an acceptable level of quality. In conventional analog systems the voice signal, also referred to as base band signal, modulates a carrier. Information can be imprinted on the carrier amplitude variations, carrier phase or frequency. The most relevant techniques employed to modulate a carrier are called amplitude modulation (AM), phase modulation (PM) and frequency modulation (FM). At the receiver the original base band signal is extracted from the modulated carrier. Disturbances in the radio transmission channel add their own influences on the carrier. This unwanted modulation caused by disturbances in the channel cause audible effects which can disturb a conversation.

In GSM, which is a digital transmission system, there is no continuous signal transmission through a selected channel. Due to the use of TDMA, information is transmitted in pulses, such that the information contents (the digital representation of the original audio signal) is coded in the time domain for transmission through the channel. At the receiver, information is recovered by decoding the digital signal so as to reproduce the original base band audio signal.

Due to a limitation of channel capacity available for transmission, it is necessary to minimize the number of bits to be transmitted. The device used to transform the voice signal in a string of bits for transmission, and that recovers the voice signal from the corresponding bit string at reception, is called a codec (coder/decoder). The voice codec is part of every mobile station assigned for voice transmission. The BTS can also perform the codec function as, for example, whenever the need arises for a link between a mobile station and a terminal for fixed telephony.

The voice codec, used in the GSM system, is a type of hybrid encoder called regular pulse excitation and long-term prediction (RPE-LTP). The voice codec produces blocks of 260 bits every 20 ms (260 bits/20 ms = 13 kbps). The bits of digitized voice are passed to the channel encoder, which processes these bits for transmission. The complementary process is done at the receiver, where a voice decoder receives blocks of 260 bits from the channel decoder and converts them into a voice signal.

Two other codecs were specified for the GSM system. One is the half-rate codec. In the GSM half-rate traffic channels, half of the time windows are free. This procedure duplicates system capacity by employing the half-rate codecs. The other codec is called an enhanced full-rate codec, which is a kind of improved full-rate codec.

E.7.4.1 Requirements for Voice Coding in GSM

The direct conversion of an analog signal into a digital representation as a sequence bits, and the corresponding reverse conversion, is nowadays a routine operation. There are many kinds of low cost analog to digital (A/D) and digital to analog (D/A) converters. However those are not the only tasks required to convert analog base band signals into digital signals and vice-versa. In GSM, voice coding is a key issue and must comply with the following requirements.

- The redundancy inherent to the base band voice signal must be significantly reduced. If this redundancy is properly removed, there will be significant a time left for other uses of the channel. The process of redundancy removal from the voice signal attempts at leaving a minimum of information required to allow recovery of this signal at the receiver. Redundancy removal is essential in this case since the capacity of transmitting data in a cellular system is restricted;
- The quality of voice transmission in GSM under good conditions, i.e., conditions during which there are no disturbances enough to cause transmission bit errors, must be at least equal to that expected from conventional cellular systems in similar conditions;
- Pauses in the course of normal conversations must be detected, so that transmission could be halted (optionally) during periods. This characteristic allows traffic reduction in the air interface, allows interference reduction among cells and lengthens battery life time in portable terminals. This function is called discontinued transmission (DTX).

E.7.4.2 A/D Conversion in GSM

In GSM, as in many other systems, the sounds produced by human voice are converted to an electrical signal through a microphone. In order to digitize this signal it is first of all sampled. Initially the voice signal is high-pass filtered, so that only frequency components below 4 kHz are kept. Base band voice signals in telephony are limited in frequency to the minimum bandwidth (300 Hz to 3.4 kHz) sufficient to allow non-ambiguous voice recognition. After filtering the signal is sampled. At every 125 μs, a sample is extracted from the analog signal and is quantized to form a digital word having 13 bits/sample. The 125 μs sampling interval results from the use of a sampling rate of 8,000 samples per second.

The signal, quantized with 13 bits per sample, is represented by $2^{13} = 8,192$ quantization levels. An 8,000 samples per second sampling rate means that the output of the A/D converter produces a data rate of $8,000 \times 13$ bps $=$ 104 kbps. This interface in the voice coding process is called Digital Audio Interface (DAI).

The 104 kbps data rate is selected to be economically transmitted by the radio transmission interface. The voice coder must then significantly reduce its rate by removing irrelevant components from the DAI data.

E.7.4.3 Voice Codecs Specified for GSM

Full-rate voice coding in GSM employs the RPE/LTP, as mentioned earlier. It combines linear predictive coding (LPC), which exploits the short duration correlation in the voice signal, with the long-term prediction (LTP). The residual signal, which excites the vocal tract model in the receiver, consists of a set of regular pulses.

The half-rate voice coder uses a combination of CELP/VSELP codecs at a rate of 5.6 kbps. The motivation for the use of this codec is the possibility of using a time window at every two TDMA frames, i.e., to double the channel traffic capacity. Some disadvantages exist however as, for example, voice quality being inferior to that offered by a full-rate voice channels.

The improved full-rate coder is called ACELP (*algebraic code excited linear prediction coder*). It has the same rate as the RPE/LTP codec, however has superior voice quality. Its disadvantage is the increased complexity in relation to the RPE/LTP codec.

E.7.4.4 Discontinued Transmission

Detection of pauses in the voice signal is a requirement for the GSM voice codec. Whenever a conversation pause is detected, transmission is halted during the pause period. The use of this function is an option for the network. The DTX option tends to reduce interference in adjacent cells because the actual transmission time for the conversation signal is much reduced. Besides, energy consumption at portable terminals is reduced, consequently requiring smaller batteries. Pauses in normal conversations occur at a rate of roughly 50 % of the total conversation time. This means that a telephonic channel is effectively used roughly half of the time allocated for voice transmission. The DTX function includes two additional characteristics.

- Voice activity detection (VAD), in order to determine the presence or absence of voice on the microphone. Actually, it has to discriminate whether the sound detected represents voice or just noise, even when background noise is rather strong. This is not a simple task since, if the voice signal is mistakenly considered to be noise, the transmitter is switched off, provoking an undesired effect called clipping (which sounds like a click);

- The complete absence of a background sound can disturb a conversation, since it can be interpreted as a lost call. Furthermore, the users tend to talk louder whenever there is total background silence. For that reason there is a need for the presence of background noise at a minimum level during pauses in the conversation. This background noise is called *presence* noise or *comfort* noise. This is achieved by the transmission of silence information frames (SID) at each 480 ms. Once as SID frame is received, the voice decoder at the receiver has to simulate the existence of a wire connection by generating some background noise. This is the *comfort* noise which is responsible by making the system *present*.

E.7.5 Channel Coding

The RF channel is a rather rough channel for the data. For this reason it is important somehow to protect the signal during transmission. This data protection comes in the form of channel coding.

Channel coding embraces a vast area of engineering and research, being the territory of many talented scientists and engineers. There are many techniques developed for the protection of data during transmission or storage. In this context coding usually adds redundant bits to the data bits in order to protect information. This procedure makes transmission more reliable because, within certain limits, it allows the detection and possibly the correction of data corrupted by the RF path. A very simple channel coding scheme consists of segment the data stream into blocks or words, and to add a single bit to each block, which tells the receiver whether the block is correct or whether it contains an odd number of errors. This simple procedure is called parity coding. The channel coding mechanisms used in GSM are a lot more elaborate than single parity checking.

E.7.5.1 Voice Channel Coding

The voice codec in GSM operates on 20 ms voice signal blocks. After passing through the codec, 260 bit blocks of data are produced at a rate of 13 kbps. Each 260 bit block (corresponding to 20 ms) is subdivided into three classes called Ia, Ib and II. The most important class is class Ia which contains 50 bits. Next in importance comes class Ib which contains 132 bits. The least important class is class II which contains the remaining 78 bits.

Voice data are coded by a serial concatenation of a block code followed by a convolutional code. First the bits in class Ia are coded in blocks. This employs a cyclic code just for detecting errors. Three parity bits are appended to the 50 bits in class Ia, which give the decoder an indication of the occurrence or not of errors that were not previously detected or corrected. This is the first part of the coding process and its decoding occurs in the second decoding part, in the receiver, after decoding the convolutional code. The second decoding operation is responsible for detecting errors that escaped the previous stage of convolutional decoding. If the block decoder detect at least one error in the bits in class Ia, then the entire 260 bit block is discarded. Whenever a block is discarded in this manner, the voice codec is informed that a voice block was abandoned and that it is necessary to interpolate the subsequent voice data blocks. This procedure provides better voice quality than if the voice codec had to reproduce a sound directly from corrupted data.

The second part of channel coding of voice data is done with convolutional. This coding technique also appends redundant bits to the data bits in a manner to allow the decoder to detect and to correct errors within certain limits. Convolutional coding is applied to the codewords of the block code, i.e., it encodes bits in class Ib as well as bits in class Ia, including the parity bits produce by block coding as well. A convolutional code can be defined by three positive integers, n, k and K, and a $k \times n$ matrix, where n denotes the number of binary digits output by the encoder for every k bits fed at its input, and K denotes the encoder memory. The asymptotic code rate R is defined as $R = k/n$. The convolutional code employed in the serial concatenation is a rate $R = 1/2$ and has a memory $K = 5$. This memory value means that five consecutive bits are used for calculating redundant bits and that, for each data bit, one redundant bit is generated. Before the information bits are encoded, four bits are added. These extra four bits are all-zero and are used for initializing the convolutional encoder. In Chapter 7 there is a more detailed description of a convolutional encoder. In the case of the GSM system, each bit at the encoder input originates two bits at the encoder output. When encoding begins, all memory elements are assumed to contain zeros.

Figure E.10 illustrates the complete channel coding scheme for all voice bits. Bits in class II are not protected by a code due to their lower relevance. The 189 input bits become 378 bits at the output and are concatenated with the 78 bits in class II, forming finally a block 456 bits long. The number 456 is equal to 4×114, which is the number of bits transmitted in a normal burst, separated in two 57 bit sub-blocks.

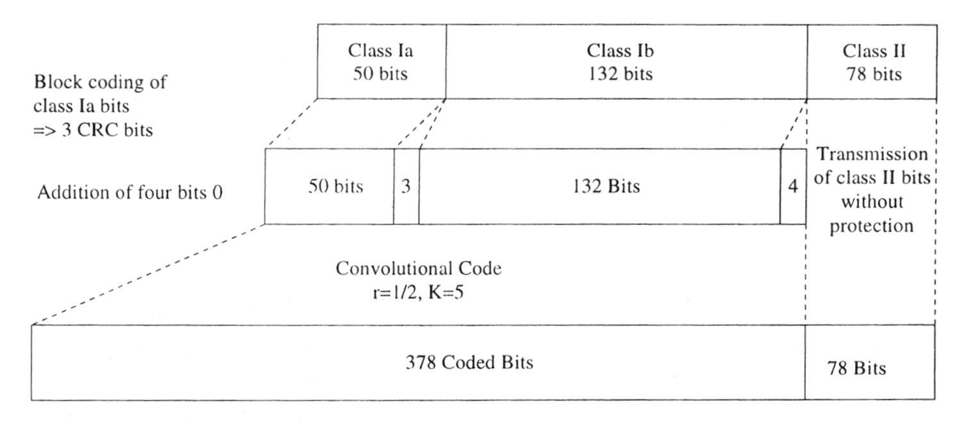

Figure E.10. Block and convolutional coding of full rate voice data.

E.7.5.2 Voice Channel Interleaving

As already mentioned, the 456 bits occupy exactly four burst time windows. If these data occupied four consecutive time windows, they would be more vulnerable to error bursts. Complete data bursts are lost regularly in radio channels. This happens when, for example, the mobile station passes through a tunnel or when some strong interference occurs. In order to avoid the big risk of loosing many consecutive data bits, consecutive data bits are scattered into more than four bursts. The 456 bit block

Coded bit index			*Position in 26 multi-frame structure*	
0	8	448	Burst even bits N	(No. 0, 4, 8, 13, 17, 21)
1	9	449	Burst even bits $N+1$	(No. 1, 5, 9, 14, 18, 22)
2	10	450	Burst even bits $N+2$	(No. 2, 6, 10, 15, 19, 23)
3	11	451	Burst even bits $N+3$	(No. 3, 7, 11, 16, 20, 24)
4	12	452	Burst odd bits $N+4$	(No. 4, 8, 13, 17, 21, 0)
5	13	453	Burst odd bits $N+5$	(No. 5, 9, 14, 18, 22, 1)
6	14	454	Burst odd bits $N+6$	(No. 6, 10, 15, 19, 23, 2)
7	15	455	Burst odd bits $N+7$	(No. 7, 11, 16, 20, 24, 3)

Table E.2. Bit reordering scheme for a TCH traffic channel.

is scattered over eight bursts in sub-blocks of 57 bits each. A sub-block is defined as the numbered bits with even index or with odd index, from the data encoded in a data burst.

The data are placed in sub-blocks in their original form; they are first reordered before being mapped to the time windows. The reordering (or permutation) is a reversible operation, so that the bits can again be placed in their original order at the receiver. The permutation of bits reduces substantially the possibility of the loss of a group of consecutive due to error bursts bits. The spreading of errors in small groups eases the task of the convolutional code, since this code is chosen to combat more effectively random errors rather than burst errors.

The 456 bits are subdivided into eight sub-blocks as follows. The bit with index 0 is placed in sub-block 1, the bit with index 1 is placed in sub-block 2, and so on, until all eight sub-blocks are filled. The bit with index 8 is placed in sub-block 1 according to Table E.2. The first four sub-blocks are filled with odd index bits of the next four consecutive data bursts (as in Figure E.11). This process of placing bits in sub-blocks is called reordering or restructuring, while the mapping of eight data bursts into sub-blocks is called diagonal interleaving.

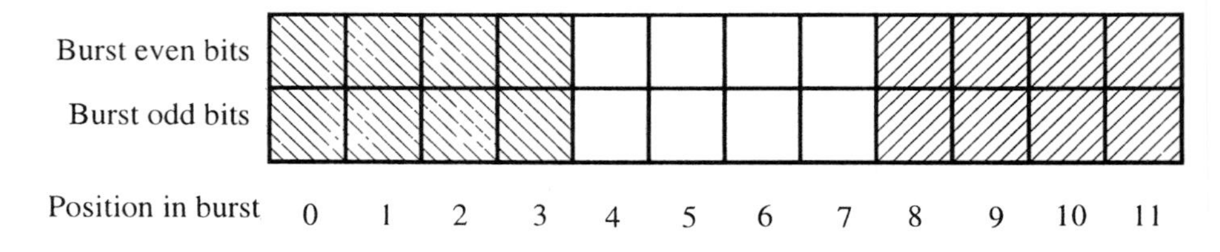

Figure E.11. Interleaving scheme for voice data.

If a traffic channel is sacrificed (stolen) by an FACCH channel, the process is indicated by the stealing flags in the data burst, in which when the first stealing flag is state 1, indicates that the even bit indexes are occupied by the FACCH channel, and the second stealing flag also in state 1, indicates that the odd bit indexes are occupied by the FACCH channel. The coding scheme for the FACCH channel is the same as that for other signaling data, i.e., different from the channel coding employed in the traffic channels. Therefore, this is the reason for the need of using stealing flags to inform the decoder that a FACCH channel is being used.

E.7.5.3 Coding and Interleaving Data Channels

The coding scheme for the data channels is rather elaborate. The reason for that is because data bits must be much more protected than the data of voice traffic channels.

For each data rate there is an associated interleaving scheme and specific parameters for the convolutional code. Since there are in total five distinct types of data channels, only the most important (TCH/F9.6) will be described.

In spite of the value 9.6 kbps being indicated for the rate in the TCH/F9.6, that value is raised to 12 kbps as a result of channel coding performed at the mobile terminal. Coding at the mobile terminal is independent of GSM, and for that reason will not be detailed here. This code is used for error detection in an environment employing wire connections.

In a channel with the rate of 9.6 kbps, the flux of bits is divided into four blocks of 60 bits each, for a total of 240 bits, which are encoded together by a convolutional code. Contrasting with the coding of voice data, the block code in this case is not applied before convolutional coding because error detection is being already performed in the terminal. In order to initiate the convolutional coding, four zero bits are appended to the 240 bits. The convolutional code parameters are the as were used for coding voice data, i.e., ($R = 1/2$, $K = 5$). The convolutional encoder accepts therefore 244 bits at the input and delivers 488 bits coded bits at the output. However 488 is greater than the 456 bits required to fill in four data bursts, and for that reason $488 - 456 = 32$ bits are removed.

The bits from user data files are more important than the user voice data bits. Therefore, the interleaving scheme for data applications needs to reach deeper and to be more complex than the interleaving used in blocks of coded voice. The blocks are spread over 22 data bursts, constituting almost a complete frame of channel traffic (Figure E.3) with its SACCH channel and its idle frame. The 456 bits are divided into 16 parts of 24 bits each ($16 \times 24 = 384$), 2 parts of 18 bits ($2 \times 18 = 36$), 2 parts of 12 bits ($2 \times 12 = 24$) and 2 parts of 6 bits ($2 \times 6 = 12$). One data burst will contain 5 or 6 blocks of consecutive data, i.e., 4 parts of 24 bits each (96 bits) added with 1 part of 18 bits ($96 + 18 = 114$), or else added with 1 part of 12 bits and 1 part of 6 bits ($96 + 12 + 6 = 114$). The blocks are spread into 22 data bursts as follows.

- Data bursts first and twenty-second transport a 6 bit block each;
- Data bursts second and twenty-first transport a 12 bit block each;
- Data bursts third and twenty transport an 18 bit block each;
- Data bursts from the fourth to the nineteenth transport a 24 bit block each.

All 456 bits are accommodated in the 22 frames of channel traffic. This structure is repeated every four data bursts in such a manner that the bits become diagonally distributed, similar to what is used in the case of voice data frames. The scheme is shown in Figure E.12. There are k data bursts shown, each one receiving n contributions of user data. Each data burst carries five or six distinct data contributions. Thus, the 22 frames can transport 5.5 information blocks. At the extreme right of the data bursts, as shown in Figure E.12, the SACCH channel (TDMA frame number 12) and the idle frame (TDMA frame number 25) retain their traditional functions and do not transport any user data. Figure E.13 shows the complete channel coding scheme for the TCH/F9.6.

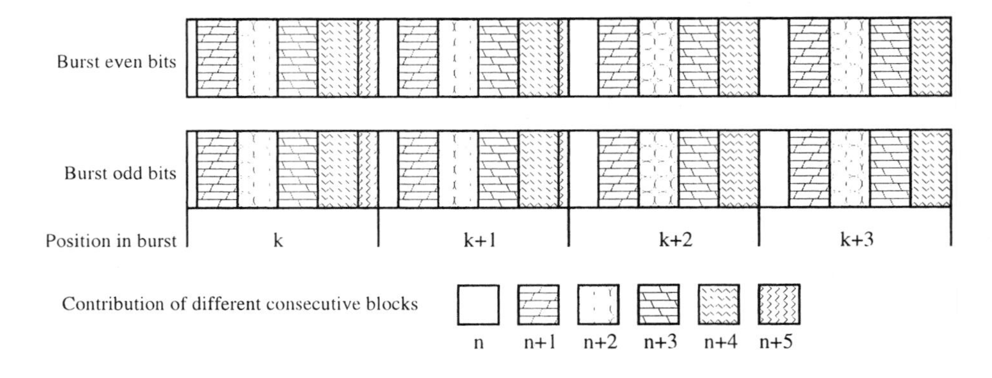

Figure E.12. Interleaving scheme for TCH/F9.6.

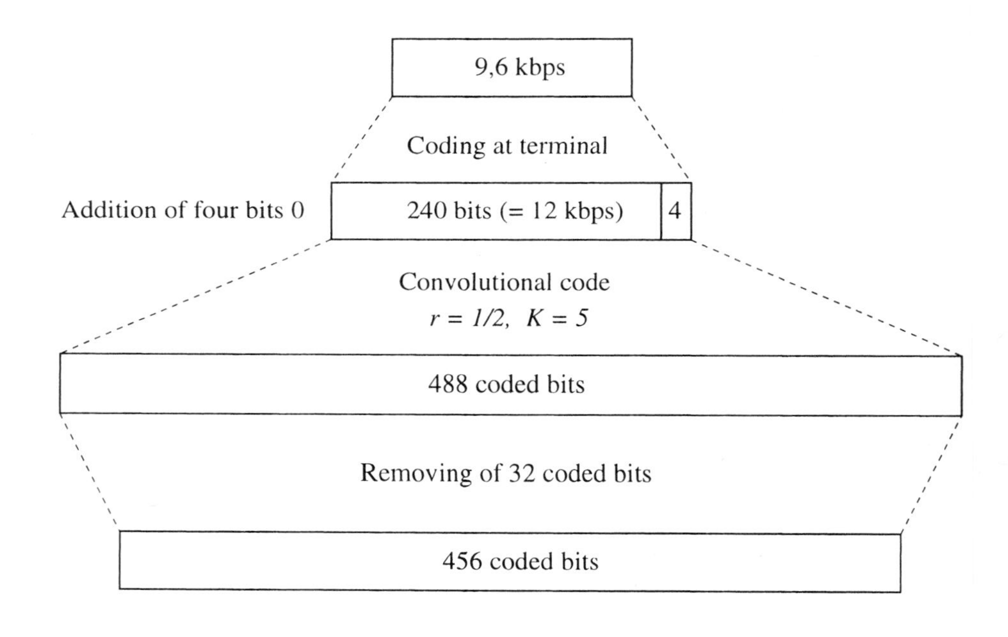

Figure E.13. Coding scheme for TCH/F9.6.

If during a data transmission session some signaling is required, the FACCH channel goes into action replacing the traffic information. The FACCH information is too important and can not wait for the slow and complicated data interleaving of depth 22. The FACCH has an interleaving depth of 8. When the FACCH replaces the user data, these are lost. Due to the small depth of the FACCH and the reduced size of its message, it is likely that the convolutional code can recover the otherwise lost user data.

E.7.5.4 Coding of Signaling Channels

Signaling data are more important than any user data that might appear in the network. This is due to its function of management of the link established between the stations involved.

Signaling contains a maximum of 184 bits, which need to be encoded. It is irrelevant whether the type of signaling information to be transmitted is mapped into a BCCH, PCH, SDCCH or SACCH channel. The format remains the same. Special formats are reserved for the SCH and the RACH channels, and the FCCH channel does not require any kind of coding.

The meaning of each one of the 184 bits is the same. Therefore, no distinction is made among them, as was done with the three classes of voice bits. The coding scheme is divided into two steps, shown in Figure E.14. The first step employs a block code, which is dedicated to the detection and correction of errors in the data bursts. These are errors that occur when a large part of a data burst, or even the complete data burst, is lost or corrupted during the passage through the channel. The block code used belongs to the family of the Fire codes. To each 184 bits, 40 parity bits are appended, resulting in a 224 bit long codeword.

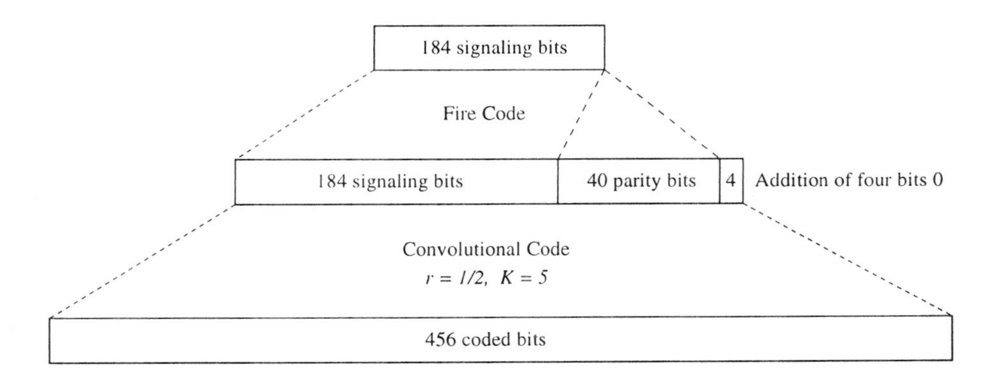

Figure E.14. Coding scheme for signaling channels.

The second step employs a convolutional code, which uses the same parameters from previous cases ($R = 1/2$, $K = 5$), and 4 zero bits are appended to the 224 bits. The convolutional code doubles the number of bits, $228 \times 2 = 456$ bits. The 456 bits fit perfectly into four data bursts, consisting of eight sub-blocks of 57 bits each. The coded data are interleaved in the four data bursts. The first four sub-blocks are placed in the locations of the even index bits of four consecutive data bursts, and the subsequent four sub-blocks are placed in the locations of the odd index bits bits of the same four consecutive data bursts, according to Table E.3. Figure E.15 shown how the data bursts are organized.

Coded bit indexes	*Position within frame structure*
0 8 448	Data burst N, even index bits
1 9 449	Data burst $N+1$, even index bits
2 10 450	Data burst $N+2$, even index bits
3 11 451	Data burst $N+3$, even index bits
4 12 452	Data burst N, odd index bits
5 13 453	Data burst $N+1$, odd index bits
6 14 454	Data burst $N+2$, odd index bits
7 15 455	Data burst $N+3$, odd index bits

Table E.3. Reordering scheme for a signaling channel.

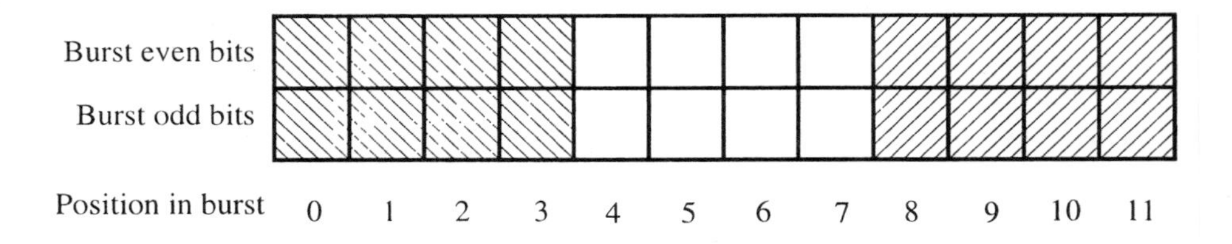

Figure E.15. Interleaving scheme for signaling data.

E.8 Cryptography

Cryptography is used to protect signaling data and user data. Cryptography is employed after the construction of all logical channels, independent of whether it is a signaling channel or a traffic channel. Cryptography is applied only to the data bursts.

Message cryptographic protection is done in two steps. First, an encrypted key is computed using algorithm A8 stored in the SIM card, together with the subscriber key and private random number provided by the network (the random number is the same used in the authentication procedure). Second, a 114 bit sequence is produced using the encrypted key, an algorithm called A5 and the TDMA frame number. An XOR operation is done involving this bit sequence and a sequence of two 57 bit data blocks from a data burst. The receiver employs the same A5 algorithm to correctly decipher the encrypted information.

E.9 Modulation

The type of modulation specified for the GSM is called Gaussian Minimum Shift Keying (GMSK), having the bandwidth by transmission interval product $BT = 0.3$, and rate 270 5/6 kbauds. The GMSK is a kind of FSK, in which frequency modulation results from the phase modulation of appropriate signals. The constant amplitude of GMSK makes it appropriate for use with high efficiency amplifiers.

The GMSK modulation is a Minimum Shift Keying (MSK) modulation passed through a Gaussian filter. Figure E.16 shows how digital data is modulated in MSK.

Digital Input		*MSK Output*	
Bit Value		*Frequency*	*Sign*
Odd Bit	*Even Bit*	*High or Low*	*+ or -*
1	1	High	+
-1	1	Low	-
1	-1	Low	+
-1	-1	High	-

Table E.4. MSK truth table.

The waveforms are aligned in phase. The data are shown in the upper part of the figure. Immediately below, the data are divided into even bits and odd bits, with the distinction that the bit durations are now doubled and there is a displacement of half a bit between them. This is just the Offset Quadrature Phase-Shift Keying (OQPSK), which guarantees soft phase transitions, of at most 90 degrees. Besides, instead of zeros and ones, the values -1 and 1 were used to represent a one and a zero, respectively, i.e., a non-return to zero (NRZ) code.

Since MSK is a form of FSK, below the even and the odd representation of bits are the high frequency and low frequency waveforms which constitute the final carrier. The MSK signal (last diagram in Figure E.16) is formed by the displacement between the low frequency and the high frequency signals according to Table E.4. At each time slot the values of even and odd bits are observed, and according to the table the MSK output is produced; a positive or a negative sign indicates whether the waveform has the same sign as the graph or an inverted sign, respectively. Figure E.17 shows the MSK frequency trajectory and Figure E.18 shows the corresponding phase trajectory.

In order to produce a GMSK signal from an MSK signal, as done earlier, it is necessary just to filter the even and the odd sequences of input bits. The filter used is a Gaussian filter with bandwidth defined as a function of the product BT. In case of the GSM system, $BT = 0.3$, for $B = 81.3$ kHz and $T = 3.7$ μs ($T = 1/270.833$).

With the use of the Gaussian filter, the GMSK softens phase transitions even further, with respect to MSK modulation. For this reason GMSK can even be called a softened MSK. This denomination makes sense since the frequency displacement in GMSK occurs with a careful phase change, as can be seen in Figure E.19. However there is an important disadvantage of GMSK with respect to MSK: intersymbol interference is larger than in MSK, contributing for an increase in the bit error rate (BER).

The implementation of a GMSK modulator can be done according to Figure E.20. The differential encoder transforms the 0 and 1 bits into 1 and -1 according to the truth table shown in Figure E.21. Differential encoding builds up the OQPSK modulation as explained earlier.

E.10 Frequency Hopping

The propagation of a radio transmission signal and its corresponding voice quality are influenced by the environment through which the signal travels. Digital transmission improves signal quality by means of error protection techniques. Other techniques are also employed in conjunction with channel coding in order to provide further im-

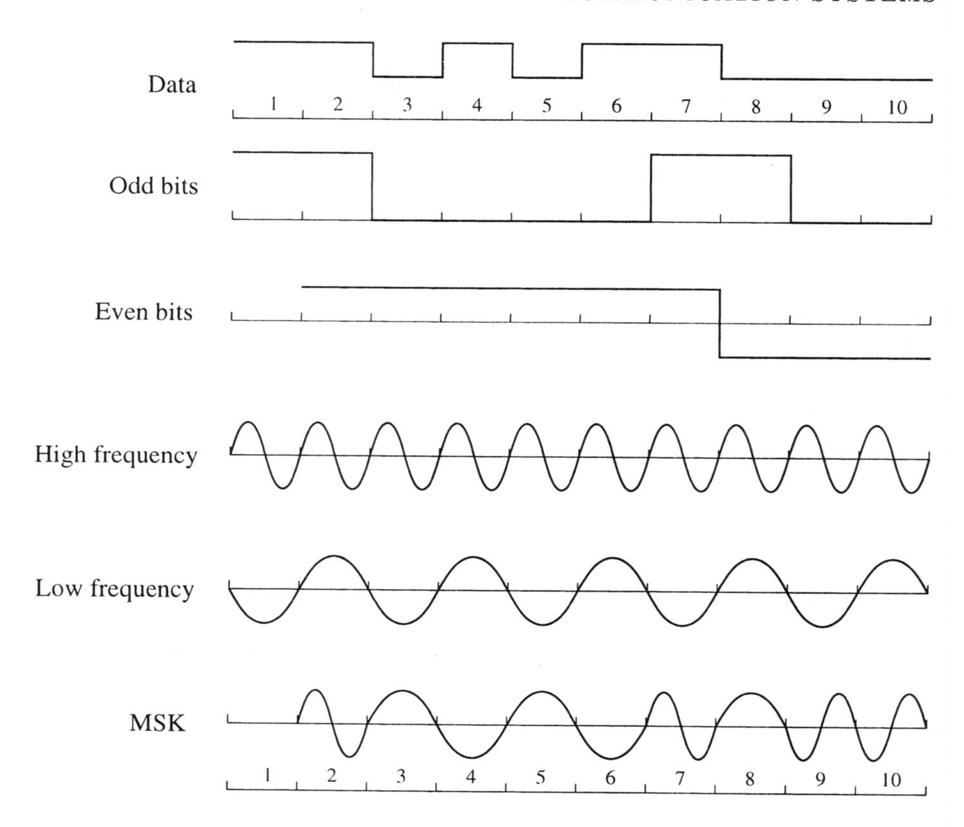

Figure E.16. MSK signal generation.

provement in signal. The radio transmission channel suffers frequency selective fading, meaning that propagation conditions may vary with time and from channel to channel.

In other words, for example, if channel 2 experiences problems when a mobile terminal passes near a big building, it could happen that channel 5 might not experience no degradation at all for the signal that it carries. In this manner, it could happen that while some channels experience signal degradation, it could be that other channels experience an improvement in signal to noise ratio (SNR) while the mobile terminal moves inside the same cell. In order to compensate for these differences in signal quality due to fading, the technique called frequency hopping was introduced. The objective of frequency hopping is to provide mobile stations with roughly the same performance in the presence of varying propagation conditions. This is achieved by periodically changing the channel operating frequencies. In this manner all mobile terminals experience, on average, similar propagation conditions.

There are two types of frequency hopping, the slow and the fast. In the slow frequency hopping (SFH), the channel operating frequency is changed at every new TDMA frame, whereas in the fast frequency hopping (FFH), the channel operating

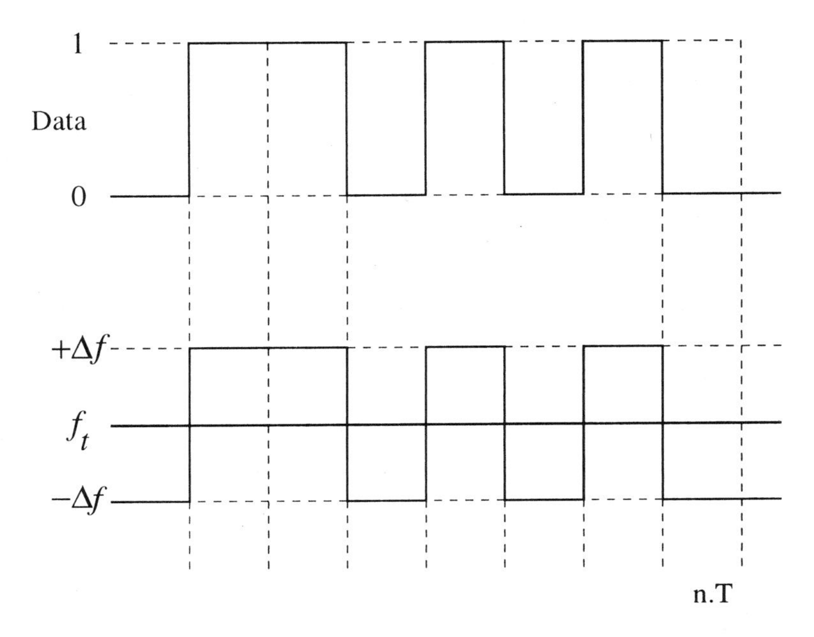

Figure E.17. Frequency trajectory of MSK.

frequency is changed many times per transmitted symbol. GSM uses only the SFH type, and therefore the channel operating frequencies are changed every 4.615 ms (duration of a TDMA frame).

The use of frequency hopping is an individual option of each cell. It is not required that a base station must have support for this function. However, a mobile station must enter the frequency hopping mode whenever the base station whenever requested by the base station. The use of frequency hopping adds frequency diversity to the system. When a mobile station approaches the border of its current cell, interference from neighboring cells increases, and in this case the base station can decide that the mobile station should enter the frequency hopping mode. The base station simply assigns a complete set of RF channels for which the mobile station can hop. In order to hop in frequency the mobile station must follow instructions provided by an algorithm. There are many algorithms for frequency hopping, one of which is cyclic hopping, where the hops are done according to a list of frequencies, repeating the process whenever the last frequency in the list is used. Another algorithm for the purpose is called (pseudo) random hopping, where the hops in frequency are done in an apparently random manner, from a list of frequencies.

E.11 GSM Services

During the design of GSM various services were defined, meant to be offered to the subscribers. It should be noticed that not all GSM services were devised from the beginning of GSM, but were gradually introduced. The initial GSM Memorandum (MoU) defined four classes for the introduction of different services:

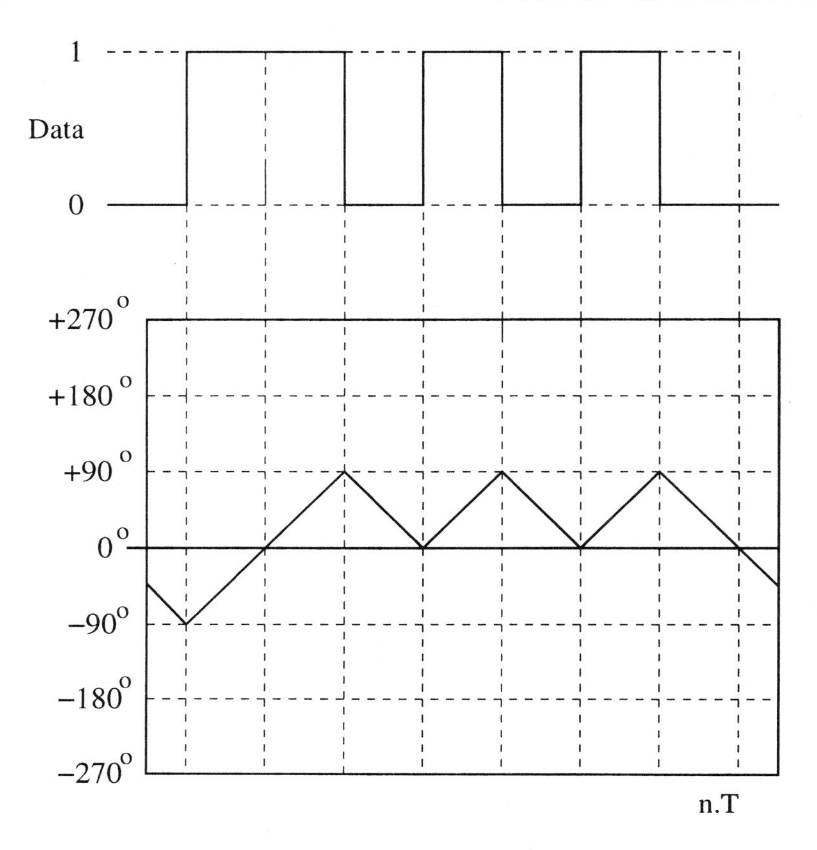

Figure E.18. Phase trajectory of MSK.

- E1: introduced when GSM service started;
- E2: introduced in the end of 1991;
- Eh: introduced in half rate channels;
- A: optional services.

There are basically two services offered by GSM: telephony (also known as teleservice) and data (also known as bearer service). The telephony services are mainly voice services which offer the subscribers the capacity (including the required terminal equipment) of communicating with other subscribers. The data services provide the necessary capacity to transmit appropriate data signals between two access points, creating an interface for the network. Three categories of service can be distinguished.

- Teleservices;
- Bearer services;
- Supplementary services.

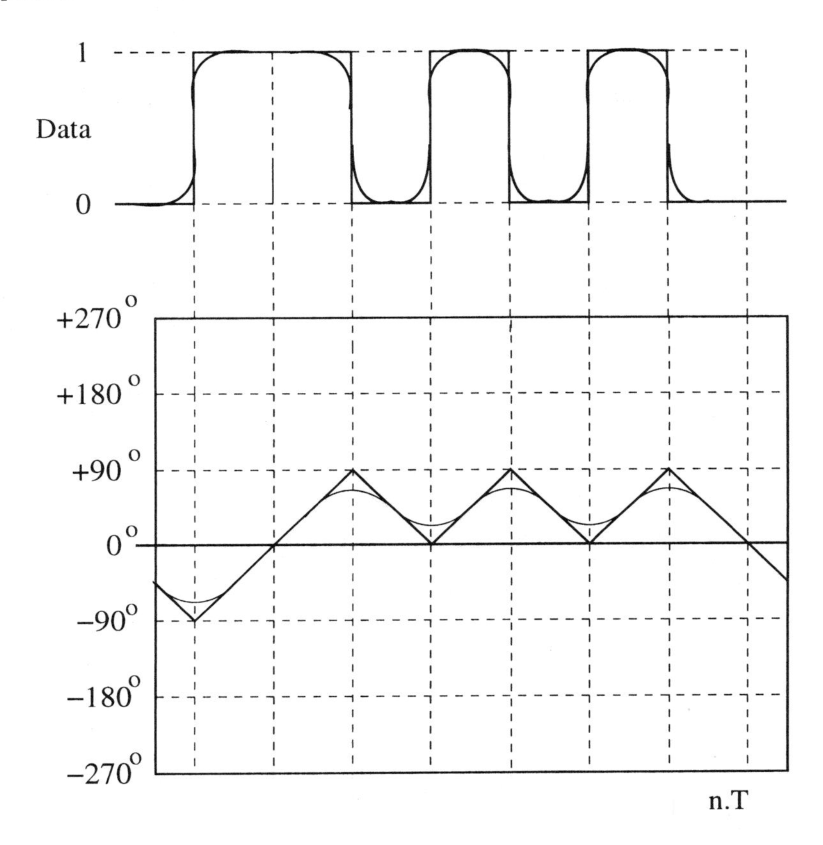

Figure E.19. Phase trajectory, MSK *versus* GMSK.

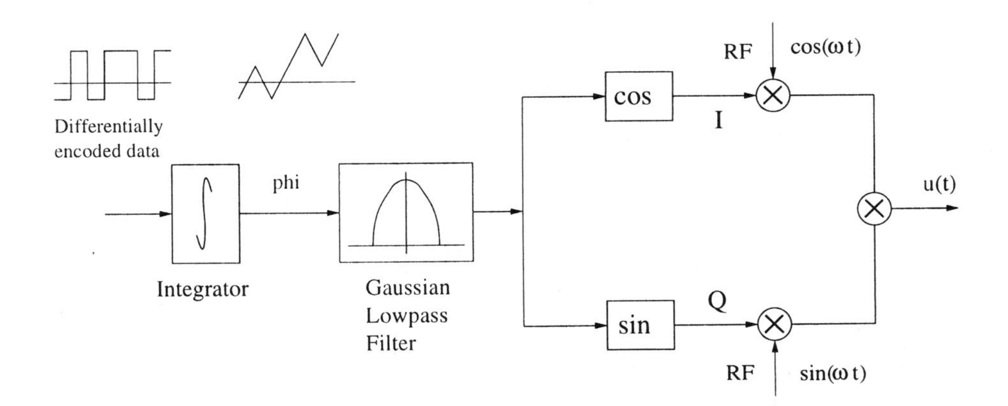

Figure E.20. GMSK Modulator.

Differential Enconding

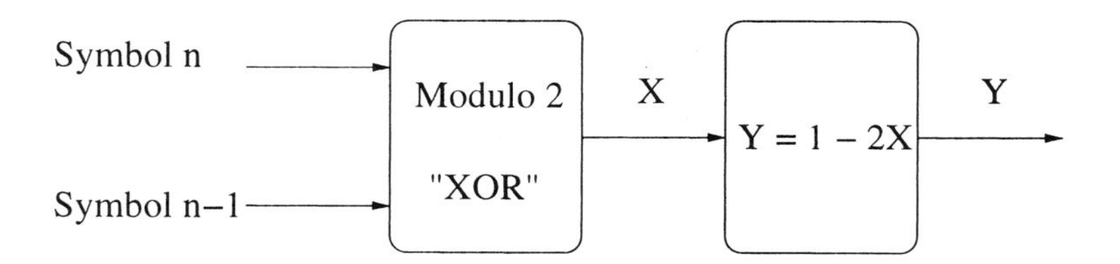

n	n−1	X	Y
0	0	0	+1
0	1	1	−1
1	0	1	−1
1	1	0	+1

Truth Table for Differential Encoding

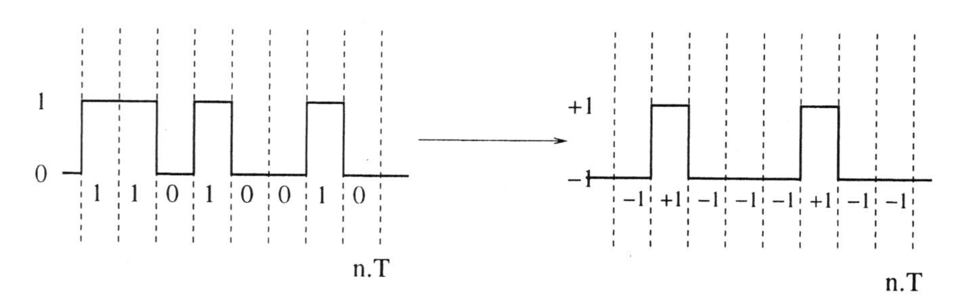

Figure E.21. Differential encoding.

E.11.1 Teleservices

Besides normal telephony (E1, Eh) and emergency calls (E1, Eh), the following teleservices are provided by GSM.

- Dual-tone multi-frequency (DTMF) – DTMF is a tone signaling scheme often used for various purposes via a telephone network, such as the remote control of a machine;
- Group III Facsimile (E1) – GSM supports CCITT group III facsimile. Following this standard, the fax machines are designed to be connected to a telephone using analog signals. A special fax converter is used in the GSM system to allow communication between analog fax machines in the network;
- Short message services (E1, E2, A) – A convenient facility in the GSM network is the short message service. A message with a maximum of 160 alphanumeric characters can be sent to/from a mobile station. This service can be seen as an advanced form of alphanumeric paging with some advantages. If the subscriber mobile unit is switched off or has left the coverage area, the message is stored and sent again to that subscriber when the mobile unit is switched on or reenters the network coverage area. This function ensures that the message is received;
- Cell broadcast (E1, E2, A) – A variation of the short message service is called cell broadcast. A message having at most 93 characters can be sent to all mobile subscribers in a certain geographic area. Typical applications include announcements of traffic jam and road accident reports;
- Voice electronic mail – This service is actually a server within the network, controlled by the subscriber. Calls can be sent to the subscriber electronic mail box and the subscriber can check them using a personal password;
- *Fax mail* – With this service the subscriber can receive fax messages in any fax machine. The messages are stored in a service center from which they can be recovered by the subscriber using a special password for the desired fax number.

E.11.2 Bearer services

Bearer services are used to carry user data. A few bearer services are listed next.

- Synchronous and asynchronous data, $300 - 9,600$ bps (E1);
- Alternate voice and data, $300 - 9,600$ bps (E1);
- Access to asynchronous PAD (packet-switched, packet assembler/disassembler), $300 - 9,600$ bps (E1);
- Access to dedicated synchronous data packets, $2,400 - 9,600$ bps (E2).

E.11.3 Supplementary services

GSM supports a set of supplementary services which can complement telephony as well as data services. A list of supplementary services is presented next.

- Call forwarding (E1). The subscriber can redirect received calls to another number if the called mobile unit is engaged (CFB), is not found (CFNRc) or if there is no reply (CFNRy). Call forwarding can also be applied unconditionally (CFU);
- Call barring.
  There are different types of call barring.

- Barring of all outgoing calls (BAOC) (E1);
- Barring of outgoing international calls (BOIC) (E1);
- Barring of outgoing international calls except those directed to the Public Switched Telephone Network (PSTN) of the country of origin (BOIC-exHC) (E1);
- Barring of all incoming calls (BAIC) (E1);
- Barring received calls when conducting roaming (A).

- Call hold (E2). This service allows the subscriber to interrupt a call (self made) and subsequently reestablish this call. The call hold service is used only in normal telephony;
- Call waiting, CW (E2). This service allows the subscriber to be notified of a second call, during a conversation. The subscriber can answer, reject, or ignore this second call. Call waiting is applied to all telecommunications services of GSM using a switched circuit connection;
- Advice of charge, AoC (E2). This service provides the user on line information about the charge level in the battery;
- Multiparty service (E2). This service allows a mobile subscriber to establish a group call, that is, a simultaneous conversation among three to six subscribers. This service applies only to normal telephony;
- Closed user groups (CUG) (A). CUGs are compared generally to a PBX. They are groups of subscribers that can only call those in the group and certain numbers outside the group;
- Calling line identification presentation/restriction (A). These services supply the called group with the number of the integrated services digital network (ISDN) of the calling group. The restriction service allows the calling group to restrict the presentation;
- Operator determined barring (A). Restriction by the operator of different services and types of call.

E.11.4 Logic Channels

As already mentioned, data in GSM are transmitted in well defined time windows, where a group of eight time windows is called a frame. The time windows constitute physical channels. The transmitted data can be traffic data or signaling data, because it is necessary to organize this data in an appropriate form. The types of organization of the data to be transmitted are called logic channels.

The concept of a logic channel is a bit distant from the physical nature of the signal and closer to the nature of the information to be transmitted. The way of transmitting information depends on the type of information to be transmitted. Different types of information can exist in the system in different logic channels. The contents of the logic channels can appear in any physical channel (frequency or time window), but once a physical channel is assigned to carry the contents of a logic channel this assignment must remain unaltered.

A logic channel transports signaling data and user data. The data, no matter which kind, are mapped into a physical channel. The way the data are mapped into the physical channel depends on the data contents. Important data have higher priority than routine data. The mapping schemes produce some channel structures which are considered combinations of structures. There are seven combinations of logic channels which can be mapped into physical channels.

The GSM system distinguishes between traffic channels, used for user data, and control channels, used for messages of network management and for some link maintenance tasks. An analogy can be made considering passengers and crew in a plane. Passengers are user data and pilots and stewardesses are control data. It is also possible to distinguish in the same analogy, the difference between a logic channel and physical channel. The way the crew and the passengers are organized constitutes two logic channels (control channel and traffic channel, respectively) e the plane constitutes the physical channel. Unless otherwise specified, in the sequel the word channel will refer to a logic channel. The logic channels used in GSM are illustrated in Figure E.22.

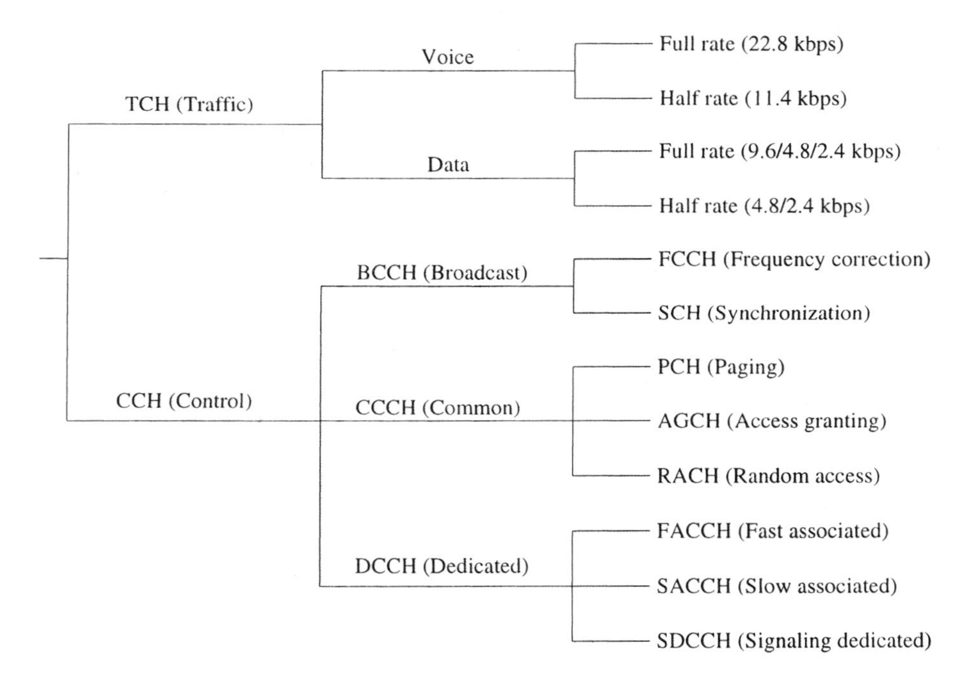

Figure E.22. Logic channels.

E.11.4.1 Traffic Channels

Traffic channels allow the user to transmit voice or data. Depending on the type, voice or data, different channels can be used, as follows.

- Full rate traffic channels (TCH/FS) are the type of channel used at the beginning of GSM to transmit voice. Their transmission rate is 13 kbps;

- Half rate traffic channels (TCH/H) were initially conceived only as an option for future use. The idea behind was to double system capacity by compressing data by a factor of two. But this gain in capacity should not compromise then quality of voice achieved with the full rate voice channels. For this reason another way of doing voice coding had to be adopted for the half rate traffic channels;
- TCH/F9.6/4.8/ 2.4. Are used for data transmission at rates 9.6/4.8/ 2.4 kbps. Depending on the equipment used on both ends of a link, it can be either fax data or computer data. The data rate used in the system depends on the capacity of the mobile station. For each data rate a different coding scheme is used, which require appropriate software within the mobile stations. There are now available transceivers that can be connected to the PCMCIA port of laptop computers, to be used for data transmission or fax;
- TCH/H4.8/2.4. Are used for transmission in a half rate traffic channel. Obviously, the implementation of this service depends on the availability of a half rate traffic channel.

E.11.4.2 Control Channels

Control channels do not transport subscriber's voice data, fax data or even computer data. Control channels carry the data needed by the network and the mobile stations to ensure that all traffic keeps reliable and efficient. Depending on their tasks, there are four distinct classes of control channels as follows. The broadcast channels, the common control channels, the dedicated control channels and the associated control channels.

The broadcast channels (BCH) are transmitted only by the base station and are responsible for providing sufficient information for the mobile station to synchronize with the network. The mobile stations never transmit a BCH. There are three types of BCH as follows.

- Broadcast control channel (BCCH). This channel informs the mobile station about specific channel parameters which are necessary to identify the network or to gain access to the network. These parameters are, among others, the location area code (LAC), the MNC (to identify the operator), the information of which frequencies can be found in neighboring cells, options of different cells and access parameters;
- Frequency correction channel (FCCH). This channel provides the system reference frequency for the mobile station. It is a logic channel mapped only to the frequency correction burst, which contains only the FCCH;
- Synchronization channel (SCH). This channel provides the key (training sequence) to the mobile station. The key is required to demodulate the information transmitted by the base station. The SCH is mapped to the synchronization burst.

The common control channels (CCCH) help in establishing a dedicated link between the mobile station and the base station. They are channels which provide tools to make calls, which can be originated in the network as well as in the mobile station. There are three types of CCCH as follows

- Random access channel (RACH). This channel is used by the mobile station to request a dedicated channel from the network. The base station never uses the RACH. The RACH is mapped to the random access burst and contains the first message sent to the base station. A measurement of the delay in the mobile station still has to be made before the link is permanently established;

- Paging channel (PCH). This channel is used by the base station to alert a mobile station (within the cell) about a call received;
- Access granting channel (AGCH). This channel is used by the base station to inform the mobile station which dedicated channel should be used. Besides, it is also informed which time advance should be used. The message in the AGCH is a reply from the base station to an RACH message from the mobile station.

The dedicated control channels (DCCH) are used for message transfers between the network and a mobile station, but they are not used for traffic. They are also used for low level signaling messages between the stations themselves. The network messages are necessary for the registration procedure or for a call setup. The low level signaling messages are used for call maintenance. Sometimes the network itself has to be involved in call maintenance.

- Signaling dedicated control channel (SDCCH). This channel is used for the transfer of signaling information between a mobile station and a base station.

In the downlink direction the base station transmits a reduced set of parameters containing system information to keep the mobile station updated about the most recent changes. These data are similar to those transmitted in the BCCH, with a few additional control parameters to command the mobile station to use a specific time advance value or another power level.

In the uplink direction the mobile station informs the measurement results performed in neighboring cells. The measurements are sent to the network with the aim of helping handover decisions. The mobile station tells the network which time advance configurations and which power level it is currently using.

The associated control channels have functions similar to those of dedicated control channels. They are used for transporting important information as well as for maintenance of a call in situations of urgency in the exchange of signaling.

- Slow associated control channel (SACCH). This channel is always used in association with a traffic channel or an SDCCH channel. If a base station assigns a traffic channel, there will always be an SACCH associated with this channel. The same is true in case an SDCCH channel is assigned. The purpose of the SACCH channel is the maintenance of the connection. The SACCH transports control parameters and measurement parameters or routine data necessary to keep the link between the mobile station and the base station;
- Fast associated control channel (FACCH). This channel can transport the same information as the SDCCH. The difference is that the SDCCH has its specific channel, while the FACCH replaces totally or partially a traffic channel. If during a call the need arises for some urgent signaling, then the FACCH appears in the place of the traffic channel. This can happen in the case of handover, in which it is necessary enough exchange of signaling between the mobile station and the base station. The FACCH is used to transmit the longer signaling. This is done by replacing traffic channels and indicating the presence of the FACCH by the use of stealing flags.

E.12 EDGE

Enhanced Data rates for Global Evolution (EDGE) is a radio-based, high-speed mobile data standard, that boosts network capacity and data rates, for both circuit and packet switching. EDGE uses 200 kHz radio channels, which are the same as current GSM channel widths. From a technical perspective, EDGE allows the GSM and GPRS network to offer a set of new radio access bearers to its core network.

EDGE is designed to improve spectral efficiency through link quality control. It requires wider transmission channel widths and features flexible time slots to mix and match the usual forms of communications, including voice, data, and video. Since it is fully based on GSM, it requires relatively small changes to the network hardware and software. The operators need not make any changes to the network structure or invest in new regulatory licenses.

EDGE introduces 8-ary Phase Shift Keying (8PSK), a linear, higher-order modulation, in addition to Gaussian Minimum Shift Keying (GMSK) allows the data transmission rates to be tripled. An 8PSK signal carries three bits per modulated symbol over the radio path, compared to a GMSK signal, which carries only one bit per symbol.

The standard GSM carrier symbol rate (270.833 ksps) is the same as with 8PSK. The burst lengths are the same as the existing GMSK Time Division Multiple Access (TDMA) structure, and the same 200 kHz nominal frequency spacing between carriers is used. While GSM uses GMSK, EDGE uses both 8PSK and GMSK. The GMSK scheme has a payload of 116 bits, while the 8PSK has a payload of 346 bits.

E.13 Problems

1 What is the main reason for the adoption of the GMSK modulation in the European GSM system?

2 Explain how the QPSK and MSK modulation systems work, presenting their basic characteristics. Present their variants employed in cellular mobile communications.

3 Which are the main characteristics of the GMSK modulation?

4 A cellular GSM system operates with bit error probability $P_b = 10^{-4}$ in the a absence of fading. What is the required signal to noise ratio for this system?

5 Consider the GSM system in the previous question. A user is cycling his bicycle at an average speed of 4,8 km/h, and simultaneously making a telephone call.

(a) What is the corresponding error probability?

(b) By how many decibels the signal power must be raised in order to keep the same error rate as before?

6 At a given moment the user in the previous question accelerates his bicycle to 14,4 km/h. What will happen to the error probability?

7 The user is approaching a base station in line of sight, which has a 20 m high antenna situated 1 km away. The telephone set is approximately 1,8 m above the ground. Assume the existence of selective Rayleigh fading and an average frequency of 900 MHz for the system.

(a) If the received signal has a power of -80 dBm, what is the power transmitted by the base station?

(b) What is the power of the interfering signal?

References

Abut, H., Gray, R. M., and Rebolledo, G. (1982). Vector quantization of speech and speech-like waveforms. *IEEE Transactions on Acoustics, Speech, and Signal Processing*, 30(3):423–435.

Aguiar Neto, B. G. (1995). *Voice Processing and Transmission (Notes).* Federal University of Paraíba, Campina Grande - PB.

Alcaim, A., Solewicz, J. A., and Moraes, J. A. (1992). Freqüência de ocorrência dos fones e listas de frases foneticamente balanceadas no português falado no Rio de Janeiro. *Revista da Sociedade Brasileira de Telecomunicações*, 7(1):23–41.

Alencar, Marcelo S. (1989). "Measurement of the Probability Density Function of Communication Signals". In *Proceedings of the IEEE Instrumentation and Measurement Technology Conference - IMTC'89*, pages 513–516, Washington, D. C.

Alencar, Marcelo S. (1993a). "A Frequency Domain Approach to the Optimization of Scalar Quantizers". In *Proceedings of the IEEE International Symposium on Information Theory, San Antonio, USA*, page 440.

Alencar, Marcelo S. (1993b). "A New Bound on the Estimation of the Probability Density Function Using Spectral Analysis". In *Proceedings of the IEEE International Symposium on Information Theory, San Antonio, USA*, page 190.

Alencar, Marcelo S. (1998a). "Estimation of the Probability Density Function Using Spectral Analysis". *Electronic Letters*, 34(2):150–151.

Alencar, Marcelo S. (1998b). "Optimization of Quantizers in the Frequency Domain". *Electronic Letters*, 34(2):155–156.

Alencar, Marcelo S. (1999). *Principles of Communications (Portuguese).* University Publishers, UFPB, João Pessoa, Brazil.

Alencar, Marcelo S. (2002). *Digital Telephony, Fourth Edition (Portuguese).* Érica Publishers Ltd., ISBN 85-7194-559-4, São Paulo, Brazil.

Alencar, Marcelo S. and Neto, Benedito G. Aguiar (1991). "Estimation of the Probability Density Function by Spectral Analysis: A Comparative Study". In *Proceedings of the Treizième Colloque sur le Traitement du Signal et des Images – GRETSI,*, pages 377–380, Juan-Les-Pins, France.

Armstrong, E. H. (1936). "A Method of Reducing Disturbances in Radio Signaling by a System of Frequency Modulation". *Proceedings of the IRE*, 24:689–674.

Ash, Robert B. (1990). *Information Theory.* Dover Publications, Inc., New York.

Assis, Mauro S. (1994). "Mobile Cellular Telephony". Technical report, Ministry of Communications, Brazil.

Atal, B. S., Cuperman, V., and Gersho, A., editors (1993). *Speech and Audio Coding for Wireless and Network Applications.* Kluwer Academic Publishers.

Barbosa, Sérgio Gonçalves Donato (2002). "Platform for Simulation of a GSM Mobile Communication System". Master's Dissertation, Federal University of Paraíba.

Baskakov, S. I. (1986). *Signals and Circuits.* Mir Publishers, Moscow, USSR.

Bayless, J. W., Campanella, S. J., and Goldberg, A. J. (1973). Voice Signals: Bit by Bit. *IEEE Spectrum*, pages 28–34.

Bellamy, John (1991). *Digital Telephony.* John Wiley & Sons, Inc., New York, USA.

Bello, Philip A. (1963). Characterization of Randomly Time-Variant Linear Channels. *IEEE Transactions on Communications Systems*, pages 360–393.

Bennett, W. R. (1948). "Spectra of Quantized Signals". *The Bell System Technical Journal*, 27:446–472.

Blachman, N. M. and McAlpine, G. A. (1969). "The Spectrum of a High-Index FM Waveform: Woodward's Theorem Revisited". *IEEE Transactions on Communications Technology*, 17(2).

Blahut, Richard E. (1990). *Digital Transmission of Information.* Addison-Wesley Publishing Co., Reading, Mass.

Blake, Ian F. (1987). *An Introduction to Applied Probability.* Robert E. Krieger Publishing Co., Malabar, Florida.

Boyer, Carl (1974). *History of Mathematics (Portuguese).* Edgard Blucher Publishers Ltd., São Paulo, Brasil.

Bultitude, Robert J. C. (1987). "Measurement, Characterization and Modeling of Indoor 800/900 MHz Radio Channels for Digital Communications". *IEEE Communications Magazine*, 25(6):5–12.

Buzo, A., Gray, A. H., Gray, R. M., and Markel, J. D. (1980). Speech coding based upon vector quantization. *IEEE Transactions on Acoustics, Speech and Signal Processing*, 28:562–574.

Carlson, Bruce A. (1975). *Communication Systems.* McGraw-Hill, Tokyo, Japan.

Carson, J. R. (1922). "Notes on the Theory of Modulation". *Proceedings of the IRE.*

Chen, O. T.-C., Sheu, B. J., and Fang, W.-C. (1994). Image compression using self-organization networks. *IEEE Transactions on Circuits and Systems for Video Technology*, 4(5):480–489.

Chou, Wu and Gray, Robert M. (1991). "Dithering and Its Effect on Sigma-Delta and Multistage Sigma-Delta Modulation". *IEEE Transactions on Information Theory*, 37(3):500–512.

CNTr (1992). "Basics of Mobile Cellular Telephony". Apostila, Telebrás National Training Center, Brasília, Brasil.

Cox, David C. (1992). "Wireless Network Access for Personal Communications". *IEEE Communications Magazine*, 30(12):96–115.

Cox, R. V. (1995). Speech coding standards. In Kleijn, W. B. and Paliwal, K. K., editors, *Speech Coding and Synthesis*, pages 49–78. Elsevier.

de Souza, José Cavalcante (1996). *The Pre-Socratic – Fragments, Doxography and Comments (Portuguese).* Nova Cultural Publishers Ltd., São Paulo, Brasil.

Deller Jr., J. R., Proakis, J. G., and Hansen, J. H. L. (1993). *Discrete-time Processing of Speech Signals.* Macmillan Publishing Co.

Devasirvatham, Daniel M. J. (1984). "Time Delay Measurements of Wideband Radio Signals Within a Building". *Electronics Letters*, 20(23):950–951.

Dhir, Amit (2004). *The Digital Consumer Technology Handbook.* Elsevier, Burlington, USA.

Dimolitsas, S. (1991). Subjective quality quantification of digital voice communication systems. *IEE Proceedings-I*, 138(6):585–595.

Durant, Will (1996). *The History of Philosophy (Portuguese).* Nova Cultural Publishers Ltd., São Paulo, Brasil.

Electronic Industries Association (EIA) (1989). Celular systems. *Report IS-54*.

Eriksson, T., Lindén, J., and Skoglund, J. (1999). Interframe LSF quantization for noisy channels. *IEEE Transactions on Speech and Audio Processing*, 7(5):495–509.

Eskicioglu, A. M. and Fischer, P. S. (1995). Image quality measures and their performance. *IEEE Transactions on Communications*, 43(12):2959–2965.

Falconer, David D., Adachi, Fumiyuki, and Gudmundson, Bjorn (1995). "Time Division Multiple Access Methods for Wireless Personal Communications". *IEEE Communications Magazine*, 33(1):50–57.

Fechine, J. M. (2000). Recognition of vocal identity using hybrid modeling: Parametric and statistical (portuguese). *Ph.D. Thesis, Federal University of Paraíba.*

Flanagan, J., Schroeder, M., Atal, B., Crochiere, R., Jayant, N., and Tribolet, J. (1979). Speech Coding. *IEE Transactions on Communications*, pages 710–737.

Freeburg, Thomas A. (1991). "Enabling Technologies for Wireless In-Building Network Communications – Four Technical Challenges, Four Solutions". *IEEE Communications Magazine*, 29(4):58–64.

Gagliardi, Robert (1978). *Introduction to Communication Engineering.* John Wiley & Sons, New York.

Gagliardi, Robert M. (1988). *Introduction to Communications Engineering.* Wiley, New York.

Galton, Ian (1993). "Granular Quantization Noise in the First-Order Delta-Sigma Modulator". *IEEE Transactions on Information Theory*, 39(6):1944–1956.

Gersho, A. (1994). Advances in speech and audio compression. *Proceedings of the IEEE*, 82(6):900–918.

Gersho, A. and Gray, R. M. (1992). *Vector Quantization and Signal Compression.* Kluwer Academic Publishers, Boston, MA.

Gersho, Allen (1969). "Adaptive Equalization of Highly Dispersive Channels for Data Transmission". *The Bell System Technical Journal*, (1):55–71.

Gilhousen, Kein S. et al. (1991). "On the Capacity of a Cellular CDMA System". *IEEE Transactions on Vehicular Technology*, 40(2):303–312.

Gradshteyn, I. S. and Ryzhik, I. M. (1990). *Table of Integrals, Series, and Products.* Academic Press, Inc., San Diego, California.

Grass, J. and Kabal, P. (1991). Methods of improving vector-scalar quantization of LPC coefficients. *Proceedings of the IEEE International Conference on Acoustics, Speech, and Signal Processing (ICASSP'91)*, pages 657–660.

Gray, R. M. (1984). Vector quantization. *IEEE ASSP Magazine*, pages 4–29.

Gray, Robert M. (1987). "Oversampled Sigma-Delta Modulation". *IEEE Transactions on Communications*, 35(5):481–489.

Gray, Robert M. (1989). "Spectral Analysis of Quantization Noise in a Single-Loop Sigma-Delta Modulator with dc Input". *IEEE Transactions on Communications*, 37(6):588–599.

Halmos, Paul R. (1960). *Naive Set Theory.* D. Van Nostrand Company, Inc., Princeton, USA.

Hashemi, Homayoun (1991). "Principles of Digital Indoor Radio Propagation". In *IASTED International Symposium on Computers, Electronics, Communication and Control*, pages 271–273, Calgary, Canada.

Haykin, Simon (1987). *Communication Systems.* Wiley Eastern Limited, New Delhi, India.

Haykin, Simon (1988). *Digital Communications.* John Wiley and Sons, New York.

Hersent, O., Guide, D., and Petit, J.-P. (2002). *IP Telephony: Packet Based Multimidia Communication (Portuguese).* Addison Wesley.

Hsu, Hwei P. (1973). *Fourier Analysis (Portuguese).* Livros Técnicos e Científicos Publishers Ltd., Rio de Janeiro, Brasil.

James, Barry R. (1981). *Probability: An Intermediate Level Course (Portuguese).* Institute of Pure and Applied Mathematics – CNPq, Rio de Janeiro, Brasil.

Jayant, N. (1992). Signal compression: Technology targets and research directions. *IEEE Journal on Selected Areas in Communications*, 10(5):796–818.

Jayant, N., Johnston, J., and Safranek, R. (1993). Signal compression based on models of human perception. *Proceedings of the IEEE*, 81(10):1385–1422.

Jayant, N. S. and Noll, P. (1984). *Digital Coding of Waveforms.* Prentice-Hall, Englewood Cliffs, NJ.

Jeszensky, Paul Jean Etienne (2004). *Telephone Systems (Portuguese).* Manole Publishers Ltd., São Paulo, Brasil.

Karayiannis, N. B. and Pai, P.-I. (1995). Fuzzy vector quantization algorithms and their applications in image compression. *IEEE Transactions on Image Processing*, 4(9):1193–1201.

Kennedy, Robert S. (1969). *Fading Dispersive Communication Channels.* Wiley Interscience, New York.

Kim, S.-J. and Oh, Y.-H. (1999). Split vector quantization of LSF parameters with minimum of dLSF constraint. *IEEE Signal Processing Letters*, 6(9):227–229.

Kleijn, W. B. and Paliwal, K. K., editors (1995). *Speech Coding and Synthesis.* Elsevier.

Knopp, Konrad (1990). *Theory and Application of Infinite Series.* Dover Publications, Inc., New York.

Kohonen, T. (1990). The self-organizing map. *Proceedings of the IEEE*, 78(9):1464–1480.

Krishnamurthy, A. K., Ahalt, S. C., Melton, D. E., and Chen, P. (1990). Neural networks for vector quantization of speech and images. *IEEE Journal on Selected Areas in Communications*, 8(8):1449–1457.

Lafortune, Jean-François and Lecours, Michel (1990a). "Measurement and Modeling of Propagation Losses in a Building at 900 MHz". *IEEE Transactions on Vehicular Technology*, 39(2):101–108.

Lafortune, Jean-François and Lecours, Michel (1990b). Measurement and modeling of propagation losses in a building at 900 mhz. *IEEE Transactions on Vehicular Technology*, 39(2).

Laroia, R., Phamdo, N., and Farvardin, N. (1991). Robust and efficient quantization of speech LSP parameters using structured vector quantizers. *Proceedings of the IEEE International Conference on Acoustics, Speech, and Signal Processing (ICASSP'91)*, pages 641–644.

Lathi, B. P. (1988). *Modern Digital and Analog Communications Systems.* Oxford University Press.

Lathi, B. P. (1989). *Modern Digital and Analog Communication Systems.* Holt, Rinehart and Winston, Inc., Philadelphia, USA.

LeBlanc, W. P., Bhattacharya, B., Mahmoud, S. A., and Cuperman, V. (1993). Efficient search and design procedures for robust multi-stage VQ of LPC parameters for 4 kb/s speech coding. *IEEE Transactions on Speech and Audio Processing*, 1(4):373–385.

Lecours, Michel, Chouinard, Jean-Yves, Delisle, Gilles Y., and Roy, Jean (1988). "Statistical Modeling of the Received Signal Envelope in a Mobile Radio Channel". *IEEE Transactions on Vehicular Technology*, 37(4):204–212.

Lee, William C. Y. (1989). *Mobile Cellular Telecommunications Systems.* McGraw-Hill Book Company, New York, USA.

Lee, William C. Y. (1991). "Overview of Cellular CDMA". *IEEE Transactions on Vehicular Technology*, 40(2):291–302.

Leon-Garcia, Alberto (1989). *Probability and Random Processes for Electrical Engineering.* Addison-Wesley Publishing Co., Reading, Massachusetts.

Lévine, B. (1973). *Fondements Théoriques de la Radiotechnique Statistique.* Éditions de Moscou, Moscow, U.S.S.R.

Linde, Y., Buzo, A., and Gray, R. M. (1980). An algorithm for vector quantizer design. *IEEE Transactions on Communications*, 28(1):84–95.

Lipschutz, Seymour (1968). *Set Theory (Portuguese).* Ao Livro Técnico S.A., Rio de Janeiro, Brasil.

Macchi, César, Jouannaud, Jean-Pierre, and Macchi, Odile (1975). "Récepteurs Adaptatifs pour Transmission de Données a Grande Vitesse". *Annales des Télécommunications*, 30(9-10):311–330.

Madeiro, F. (1998). Vetor quantization applied to speech and image compression. *Master's Dissertation, Federal University of Paraíba.*

Markel, J. D. and Gray, A. H. (1976). *Linear Prediction of Speech.* Berlin, Germany: Springer-Verlag.

McMahon, E. L. (1964). "An Extension of Price's Theorem". *IEEE, PGIT*, 10.

Nanda, Sanjiv and Goodman, David J. (1992). *Third Generation Wireless Information Networks.* Kluver Academic Publishers, Boston.

Neal H. Shepherd, Editor (1988). "Received Signal Fading Distribution". *IEEE Transactions on Vehicular Technology*, 37(1):57–60.

Nedoma, Jiří (1957). "The Capacity of a Discrete Channel". In *Transactions of the First Prague Conference on Information Theory, Statistical Decision Functions, Random Processes*, pages 143–181, Prague, Czechoslovakia. Academia, Publishing House of the Czechoslovak Academy of Science.

Newman Jr., D. B. (1986). "FCC Authorizes Spread Spectrum". *IEEE Communications Magazine*, 24(7):46–47.

Oberhettinger, F. (1990). *Tables of Fourier Transforms and Fourier Transforms of Distributions.* Springer-Verlag, Berlin.

Paez, M. D. and Glisson, T. H. (1972). "Minimum Mean-Squared-Error Quantization in Speech PCM and DPCM Systems". *IEEE Transactions on Communications*, pages 225–230.

Pahlavan, K., Ganesh, R., and Hotaling, T. (1989). "Multipath Propagation Measurements on Manufacturing Floors at 910 MHz". *Electronics Letters*, 25(3):225–227.

Paliwal, K. K. and Atal, B. S. (1993). Efficient vector quantization of LPC parameters at 24 bits/frame. *IEEE Transactions on Speech and Audio Processing*, 1(1):3–14.

Pan, J. S., McInnes, F. R., and Jack, M. A. (1995). VQ codebook design using genetic algorithms. *Electronics Letters*, 31(17):1418–1419.

Panter, Philip F. (1972). *Communications Systems Design: Line-of-sight and Troposcatter Systems.* McGraw-Hill Book Company, New York.

Papoulis, A. (1981). *Probability, Random Variables, and Stochastic Processes.* McGraw-Hill, Tokyo.

Papoulis, A. (1983a). *Signal Analysis.* McGraw-Hill, Tokyo.

Papoulis, Athanasios (1983b). "Random Modulation: a Review". *IEEE Transactions on Accoustics, Speech and Signal Processing*, 31(1):96–105.

Papoulis, Athanasios (1991). *Probability, Random Variables and Stochastic Processes.* McGraw-Hill, Inc.

Pickholtz, Raymond L., Schilling, Donald L., and Milstein, Laurence B. (1982). "Theory of Spread–Spectrum Communications – A Tutorial". *IEEE Transactions on Communications*, COM. 30(5):855–884.

Price, R. (1958). "A Useful Theorem for Non-Linear Devices Having Gaussian Inputs". *IRE, PGIT*, 46(4).

Proakis, John G. (1990). *Digital Communications.* McGraw-Hill Book Company, New York.

Qualcomm (1992). *The CDMA Network Engineering Handbook*, volume 1. Qualcomm Incorporated, 10555 Sorrento Valley Road, San Diego, California, USA.

Qureshi, Shahid U. H. (1985). "Adaptive Equalization". *Proceedings of the IEEE*, 73(9):1349–1387.

Rabiner, L. R. and Schafer, R. W. (1978). *Digital Processing of Speech Signals.* Prentice-Hall, Upper Saddle River, New Jersey.

Raith, K. and Uddenfeldt, J. (1991). "Capacity of Digital Cellular TDMA System". *IEEE Transactions on Vehicular Technology*, 40(2):323–332.

Ramachandran, R. P., Sondhi, M. M., Seshadri, N., and Atal, B. S. (1995). A two codebook format for robust quantization of line spectral frequencies. *IEEE Transactions on Speech and Audio Processing*, 3(3):157–168.

Ramamurthi, B. and Gersho, A. (1986). Classified vector quantization of images. *IEEE Transactions on Communications*, 34(11):1105–1115.

Rao, K. K., Bojkovic, Z. S., and Milovanovic, D. A. (2002). *Multimedia Communication Systems: Techniques, Standards and Networks.* Printice Hall, New Jersey.

Rappaport, Theodore S. (1989). "Indoor Radio Communications for Factories of the Future". *IEEE Communications Magazine*, pages 15–24.

Rhee, Man Young (1998). *CDMA – Celular Mobile Comunication & Network Security.* Prentice Hall PTR. Korea.

Saleh, Adel A. M. and Valenzuela, Reinaldo A. (1987). "A Statistical Model for Indoor Multipath Propagation". *IEEE Journal on Selected Areas in Communications*, 5(2):128–137.

Scholtz, Robert A. (1982). "The Origins of Spread-Spectrum Communications". *IEEE Transactions on Communications*, COM. 30(5):822–854.

Schoroeder, M. R. and Atal, B. S. (1985). Code-excited linear prediction (CELP): High-quality speech at low bit rates. In *Proceedings of the International Conference on Acoustics, Speech and Signal Processing*, pages 937–940, Tampa, USA.

Schwartz, Mischa (1970). *Information Transmission, Modulation, and Noise.* McGraw-Hill, New York.

Schwartz, Mischa, Bennett, William, and Stein, Seymour (1966). *Communication Systems and Techniques.* McGraw-Hill, New York.

Shamai, Shlomo (1994). "Information Rates by Oversampling the Sign of a Bandlimited Process". *IEEE Transactions on Information Theory*, 40(4):1230–1236.

Shannon, Claude E. (1948). "The Philosophy of PCM". *Proceedings of the IRE*, 36:1324–1331.

Shoham, Y. (1987). Vector predictive quantization of spectral parameters for low rate speech coding. In *Proceedings of the International Conference on Acoustics, Speech and Signal Processing*, pages 2181–2184, Dallas, TX.

Soong, F. K. and Juang, B.-H. (1993). Optimal quantization of LSP parameters. *IEEE Transactions on Speech and Audio Processing*, 1(1):15–24.

Soong, F. K., Rosenberg, A. E., Juang, B.-H., and Rabiner, L. R. (1987). A vector quantization approach to speaker recognition. *AT&T Technical Journal*, 66(2):14–26.

Spanias, A. S. (1994). Speech coding: A tutorial review. *Proceedings of the IEEE*, 82(10):1541–1582.

Spiegel, Murray R. (1976). *Análise de Fourier*. McGraw-Hill do Brasil, Ltda., São Paulo.

Sripad, A. B. and Snyder, D. L. (1977). "A Necessary and Sufficient Condition for Quantization Errors to be Uniform and White". *IEEE Transactions on Accoustics, Speech and Signal Processing*, 25(5).

Steele, R. (1993). "Speech Codecs for Personal Communications". *IEEE Communications Magazine*, 31(11):76–83.

Stremler, Ferrel G. (1982). *Introduction to Communication Systems*. Addison-Wesley Publishing Company, Reading, USA.

Sugamura, N. and Farvardin, N. (1988). Quantizer design in LSP speech analysis-synthesis. *IEEE Journal on Selected Areas in Communications*, 6(2):432–440.

Taub, H. and Schilling, D. L. (1971). *Principles of Communication Systems*. McGraw-Hill, Tokyo.

Thom, D. (1991). "Characterization of Indoor Wireless Channel in the Presence of Multipath Fading". Report 1, Department of Electrical and Computer Engineering, University of Waterloo, Waterloo, Canada.

Valenzuela, Reinaldo A. (1989). "Performance of Adaptive Equalization for Indoor Radio Communications". *IEEE Transactions on Communications*, 37(3):291–293.

Viterbi, Andrew J. (1985). "When Not to Spread Spectrum – A Sequel". *IEEE Communications Magazine*, 23(4):12–17.

Woodward, P. M. (1952). "The Spectrum of Random Frequency Modulation". Memo. 666, Telecommunications Research Establishment, Great Malvern, England.

Wozencraft, J. M. and Jacobs, I. M. (1965). *Principles of Communication Engineering*. John Wiley & Sons, New York.

Wylie, C. R. (1966). *Advanced Engineering Mathematics*. McGraw-Hill Book Company, London.

Yacoub, Michel Daoud (1993). *Foundations of Mobile Radio Engineering*. CRC Press, Boca Raton, USA.

Yacoub, Michel Daoud (2002). *Wireless Technology – Protocols, Standards, and Techniques*. CRC Press, Boca Raton, USA.

Yegani, Parviz and Mcgillen, Clare D. (1991). "A Statistical Model for the Factory Radio Channel". *IEEE Transactions on Communications*, 29(10):1445–1454.

Zamir, Ram and Feder, Meir (1995). "Rate-Distortion Performance in Coding Bandlimited Sources by Sampling and Dithered Quantization". *IEEE Transactions on Information Theory*, 41(1):141–154.

Zeger, K., Vaisey, J., and Gersho, A. (1992). Globally optimal vector quantizer design by stochastic relaxation. *IEEE Transactions on Signal Processing*, 40(2):310–322.

Zumpano, Antônio and de Lima, Bernardo Nunes Borges (2004). "a medida do acaso". *Ciência Hoje*, 34(201):76–77.

About the Authors

Marcelo Sampaio de Alencar was born in Serrita, Brazil in 1957. He received his Bachelor Degree in Electrical Engineering, from Universidade Federal de Pernambuco (UFPE), Brazil, 1980, his Master Degree in Electrical Engineering, from Universidade Federal da Paraíba (UFPB), Brazil, 1988 and his Ph.D. from University of Waterloo, Department of Electrical and Computer Engineering, Canada, 1993.

He is a member of the Board of Directors of the Brazilian Telecommunications Society (SBrT), he is a member of the Brazilian Microwave Society (SBMO) and a Senior Member of the Institute of Electrical and Electronics Engineers (IEEE). He is a reviewer of the IEEE Transactions on Vehicular Technology, IEEE Transactions on Communications, Wireless Personal Communications and the Brazilian Telecommunications Journal (SBrT). He is a member of the Sister Society Board (IEEE Communications Society) (ComSoc) and Liaison to Latin America Societies.

During 1982-1984 he was Assistant Professor at the Department of Electrical Engineering at Faculdade de Engenharia de Joinville, Universidade para o Desenvolvimento do Estado de Santa Catarina (UDESC), Brazil. He is currently Vice-President Foreign Affairs of SBrT and President of the Institute for Advanced Studies in Communications (Iecom).

From 1984 to he was with the Department of Electrical Engineering, Universidade Federal da Paraiba, Brazil. Since 2003 he has been with the Department of Electrical Engineering, Universidade Federal de Campina Grande, Brazil, where he holds a position as Professor Titular (Full Professor). In 1989 he worked as a Consultant for the Regional Engineering Division of Empresa Brasileira de Telecomunicações (MCI-Embratel), Brazil. He was member of the Advisory or Technical Committee of several conferences, including WCNC, ICC, SBT, SBMO, PIMRC, SPAWC, IMOC, Globecom and ITS.

He was member of the Advisory Committee of the Center of Science and Technology, UFPB, in charge of University-Industry interaction and manager of the agreement between Universidade Federal da Paraiba and California State University (CSU), Chico. He is also in charge of the contracts between UFPB and several companies, including Embratel, Empresa de Telecomunicacoes do Rio Grande do Norte (TELERN), Companhia Hidro-elétrica do São Francisco (CHESF), Siemens and Telecom Italia Mobile (TIM).

Marcelo Alencar has been consulting for several companies and agencies, over the past 20 years, incluing MCI-Embratel, Telemar, Telecom Italia, Contol, Chesf, Siemens, Telpa, Telpe, Oi, CNPq, Capes, Anatel and Fapesp.

His main research interests include: Mobile Communications, Information Theory, Communications Theory, Signal Processing, Optical Communications and Biological Effects of Radiation. Areas in which he has over 150 papers published. He is a columnist for the newspaper Jornal do Commercio, in Brazil, since April, 2000. His biography is included in the following publications: Who's Who in the World and Who's Who in Science and Engineering, by Marquis Who's Who, New Providence, USA, and Outstanding People of the 21st Century, by the International Biographical Centre, Cambridge, England.

Marcelo Alencar is recipient of a grant from the IEEE Foundation and has authored four books: Digital Cellular Telephony (2004), Communication Systems (2001) and Digital Telephony (1998), Editora Erica Ltda., and Principles of Communications (1999), Editora Universitária (in Portuguese), and a chapter in the book Communications, Information and Network Security, Kluwer Academic Publishers, 2003, Boston, United States.

Valdemar C. da Rocha, Jr. was born in Jaboatão, Pernambuco, Brazil, on August 27, 1947. He received the B.Sc. degree in Electrical/Electronics Engineering from the Escola Politecnica, Recife, Brazil, in 1970, and the Ph.D. degree in Electronics from the University of Kent at Canterbury, England, in 1976.

In 1976 he joined the faculty of the Federal University of Pernambuco, Recife, Brazil, as an Associate Professor and founded its Electrical Engineering Postgraduate Programme. From 1992 to 1996 he was Head of the Department of Electronics and Systems and in 1993 became Professor of Telecommunications.

He has often been a consultant to both the Brazilian Ministry of Education and the Ministry of Science and Technology on postgraduate education and research in electrical engineering. For two terms (1993-

1995 and 1999-2001) he has been the Chairman of the Electrical Engineering Committee in the Brazilian National Council for Scientific and Technological Development.

During 1990-1992, he was a Guest Professor at the Swiss Federal Institute of Technology-Zurich, Institute for Signal and Information Processing.

He has been involved in the organisation of conferences in Brazil and abroad, and was a co-organizer with Jim Massey of the Cryptography session of the 1992 IEEE Information Theory Workshop in Salvador, Bahia, Brazil. He was a Technical Co-Chair with Marcelo Alencar of the 2002 International Telecommunications Symposium, Natal, Brazil. He has been many times on the Technical Committee and a guest speaker at the International Symposium on Communication Theory and Applications, held in Ambleside, UK, every two years.

He is a founding member and current President (2004-2006) of the Brazilian Telecommunication Society. He is also a Senior Member of the IEEE Communication Society (1977), the IEEE Information Theory Society (1981), a member of the Brazilian Society of Applied and Computational Mathematics (1982), and a Fellow of the Institute of Mathematics and its Applications (1992, England). His area of research interest is applied digital information theory, including error-correcting codes and cryptography.

Index